Analytic Methods in Arithmetic Geometry

CONTEMPORARY
MATHEMATICS

740

Centre de Recherches Mathématiques Proceedings

CRM

Analytic Methods in Arithmetic Geometry

Arizona Winter School 2016
Analytic Methods in Arithmetic Geometry
March 12–16, 2016
The University of Arizona, Tucson, AZ

Alina Bucur
David Zureick-Brown
Editors

AMS AMERICAN MATHEMATICAL SOCIETY

2010 *Mathematics Subject Classification*. Primary 11G05,11R45, 20D60, 05C25, 11L03, 11T23, 19F217, 11G10, 11M50, 14G10.

Library of Congress Cataloging-in-Publication Data

Cataloging-in-Publication Data has been applied for by the AMS.
See http://www.loc.gov/publish/cip/.

Contemporary Mathematics ISSN: 0271-4132 (print); ISSN: 1098-3627 (online)
DOI: https://doi.org/10.1090/conm/740

Color graphic policy. Any graphics created in color will be rendered in grayscale for the printed version unless color printing is authorized by the Publisher. In general, color graphics will appear in color in the online version.

∞ The paper used in this book is acid-free and falls within the guidelines
established to ensure permanence and durability.
Visit the AMS home page at `https://www.ams.org/`

10 9 8 7 6 5 4 3 2 1 24 23 22 21 20 19

Contents

Preface

This volume contains notes which accompanied the lectures at the nineteenth annual Arizona Winter School, held March 12–16, 2016, at the University of Arizona in Tucson. The Arizona Winter School is an intensive five-day meeting, each year organized around a different central topic in arithmetic geometry, featuring several courses by leading and emerging experts ("an annual pilgrimage," in the words of one participant). The Winter School is the main activity of the Southwest Center for Arithmetic Geometry, which was founded in 1997 by a group of seven mathematicians working in the southwest United States, and which has been supported since that time by the National Science Foundation; in recent years it has been organized in partnership with the Clay Mathematics Institute.

The special character of the Arizona Winter School comes from its format. Each speaker proposes a project, and a month before the Winter School begins, the speaker is assigned a group of graduate students who work on the project. The speakers also provide lecture notes and a bibliography. During the actual school the speaker and and his or her group of students work every evening on the assigned project. On the last day of the workshop, the students from each group present their work to the whole school. The result is a particularly intense and focused five days of mathematical activity (for the students and speakers alike).

The topic of the Winter School in 2016 was Analytic Methods in Arithmetic Geometry, and the speakers were Alina Carmen Cojocaru, Harald Andrés Helfgott, Philippe Michel, and Andrew Sutherland. Étienne Fouvry, Emmanuel Kowalski, and Will Sawin join Michel as co-authors.

We thank the authors for their hard work before, during, and after the Winter School. The anonymous reviewers made numerous valuable comments, and we thank them for their careful reading of this manuscript. We also thank the IAS School of Mathematics and Simons Foundation (#524015) for their support of Alina Bucur, and NSF (grant DMS-1555048) for their support of David Zureick-Brown while editing this volume. Finally, we are indebted to the other members (past and present) of the Southwest Center; it is thanks to their efforts that the Winter School exists in its present form.

Alina Bucur
David Zureick-Brown

Contemporary Mathematics
Volume **740**, 2019
https://doi.org/10.1090/conm/740/14901

Primes, elliptic curves and cyclic groups

Alina Carmen Cojocaru

with an appendix by Alina Carmen Cojocaru, Matthew Fitzpatrick, Thomas Insley, and Hakan Yilmaz

ABSTRACT. Given an elliptic curve defined over the field of rational numbers, what is the frequency with which its reduction modulo a prime gives rise to a cyclic group? Guided by this question, we survey results (and their methods of proof) about rational primes viewed in the context of elliptic curves.

CONTENTS

The author's work on this material was partially supported by the Simons Collaboration Grant under Award No. 318454. The work of the authors of the appendix was partially supported by the National Science Foundation RTG grant under agreement No. DMS-1246844. The computations summarized in the appendix were performed in the Mathematical Computing Laboratory at the University of Illinois at Chicago.

1. Introduction

Since the beginning of the 20th century, it has been both stimulating and rewarding to explore analogies between the group of units $k^\times$ of a finite field k and the group of points $E(k)$ of an elliptic curve E/k defined over k. In this paper we focus on $k = \mathbb{F}_p$, the finite field with p elements, with p denoting a rational prime, and we overview results about the group of points $\overline{E}(\mathbb{F}_p)$ of the reduction modulo p of an elliptic curve $E/\mathbb{Q}$, denoted $\overline{E}/\mathbb{F}_p$. Specifically, we investigate the *group structure* of $\overline{E}(\mathbb{F}_p)$ not solely for one given prime p, but also as a function of p. Moreover, upon recalling that $\mathbb{F}_p^\times$ is a cyclic group, we pursue this investigation of $\overline{E}(\mathbb{F}_p)$ guided by:

QUESTION 1. Given an elliptic curve $E/\mathbb{Q}$, how often is the group $\overline{E}(\mathbb{F}_p)$ cyclic?

We start the paper with an introduction into the realms of primes and of elliptic curves. In particular, in Section 2 we recall basic properties of rational primes, while in Section 3 we recall basic properties of elliptic curves, on which we expand in Sections 4 - 8. In Section 9 we bring together the two realms and present a strategy towards answering Question 1; based on this strategy, we derive related theorems in Sections 10 - 11. In the remaining sections we explore variations of Question 1, as follows: in Section 12 we pursue an average version of Question 1; in Sections 13 - 14 we discuss the questions of how often $\left|\overline{E}(\mathbb{F}_p)\right|$ equals a prime and how often $\left|\overline{E}(\mathbb{F}_p)\right|$ equals the prime p itself; finally, in Section 15 we discuss, briefly, function field versions of Question 1 in the settings of elliptic curves over $\mathbb{F}_q(T)$ and of Drinfeld $\mathbb{F}_q[T]$-modules over $\mathbb{F}_q(T)$. We conclude the paper with a few remarks about other variations of the cyclicity section (Section 16) and with computational data supporting conjectures discussed in the previous sections (Appendix).

NOTATION. Throughout the paper, the letters p and ℓ are used to denote rational primes; the letters k, m, and n are used to denote positive integers; the letters x, y, and t are used to denote positive real numbers. The other notation to be followed is either standard (e.g., $\mathrm{O}(\cdot)$, $\mathrm{o}(\cdot)$, $\ll$, $\gg$, $\sim$, $\asymp$, $\simeq$) or is introduced explicitly when used for the first time.

2. Primes

A fundamental problem in number theory is that of understanding the integers and, in particular, the primes. For instance, how many primes are there?

Around 300 BC, Euclid proved that there are infinitely many primes. In fact, Euclid's argument leads to the explicit lower bound $\pi(x) \gg \log\log x$ for the prime counting function

$$\pi(x) := \#\{p \leq x : p \text{ prime}\}.$$

We may ask further, how many primes are there in the interval $[2, x)$, or what is the behaviour of the function $\pi(x)$, as $x \to \infty$?

Around the 1850s, about two millennia after Euclid, using ingenious elementary arguments, Chebysheff proved that $\pi(x)$ is bounded, from above and below, by constant multiples of $\frac{x}{\log x}$. A few decades later, following Riemann's groundbreaking insights on the Riemann zeta function which were published in 1859, Hadamard and de la Vallée Poussin proved, independently, the asymptotic growth of $\pi(x)$, which had been conjectured by Legendre and Gauss in the late 1700s:

THEOREM 2. *(The Prime Number Theorem, 1896)*

$$\pi(x) \sim \frac{x}{\log x} \sim \int_2^x \frac{dt}{\log t}.$$

Thanks to the Prime Number Theorem, asymptotic formulae for other prime counting functions may be derived, including:

$$\sum_{p \leq x} \frac{1}{\log p} \sim \frac{x}{(\log x)^2} \sim \int_2^x \frac{1}{\log t} \cdot \frac{dt}{\log t}, \tag{1}$$

$$\sum_{p \leq x} \frac{1}{2\sqrt{p}} \sim \frac{\sqrt{x}}{\log x} \sim \int_2^x \frac{1}{2\sqrt{t}} \cdot \frac{dt}{\log t}. \tag{2}$$

This is, by no means, the end of the study of primes. Not only there are infinitely many of them, but infinitely many of them (seem to) appear in interesting sequences. For example, Euclid-type arguments may be used to prove that infinitely many primes lie in *certain* arithmetic progressions. Furthermore, analytic arguments, introduced by Dirichlet between 1837-1839, may be used to prove that infinitely many primes lie in *all* (admissible) arithmetic progressions. Explicitly, following de la Vallée Poussin's work on primes, Dirichlet's result may be rephrased as:

THEOREM 3. *(Dirichlet's Theorem for Primes in Arithmetic Progressions) For any coprime integers a, m with $m \geq 1$, we have*

$$\pi(x, m, a) := \#\{p \leq x : p \equiv a (\operatorname{mod} m)\} \sim \frac{1}{\phi(m)}\, \pi(x).$$

Here, $\phi(\cdot)$ denotes the Euler function.

Theorem 3 is only a particular case of the more general Chebotarev Density Theorem, proven by Chebotarev in the 1920s, which, in its simplest form, states:

THEOREM 4. *(The Chebotarev Density Theorem, 1922) For any finite, Galois extension $K/\mathbb{Q}$ and for any conjugacy class $\mathcal{C} \subseteq \operatorname{Gal}(K/\mathbb{Q})$, we have*

$$\pi_{\mathcal{C}}(x, K/\mathbb{Q}) := \#\left\{p \leq x : \left(\frac{K/\mathbb{Q}}{p}\right) = \mathcal{C}\right\} \sim \frac{|\mathcal{C}|}{[K : \mathbb{Q}]}\, \pi(x), \tag{3}$$

where $\left(\frac{K/\mathbb{Q}}{p}\right)$ is the Artin symbol at p in the extension $K/\mathbb{Q}$.

Theorem 3 can be recovered from (3) by taking $K = \mathbb{Q}(\zeta_m)$, with ζ_m a primitive m-th root of unity, and by remarking that, since $\operatorname{Gal}(\mathbb{Q}(\zeta_m)/\mathbb{Q}) \simeq (\mathbb{Z}/m\mathbb{Z})^\times$, any conjugacy class $\mathcal{C} \subseteq \operatorname{Gal}(\mathbb{Q}(\zeta_m)/\mathbb{Q})$ is a singleton set, uniquely determined by $a (\operatorname{mod} m)$ for some $a \in \mathbb{Z}$ with $\gcd(a, m) = 1$.

Other sets of primes, conjectured to be infinite, have been the focus of celebrated conjectures, such as the following three.

CONJECTURE 5. *(Artin's Primitive Root Conjecture, 1927) For any non-zero integer a, not a unit and not a square, there exists a constant $C_{Artin}(a) > 0$ such that*

$$\#\{p \leq x : \mathbb{F}_p^\times = \langle a (\operatorname{mod} p)\rangle\} \sim C_{Artin}(a) \int_2^x \frac{dt}{\log t}. \tag{4}$$

Moreover,

$$C_{Artin}(a) := c_a \prod_{\ell} \left(1 - \frac{1}{\left[\mathbb{Q}\left(\zeta_\ell, a^{\frac{1}{\ell}}\right) : \mathbb{Q}\right]}\right), \tag{5}$$

where, upon writing the integer a uniquely as

$$a = a_{sf}\ a_{non\text{-}sf}^2$$

for some integers a_{sf}, $a_{non\text{-}sf}$ with a_{sf} squarefree, we have

$$c_a := \begin{cases} 1 & \text{if } a_{non\text{-}sf} \not\equiv 1 (\text{mod}\, 4), \\ 1 - \mu(a_{non\text{-}sf}) \prod_{\ell \mid a} \frac{1}{\left[\mathbb{Q}\left(\zeta_\ell, a^{\frac{1}{\ell}}\right):\mathbb{Q}\right]-1} & \text{if } a_{non\text{-}sf} \equiv 1 (\text{mod}\, 4). \end{cases}$$

Here, $\mu(\cdot)$ denotes the Möbius function.

CONJECTURE 6. *(Hardy-Littlewood Twin Prime Conjecture, 1923)*
For any non-zero integer a, there exists a constant $\mathfrak{S}(a) \geq 0$ such that

$$\#\{p \leq x : p + a \text{ is prime}\rangle\} \sim \mathfrak{S}(a) \int_2^x \frac{1}{\log t} \cdot \frac{dt}{\log t}. \tag{6}$$

Moreover,

$$\mathfrak{S}(a) := \begin{cases} 2 \prod_{\ell \neq 2} \frac{\ell(\ell-2)}{\ell-1} \prod_{\ell \mid a} \frac{\ell-1}{\ell-2} & \text{if } a \text{ is even}, \\ 0 & \text{if } a \text{ is odd}. \end{cases} \tag{7}$$

CONJECTURE 7. *(Hardy-Littlewood Quadratic Polynomial Conjecture, 1923)*
For any integers a, b, c, with $a > 0$ and $D := b^2 - 4ac$ not a square, there exists a constant $\mathfrak{S}(a, b, c) \geq 0$ such that

$$\#\{p \leq x : p = an^2 + bn + c \text{ for some integer } n\} \sim \mathfrak{S}(a, b, c) \int_2^x \frac{1}{\sqrt{t}} \cdot \frac{dt}{\log t}. \tag{8}$$

Moreover,
(9)

$$\mathfrak{S}(a, b, c) := \begin{cases} \frac{\gcd(2, a+b)}{\sqrt{a}} \prod_{\substack{\ell \neq 2 \\ \ell \mid a, \ell \mid b}} \frac{\ell}{\ell-1} \prod_{\substack{\ell \neq 2 \\ \ell \nmid a}} \left(1 - \frac{\left(\frac{D}{\ell}\right)}{\ell-1}\right) & \text{if } 2 \nmid \gcd(a+b, c), \\ 0 & \text{if } 2 \mid \gcd(a+b, c). \end{cases}$$

Here, $\left(\frac{\cdot}{\ell}\right)$ denotes the quadratic symbol modulo ℓ.

While Conjectures 5 - 7 are still open, significant progress has been made towards each of them by using the theory of sieves, a branch of number theory which was started by Brun in the second decade of the 1900s and which continues to grow. We refer the reader to [**HaRi**] for a sound introduction to the methods towards such progress and to [**CoMu05**], [**Grv**], [**Mo**], [**Sa**] and [**So**] for related more recent works.

The study of the above three conjectures reveals the importance of the study of primes in arithmetic progressions pursued not only for one modulus, but also for varying moduli. In turn, this latter study reveals the importance of the error terms in Dirichlet's Theorem 3.

In this direction, analytic methods pertaining to zeta and L-functions, as well as to sieves, have been successfully used to prove the following results.

THEOREM 8. *(Conditional Effective Dirichlet's Theorem)*
For any $x > 0$ and any coprime integers a, m with $1 \leq m \leq \frac{x^{\frac{1}{2}}}{(\log x)^3}$, the Generalized Riemann Hypothesis (GRH for short) for Dirichlet L-functions is equivalent to

$$\pi(x, m, a) = \frac{1}{\phi(m)} \pi(x) + \mathrm{O}\left(x^{\frac{1}{2}} \log(mx)\right).$$

THEOREM 9. *(The Siegel-Walfisz Theorem, 1936)*
For any $A > 0$, there exists a constant $C(A) > 0$ such that, for any $x > 0$ and any coprime integers a, m with $1 \leq m \leq (\log x)^A$, we have

$$\pi(x, m, a) = \frac{1}{\phi(m)} \pi(x) + \mathrm{O}\left(x \exp\left(-C(A)\sqrt{\log x}\right)\right).$$

THEOREM 10. *(The Brun-Titchmarsh Theorem, 1930)*
For any $x > 0$, any $\varepsilon > 0$, and any coprime integers a, m with $1 \leq m \leq x^{1-\varepsilon}$, we have

$$\pi(x, m, a) \ll_{\varepsilon} \frac{x}{\phi(m) \log \frac{x}{m}}.$$

THEOREM 11. *(The Barban-Davenport-Halberstam Theorem, 1963-1966)*
For any $x > 0$, $A > 0$, and $Q > 0$ with $\frac{x}{(\log x)^A} \leq Q \leq x$, we have

$$\sum_{m \leq Q} \sum_{\substack{1 \leq a \leq m \\ \gcd(a,m)=1}} \left| \pi(x, m, a) - \frac{1}{\phi(m)} \pi(x) \right|^2 \leq Q\, x\, \log x.$$

THEOREM 12. *(The Bombieri-Vinogradov Theorem, 1965)*
For any $A > 0$, there exists $B = B(A) > 0$ such that

$$\sum_{m \leq \frac{x^{\frac{1}{2}}}{(\log x)^B}} \max_{y \leq x} \max_{\gcd(a,m)=1} \left| \pi(y, m, a) - \frac{1}{\phi(m)} \pi(y) \right| \ll_A \frac{x}{(\log x)^A}.$$

We refer the reader to [**CoMu05**], [**Dav**] and [**MoVa**] for proofs of these results; for now, we only highlight an important character sum estimate used in these proofs:

THEOREM 13. *(The large sieve inequality)*
For any $M, N, Q > 0$ and $(a_n)_n \subseteq \mathbb{C}$, we have

$$\sum_{m \leq Q} \frac{m}{\phi(m)} \sum_{\substack{\chi (\mathrm{mod}\, m) \\ \chi \text{ primitive}}} \left| \sum_{M < n \leq M+N} a_n\, \chi(n) \right|^2 \leq (N + 3Q^2) \sum_{M < n \leq M+N} |a_n|^2,$$

where χ denotes an arbitrary Dirichlet character modulo m.

We also record an application of Theorems 9 - 12, due to Indlekofer, Wehmeier and Lucht, which shall be used by us in Section 12:

PROPOSITION 14. *(* [**InWeLu**, Thm. 3, p. 1957]*)*
For any $A > 0$ and $x > 0$, we have

$$\sum_{p \leq x} \prod_{\ell | p-1} \left(1 - \frac{1}{\ell(\ell^2 - 1)}\right) = \prod_{\ell} \left(1 - \frac{1}{\ell(\ell - 1)^2(\ell + 1)}\right) \cdot \pi(x) + \mathrm{O}_A\left(\frac{x}{(\log x)^A}\right).$$

Moreover, we record a variation of Theorem 11 related to twin primes, due to Balog, Cojocaru and David, which we shall refer to in Section 13:

THEOREM 15. *(*[**BaCoDa**, Thm. 4, p. 4]*)*
For any $A > 0$, there exists $B(A) > 0$ such that, for any $B > B(A)$ and for any $x > 0$, $\varepsilon > 0$, $R > 0$, $Q > 0$, $X > 0$, $Y > 0$ satisfying

$$x^{\frac{1}{3}+\varepsilon} \leq R \leq x, \ \ Q \leq \frac{x}{(\log x)^B}, \ \ 2 \leq X + Y \leq x,$$

we have

$$\sum_{0<|r|\leq R} \sum_{m \leq Q} \sum_{a (\mathrm{mod}\, m)} \left| \sum_{\substack{X<p\leq X+Y \\ p \equiv a (\mathrm{mod}\, m) \\ p-p'=r}} \log p \cdot \log p' - \mathfrak{S}(r,m,a)Y \right|^2 \ll \frac{Rx^2}{(\log x)^A},$$

where p and p' denote rational primes, $\mathfrak{S}(rm)$ is as in (7), and

$$\mathfrak{S}(r,m,a) \ := \ \begin{cases} \frac{1}{\phi(m)}\mathfrak{S}(rm) & \text{if } 2 \mid r, \ \gcd(a,m) = \gcd(a-r,m) = 1, \\ 0 & \text{otherwise.} \end{cases}$$

The proof of this result relies on sieve methods, as well as on yet another analytic technique known as *the circle method.*

An immediate question to ask is whether statements similar to Theorems 8 - 12 hold in the general setting of the Chebotarev Density Theorem. Several such results exist, including the following generalizations of Theorems 8 - 9 proved by Lagarias and Odlyzko using analytic methods in algebraic number theory:

THEOREM 16. *(Effective Chebotarev Density Theorem, 1977* [**LaOd**]*)*
For any finite, Galois extension $K/\mathbb{Q}$ and for any conjugacy class $\mathcal{C} \subseteq \mathrm{Gal}(K/\mathbb{Q})$, we have:

(i) *for any $\frac{1}{2} \leq \delta < 1$, the δ-quasi-GRH for the Dedekind zeta function ζ_K of K (that is, the hypothesis that ζ_K admits a zero-free region of $\mathrm{Re}(s) > \delta$) is equivalent to the effective asymptotic formula*

$$\pi_{\mathcal{C}}(x, K/\mathbb{Q}) = \frac{|\mathcal{C}|}{[K:\mathbb{Q}]}\pi(x) + \mathrm{O}\left(|\mathcal{C}| \, x^{\delta} \left(\frac{\log|\mathrm{disc}(K/\mathbb{Q})|}{[K:\mathbb{Q}]} + \log x\right)\right);$$

(ii) *there exist positive absolute constants C_1 and C_2 such that, for any $x > 0$ satisfying*

$$\sqrt{\frac{\log x}{[K:\mathbb{Q}]}} \geq C_2 \max\left(\log|\mathrm{disc}(K/\mathbb{Q})|, |\mathrm{disc}(K/\mathbb{Q})|^{\frac{1}{[K:\mathbb{Q}]}}\right),$$

we have

$$\pi_{\mathcal{C}}(x, K/\mathbb{Q}) = \frac{|\mathcal{C}|}{[K:\mathbb{Q}]}\pi(x) + \mathrm{O}\left(|\tilde{\mathcal{C}}| \, x \exp\left(-C_1\sqrt{\frac{\log x}{[K:\mathbb{Q}]}}\right)\right).$$

Here, $\tilde{\mathcal{C}}$ denotes the set of conjugacy classes contained in $\mathcal{C}$.

Note that the formulation of part (i) above can be deduced from the version of the Chebotarev Density Theorem given in [**Se81**, p. 133].

In the context of the Chebotarev Density Theorem 4, the question of understanding $\pi_{\mathcal{C}}(x, K/\mathbb{Q})$ for ranges larger than the ones of Theorem 16 is mostly open.

In Section 10, we will present some answers to this question when K belongs to the family of division fields defined by an elliptic curve (see Theorems 65 and 66).

We conclude here our overview of the study of primes in the classical setting and shift our focus to elliptic curves. We refer the reader to [**Ap**], [**CoMu05**], [**Dav**], [**FrIw**], [**HaRi**], [**HaWr**], [**IwKo**], [**Te**] for proofs of the above results and for original references. Starting with Section 10, we will echo Conjectures 5 - 7 in the context of elliptic curves, as motivated by Question 1, and we will refer to several of the results of this section when summarizing the progress made towards answering Question 1 and its variations.

3. Elliptic curves: generalities

In what follows we review basic properties of elliptic curves over arbitrary fields. We expand on these properties over several ensuing sections (Sections 3 - 8), after which we start pursuing explicitly Question 1. For a thorough introduction to the theory of elliptic curves, including proofs and original references, we refer the reader to [**Si**] and [**Was**]; for properties not covered in these texts, we provide references ourselves.

DEFINITION 17. An **elliptic curve** E over a field K is a smooth, projective curve, defined over K, of genus 1, and having a fixed K-rational point, typically denoted $\mathcal{O}$. The set of K-rational points of E is denoted $E(K)$.

It can be proved that, when $\operatorname{char} K \neq 2, 3$, an elliptic curve E/K is defined by a **Weierstrass equation**

$$E_{a,b} : y^2 = x^3 + ax + b, \tag{10}$$

with $a, b \in K$ and with **discriminant**

$$\Delta_E = \Delta_{a,b} := -16\left(4a^3 + 27b^2\right) \neq 0. \tag{11}$$

Moreover, it can be proved that, when $K = \mathbb{Q}$, the coefficients a, b may be chosen to be in $\mathbb{Z}$.

The quantity

$$j_E = j_{a,b} := -1728\frac{4a^3}{\Delta_{a,b}} \in K, \tag{12}$$

defined by the coefficients of (10), is called the j-**invariant** of the curve. As we will recall in Proposition 19 below, it is an invariant for the $\overline{\mathbb{Q}}$-isomorphism class of E, where $\overline{\mathbb{Q}}$ denotes an algebraic closure of $\mathbb{Q}$.

When the elliptic curve is described by a Weierstrass equation such as (10), the fixed K-rational point $\mathcal{O} \in E(K)$ is the projective point $[0 : 1 : 0]$; henceforth, we refer to $\mathcal{O}$ as the **point at infinity** of the elliptic curve.

DEFINITION 18. Given two elliptic curves E/K, E'/K, and given a field extension $K \subseteq L$, an L-**morphism** between E/K and E'/K is a curve morphism $E \longrightarrow E'$, defined over L, which preserves the fixed rational points $\mathcal{O}$ of the curves. We denote by $\operatorname{End}_L(E)$ and $\operatorname{Aut}_L(E)$ the rings of L-endomorphisms and, respectively, of L-automorphisms, of the elliptic curve E.

As mentioned above, the j-invariant of an elliptic curve encodes information about the $\overline{K}$-isomorphism class of the curve, where $\overline{K}$ denotes an algebraic closure of K:

PROPOSITION 19. *Let K be a field with* $\operatorname{char} K \neq 2, 3$. *For any two elliptic curves $E_{a,b}/K$, $E_{a',b'}/K$ with Weiestrass equations $y^2 = x^3 + ax + b$, respectively $y^2 = x^3 + a'x + b'$, we have that*

$$E_{a,b} \simeq_{\overline{K}} E_{a',b'} \ \Leftrightarrow\ j_{a,b} = j_{a',b'};$$

equivalently,

$$E_{a,b} \simeq_{\overline{K}} E_{a',b'} \ \Leftrightarrow\ \exists\, u \in \overline{K}^{\times} \text{ such that } a = u^4 a' \text{ and } b = u^6 b'. \tag{13}$$

Furthermore, the isomorphism between $E_{a,b}$ and $E_{a',b'}$ can be defined over $K(u)$.

COROLLARY 20. *For any $p \geq 5$ and for any elliptic curve $E_{a,b}/\mathbb{F}_p$, we have*

$$\#\left\{E_{a',b'}/\mathbb{F}_p \text{ elliptic curve} : E_{a',b'} \simeq_{\mathbb{F}_p} E_{a,b}\right\} = \frac{p-1}{\left|\operatorname{Aut}_{\overline{\mathbb{F}}_p}(E_{a,b})\right|}.$$

One of the first remarkable properties of an elliptic curve concerns the algebraic structure of its set of points:

THEOREM 21. *(Poincaré's Theorem, 1901)*
For any field K and for any elliptic curve E/K, the set of K-rational points $E(K)$ is endowed with an additive law defined through the chord-tangent method; with respect to this law, $E(K)$ is an abelian group.

PROPOSITION 22. *For any field K, for any two elliptic curves E_1/K, E_2/K, and for any K-morphism $f : E_1 \longrightarrow E_2$, we have*

$$f(P+Q) = f(P) + f(Q) \ \ \forall P, Q \in E_1(K).$$

The algebraic structure of the group of points of an elliptic curve is related to the algebraic structure of the ring of endomorphisms of the curve:

THEOREM 23. *(Deuring's Endomorphism Ring Classification Theorem, 1941)*
For any field K and for any elliptic curve E/K, the ring $\operatorname{End}_{\overline{K}}(E)$ *is isomorphic either to $\mathbb{Z}$, or to an order in an imaginary quadratic field, or to an order in a quaternion algebra. Moreover,*

(i) *if* $\operatorname{char} K = 0$, *only the first two possibilities occur, in which case we say that E/K is* **without complex multiplication** *(without CM, for short) and, respectively,* **with complex multiplication** *(with CM, for short);*
(ii) *if* $\operatorname{char} K > 0$, *then only the latter two possibilities occur, in which case we say that E/K is* **ordinary** *and, respectively,* **supersingular**.

The classification of the endomorphism ring of an elliptic curve naturally calls for the classification of the automorphism ring, which we now recall:

THEOREM 24. *(Deuring's Automorphism Ring Classification Theorem, 1941)*
For any field K with $\operatorname{char} K \neq 2, 3$ *and for any elliptic curve E/K, there exists a* $\operatorname{Gal}(\overline{K}/K)$*-module isomorphism*

$$\operatorname{Aut}_{\overline{K}}(E) \simeq \mu_n,$$

where

$$n := \begin{cases} 6 & \text{if } j_E = 0, \\ 4 & \text{if } j_E = 1728, \\ 2 & \text{if } j_E \neq 0, 1728, \end{cases}$$

and where $\mu_n < \mathbb{C}^\times$ *denotes the group of n-th roots of unity in the complex plane. In particular, if* $p \geq 5$, *if* $K = \mathbb{F}_p$, *and if* $E = E_{a,b}$ *is defined by* (10) *for some residue classes* $a(\bmod p)$, $b(\bmod p)$, *then*

$$\left|\mathrm{Aut}_{\overline{\mathbb{F}}_p}(E)\right| = \begin{cases} 6 & \text{if } p \mid a \text{ and } p \equiv 1(\bmod 3), \\ 4 & \text{if } p \mid b \text{ and } p \equiv 1(\bmod 4), \\ 2 & \text{otherwise.} \end{cases}$$

We conclude here our introduction to elliptic curves over arbitrary fields and make the following convention:

General setting for the remainder of the paper. *Henceforth, if not otherwise stated, our setting is that of an elliptic curve* $E/\mathbb{Q}$ *defined by a Weiestrass equation* (10) *with integer coefficients, whose reduction modulo a prime* $p \nmid \Delta_E$ *we denote by* $\overline{E}/\mathbb{F}_p$.

4. Elliptic curves over $\mathbb{Q}$: group structure

In this section we give an overview of the basic properties of the group of points $E(\mathbb{Q})$ of an elliptic curve $E/\mathbb{Q}$. To start, what is the group structure of $E(\mathbb{Q})$?

THEOREM 25. *(Mordell's Theorem, 1922)*
For any elliptic curve $E/\mathbb{Q}$, *its group of* $\mathbb{Q}$*-rational points is finitely generated, that is,*

$$E(\mathbb{Q}) \simeq \mathbb{Z}^r \oplus E(\mathbb{Q})_{tors},$$

where $r = r_{alg}(E)$ *is some non-negative integer, called the* **algebraic rank of** $E/\mathbb{Q}$, *and* $E(\mathbb{Q})_{tors}$ *is the group of points of finite order in* $E(\mathbb{Q})$, *called the* **torsion subgroup of** $E(\mathbb{Q})$.

Several results proven over the course of the 20th century have led to the complete classification of the structure of the torsion subgroup, as recalled below.

THEOREM 26. *(Rational Torsion Classification Theorem, 1977-1978)*
For any elliptic curve $E/\mathbb{Q}$, *the following properties hold.*

(i) *(Mazur* [**Ma77**], [**Ma78**]*)*
The torsion subgroup $E(\mathbb{Q})_{tors}$ *is isomorphic to one of the groups:*

$$\{\mathcal{O}\}, \mathbb{Z}/2\mathbb{Z}, \mathbb{Z}/3\mathbb{Z}, \mathbb{Z}/4\mathbb{Z}, \mathbb{Z}/5\mathbb{Z}, \mathbb{Z}/6\mathbb{Z}, \mathbb{Z}/7\mathbb{Z}, \mathbb{Z}/8\mathbb{Z}, \mathbb{Z}/9\mathbb{Z}, \mathbb{Z}/10\mathbb{Z}, \mathbb{Z}/12\mathbb{Z},$$

$$\mathbb{Z}/2\mathbb{Z} \times \mathbb{Z}/2\mathbb{Z}, \mathbb{Z}/2\mathbb{Z} \times \mathbb{Z}/4\mathbb{Z}, \mathbb{Z}/2\mathbb{Z} \times \mathbb{Z}/6\mathbb{Z}, \mathbb{Z}/2\mathbb{Z} \times \mathbb{Z}/8\mathbb{Z}.$$

(ii) *(Olson* [**Ol**]*)*
If $\mathrm{End}_{\overline{\mathbb{Q}}}(E) \not\simeq \mathbb{Z}$, *then the torsion subgroup* $E(\mathbb{Q})_{tors}$ *is isomorphic to one of the groups:*

$$\{\mathcal{O}\}, \mathbb{Z}/2\mathbb{Z}, \mathbb{Z}/3\mathbb{Z}, \mathbb{Z}/4\mathbb{Z}, \mathbb{Z}/6\mathbb{Z}, \mathbb{Z}/2\mathbb{Z} \times \mathbb{Z}/2\mathbb{Z}.$$

Moreover, each of the groups listed above occurs infinitely often.

In practice, the torsion subgroup $E(\mathbb{Q})_{\mathrm{tors}}$ can be determined relatively quickly by combining Theorem 26 with the following two results.

THEOREM 27. *(Nagell-Lutz Rational Torsion Criterion, 1935-1937)*
For any elliptic curve $E/\mathbb{Q}$, *with Weierstrass equation* (10), *and for any point* $P \in E(\mathbb{Q})_{tors} \backslash \{\mathcal{O}\}$, *whose coordinates we denote by* $(x(P), y(P))$, *we have*

$$x(P), y(P) \in \mathbb{Z}$$

and

$$\textit{either} \quad 2P = \mathcal{O}, \quad \textit{or} \quad y(P)^2 \mid 4a^3 + 27b^2.$$

THEOREM 28. *(Reduction Modulo p Theorem)*
For any elliptic curve $E/\mathbb{Q}$ with Weierstrass equation (10) *and for any prime $p \nmid \Delta_E$, define the reduction map*

$$E(\mathbb{Q})_{tors} \longrightarrow \overline{E}(\mathbb{F}_p)$$

$$\mathcal{O} \mapsto \mathcal{O},$$

$$P = (x(P), y(P)) \mapsto \overline{P} = (x(P)(\bmod p), y(P)(\bmod p)).$$

If $p \nmid 2\Delta_E$, then this map is an injective group homomorphism.

In summary, the torsion subgroup $E(\mathbb{Q})_{\text{tors}}$ is well understood: it is completely classified and it can be determined algorithmically.

In contrast, the algebraic rank $r_{\text{alg}}(E)$ remains enigmatic. In practice, for an elliptic curve $E/\mathbb{Q}$ defined by the Weierstrass equation (10) with a, b moderate in size, there do exist algorithms for computing its algebraic rank; nevertheless, to ensure that the algorithms always terminate is an open problem connected to the Birch & Swinnerton-Dyer Conjecture from the 1960s [**BiSD**]. Briefly, this conjecture focuses on the sum

$$\sum_p \frac{|\overline{E}(\mathbb{F}_p)|}{p},$$

which is related to the behaviour of the logarithmic derivative of the Hasse-Weil zeta function of E at $s = 1$; in turn, this derivative is related to the value of the L-function $L(E, s)$ of E at $s = 1$ and, in particular, to the integer $r_{\text{an}}(E) := \text{ord}_{s=1} L(E, s)$, called the **analytic rank** of E. The Birch & Swinnerton-Dyer Conjecture predicts that $r_{\text{alg}}(E) = r_{\text{an}}(E)$. While still open, it has been the focus of significant research on both the algebraic and analytic sides of arithmetic geometry. For more on the status of this conjecture and developments about ranks of elliptic curves, see [**Poo**], [**RuSi02**] and [**Wi**].

To conclude, we remark that, on one hand, the *structures* of *finitely many* groups $\overline{E}(\mathbb{F}_p)$ are sufficient to determine the torsion subgroup $E(\mathbb{Q})_{\text{tors}}$; on the other hand, the *orders* of *infinitely many* groups $\overline{E}(\mathbb{F}_p)$ are helpful to determine the rank $r_{\text{alg}}(E)$ of $E/\mathbb{Q}$ (albeit conjecturally). Two questions emerge:

QUESTION 29. Given an elliptic curve $E/\mathbb{Q}$ and a prime $p \nmid \Delta_E$, what is the group order of $\overline{E}(\mathbb{F}_p)$?

QUESTION 30. Given an elliptic curve $E/\mathbb{Q}$ and a prime $p \nmid \Delta_E$, what is the group structure of $\overline{E}(\mathbb{F}_p)$?

We will devote Section 8 to answering these two questions. The intermediate Sections 5 - 7 will prepare some further background about elliptic curves.

5. Elliptic curves over $\mathbb{Q}$: division fields

In this section we summarize the main properties of the field extensions of $\mathbb{Q}$ defined by the torsion points of an elliptic curve $E/\mathbb{Q}$. These fields, called the **division fields** of E, are essential to our investigations of Question 1.

For every integer $m \geq 1$, we let $E[m]$ be the **group of m-division points** of $E\left(\overline{\mathbb{Q}}\right)$, i.e.

$$E[m] := \left\{P \in E\left(\overline{\mathbb{Q}}\right) : mP = \mathcal{O}\right\}.$$

This is a free $\mathbb{Z}/m\mathbb{Z}$-module of rank 2, acted on by the absolute Galois group

$$G_{\mathbb{Q}} := \operatorname{Gal}\left(\overline{\mathbb{Q}}/\mathbb{Q}\right).$$

The group action gives rise to a Galois representation

$$\varphi_{E,m} : G_{\mathbb{Q}} \longrightarrow \mathrm{GL}_2(\mathbb{Z}/m\mathbb{Z}),$$

defined by restricting an arbitrary $\sigma \in G_{\mathbb{Q}}$ to $E[m]$ and by composing this restriction with an isomorphism

$$\operatorname{Aut}(E[m]) \simeq \mathrm{GL}_2(\mathbb{Z}/m\mathbb{Z}).$$

By taking the inverse limit over all integers $m \geq 1$, ordered by divisibility, and by choosing bases compatibly, from the representations $\varphi_{E,m}$ we obtain a continuous Galois representation

$$\varphi_E : G_{\mathbb{Q}} \longrightarrow \mathrm{GL}_2(\hat{\mathbb{Z}})$$

and its projections

$$\varphi_{E,m^\infty} : G_{\mathbb{Q}} \longrightarrow \mathrm{GL}_2(\mathbb{Z}_m).$$

Here, $\hat{\mathbb{Z}}$ denotes the inverse limit over all m of the rings $\mathbb{Z}/m\mathbb{Z}$. Upon using the isomorphism $\hat{\mathbb{Z}} \simeq \prod_\ell \mathbb{Z}_\ell$ given by the Chinese Remainder Theorem, $\mathbb{Z}_m$ denotes the quotient ring of $\hat{\mathbb{Z}}$ corresponding to $\prod_{\ell | m} \mathbb{Z}_\ell$.

Note that, in the language of these representations, we have

$$\mathbb{Q}(E[m]) = \overline{\mathbb{Q}}^{\operatorname{Ker}\varphi_{E,m}}$$

and

$$\mathbb{Q}(E_{\text{tors}}) := \bigcup_{m \geq 1} \mathbb{Q}(E[m]) = \overline{\mathbb{Q}}^{\operatorname{Ker}\varphi_E}.$$

The ramification of the extension $\mathbb{Q}(E[m])/\mathbb{Q}$ is controlled by m and by the discriminant of the curve:

THEOREM 31. *(The Néron-Ogg-Shafarevich Criterion, 1964-1967)*
For any elliptic curve $E/\mathbb{Q}$ with Weierstrass equation (10) *and for any integer $m \geq 1$, if p is a prime which ramifies in $\mathbb{Q}(E[m])/\mathbb{Q}$, then $p \mid m\Delta_E$.*

The extension $\mathbb{Q}(E[m])/\mathbb{Q}$ has several remarkable arithmetic properties, some of which arise as consequences to the existence of a pairing $E[m] \times E[m] \longrightarrow \mu_m$, called the Weil pairing on $E/\mathbb{Q}$ (see [**Si**, p. 96] for the definition), such as the following:

THEOREM 32. *For any elliptic curve $E/\mathbb{Q}$ with Weierstrass equation* (10) *and for any integer $m \geq 1$, we have*

$$\mathbb{Q}(\zeta_m) \subseteq \mathbb{Q}(E[m]).$$

Consequently, if $p \nmid m\Delta_E$ is a prime which splits completely in $\mathbb{Q}(E[m])$, then $p \equiv 1 (\operatorname{mod} m)$.

The structure of $\mathrm{End}_{\overline{\mathbb{Q}}}(E)$ impacts the Galois properties of the extension $\mathbb{Q}(E[m])/\mathbb{Q}$. Before stating these properties, let us quickly revisit elliptic curves with complex multiplication:

THEOREM 33. *(Classification Theorem of the j-invariants of CM elliptic curves over $\mathbb{Q}$)*

For any elliptic curve $E/\mathbb{Q}$ such that $\mathrm{End}_{\overline{\mathbb{Q}}}(E) \not\simeq \mathbb{Z}$, the imaginary quadratic order $\mathrm{End}_{\overline{\mathbb{Q}}}(E)$ has class number 1. Consequently, $\mathrm{End}_{\overline{\mathbb{Q}}}(E)$ is isomorphic to one of the following thirteen orders (listed in decreasing order of their discriminants):

$$\mathbb{Z}+\mathbb{Z}\left[\frac{1+\sqrt{-3}}{2}\right],\ \mathbb{Z}+2\mathbb{Z}\left[\frac{1+\sqrt{-3}}{2}\right],\ \mathbb{Z}+3\mathbb{Z}\left[\frac{1+\sqrt{-3}}{2}\right],\ \mathbb{Z}+\mathbb{Z}\left[\sqrt{-1}\right],\ \mathbb{Z}+2\mathbb{Z}\left[\sqrt{-1}\right],$$

$$\mathbb{Z}+\mathbb{Z}\left[\frac{1+\sqrt{-7}}{2}\right],\ \mathbb{Z}+2\mathbb{Z}\left[\frac{1+\sqrt{-7}}{2}\right],\ \mathbb{Z}+\mathbb{Z}\left[\sqrt{-2}\right],\ \mathbb{Z}+\mathbb{Z}\left[\frac{1+\sqrt{-11}}{2}\right],$$

$$\mathbb{Z}+\mathbb{Z}\left[\frac{1+\sqrt{-19}}{2}\right],\ \mathbb{Z}+\mathbb{Z}\left[\frac{1+\sqrt{-43}}{2}\right],\ \mathbb{Z}+\mathbb{Z}\left[\frac{1+\sqrt{-67}}{2}\right],\ \mathbb{Z}+\mathbb{Z}\left[\frac{1+\sqrt{-163}}{2}\right].$$

Moreover, the j-invariant j_E of the elliptic curve is one of the following thirteen integers (listed in the order matching the quadratic orders above):

$$0, 2^4\cdot 3^3\cdot 5^3,\ -2^{15}\cdot 3\cdot 5^3,\ 2^6\cdot 3^3,\ 2^3\cdot 3^3\cdot 11^3,$$
$$-3^3\cdot 5^3,\ 3^3\cdot 5^3\cdot 17^3, 2^6\cdot 5^3,\ -2^{15},$$
$$-2^{15}\cdot 3^3,\ -2^{18}\cdot 3^3\cdot 5^3,\ -2^{15}\cdot 3^3\cdot 5^3\cdot 11^3,\ -2^{18}\cdot 3^3\cdot 5^3\cdot 23^3\cdot 29^3.$$

In the setting of Theorem 33, we introduce the auxiliary notation

$$\begin{aligned} O &:= \mathrm{End}_{\overline{\mathbb{Q}}} E),\\ \widehat{O} &:= \varprojlim_m O/mO,\\ K &:= O\otimes_{\mathbb{Z}}\mathbb{Q},\\ G_K &:= \mathrm{Gal}\left(\overline{K}/K\right), \end{aligned}$$

where the inverse limit is over integers $m \geq 1$ ordered by divisibility.

THEOREM 34. *(Open Image Theorem for CM Elliptic Curves, 1955* [**We55**], [**We55bis**]*)*

For any elliptic curve $E/\mathbb{Q}$ such that $O := \mathrm{End}_{\overline{\mathbb{Q}}}(E) \not\simeq \mathbb{Z}$, with the above notation, we have:

(i) *$\mathbb{Q}(E_{tors})$ is a free $\widehat{O}$-module of rank 1, acted on by G_K;*
(ii) *the representation*

$$\varphi_E|_{G_K} : G_K \longrightarrow \mathrm{GL}_1\left(\widehat{O}\right) = \widehat{O}^\times \tag{14}$$

has open image, that is,

$$\left|\widehat{O}^\times : \varphi_E|_{G_K}(G_K)\right| < \infty.$$

In particular, there exists a smallest integer $m_E \geq 1$ such that, for each $m \geq 1$,

$$\mathrm{Gal}(K(E[m])/K) \simeq \mathrm{pr}^{-1}\left(\mathrm{Gal}(K(E[\gcd(m, m_E)])/K)\right),$$

where $\mathrm{pr} : (O/mO)^\times \longrightarrow (O/\gcd(m, m_E)\, O)^\times$ is the natural projection.

COROLLARY 35. *For any elliptic curve $E/\mathbb{Q}$ such that $O := \mathrm{End}_{\overline{\mathbb{Q}}}(E) \not\simeq \mathbb{Z}$ and for any integer $m \geq 1$, written uniquely as $m = m_1 m_2$ for some integers m_1, m_2 with* $\gcd(m_1, m_E) = 1$ *and $m_2 \mid m_E^\infty$, we have*

$$\mathrm{Gal}(K(E[m])/K) \simeq (O/m_1 O)^\times \times H_{m_2}$$

for some $H_{m_2} \leq (O/m_2 O)^\times$.

The situation for elliptic curves without complex multiplication is very different:

THEOREM 36. *(Open Image Theorem for non-CM Elliptic Curves , 1972* [**Se72**]*)* *For any elliptic curve $E/\mathbb{Q}$ such that $\mathrm{End}_{\overline{\mathbb{Q}}}(E) \simeq \mathbb{Z}$, the representation φ_E has open image, that is,*

$$\left|\mathrm{GL}_2\left(\hat{\mathbb{Z}}\right) : \varphi_E(G_{\mathbb{Q}})\right| < \infty.$$

In particular, there exists a smallest integer $m_E \geq 1$ such that

$$\varphi_E(G_{\mathbb{Q}}) = \mathrm{pr}^{-1}(\varphi_{E,m_E}(G_{\mathbb{Q}})),$$

where $\mathrm{pr} : \mathrm{GL}_2\left(\hat{\mathbb{Z}}\right) \longrightarrow \mathrm{GL}_2(\mathbb{Z}/m_E\mathbb{Z})$ is the natural projection.

COROLLARY 37. *For any elliptic curve $E/\mathbb{Q}$ such that $\mathrm{End}_{\overline{\mathbb{Q}}}(E) \simeq \mathbb{Z}$ and for any integer $m \geq 1$, written uniquely as $m = m_1 m_2$ for some integers m_1, m_2 with* $\gcd(m_1, m_E) = 1$ *and $m_2 \mid m_E^\infty$, we have*

$$\mathrm{Gal}(\mathbb{Q}(E[m])/\mathbb{Q}) \simeq \mathrm{GL}_2(\mathbb{Z}/m_1\mathbb{Z}) \times H_{m_2}$$

for some $H_{m_2} \leq \mathrm{GL}_2(\mathbb{Z}/m_2\mathbb{Z})$.

A useful consequence to these theorems is an estimate for the degree of $\mathbb{Q}(E[m])/\mathbb{Q}$, whose proof we leave to the reader as an exercise (see also [**Br**] for related work):

PROPOSITION 38. *For any elliptic curve $E/\mathbb{Q}$ and for any integer $m \geq 1$, we have*

$$\frac{m^{\frac{4}{\gamma}}}{\log\log m} \ll_E [\mathbb{Q}(E[m]) : \mathbb{Q}] \ll m^{\frac{4}{\gamma}},$$

where

$$\gamma := \begin{cases} 1 & \text{if } \mathrm{End}_{\overline{\mathbb{Q}}}(E) \simeq \mathbb{Z}, \\ 2 & \text{if } \mathrm{End}_{\overline{\mathbb{Q}}}(E) \not\simeq \mathbb{Z}. \end{cases}$$

In connection with Proposition 38, let us remark that, in the theory of elliptic curves over $\mathbb{Q}$, a special role is played by the curves $E/\mathbb{Q}$ whose extensions $\mathbb{Q}(E[m])/\mathbb{Q}$ have maximal degrees, i.e. whose representation φ_E has maximal image $\varphi_E(G_{\mathbb{Q}})$. We devote the next section to a brief overview of this topic.

6. Elliptic curves over $\mathbb{Q}$: maximal Galois representations

It was observed by Serre about five decades ago that no elliptic curve $E/\mathbb{Q}$ satisfies $|\mathrm{GL}_2(\mathbb{Z}/m\mathbb{Z}) : \varphi_{E,m}(G_{\mathbb{Q}})| = 1$ for all integers m. Instead, it could happen that

$$|\mathrm{GL}_2(\mathbb{Z}/m\mathbb{Z}) : \varphi_{E,m}(G_{\mathbb{Q}})| \in \{1, 2\} \quad \forall m \geq 1, \tag{15}$$

a property which is captured by the following definition:

DEFINITION 39. An elliptic curve $E/\mathbb{Q}$ is called a **Serre curve** if

$$\left|\mathrm{GL}_2(\hat{\mathbb{Z}}) : \varphi_E(G_{\mathbb{Q}})\right| = 2.$$

In this section, we take a quick look at the underlying reason behind Serre's observation and give an overview of the main properties of Serre curves.

Denoting by Δ_{sf} the squarefree part of the discriminant Δ_E of any Weierstrass model for $E/\mathbb{Q}$, we obtain the field embeddings

$$\mathbb{Q}\left(\sqrt{\Delta_E}\right) \subseteq \mathbb{Q}(E[2]), \qquad \mathbb{Q}\left(\sqrt{\Delta_E}\right) \subseteq \mathbb{Q}\left(\zeta_{|d_E|}\right) \subseteq \mathbb{Q}\left(E[|d_E|]\right), \tag{16}$$

where

$$d_E := \operatorname{disc}\left(\mathbb{Q}\left(\sqrt{\Delta_E}\right)/\mathbb{Q}\right) = \begin{cases} \Delta_{\text{sf}} & \text{if } \Delta_{\text{sf}} \equiv 1 (\operatorname{mod} 4), \\ 4\Delta_{\text{sf}} & \text{otherwise.} \end{cases}$$

Note that the existence of an integer d_E satisfying (16) is guaranteed by the Kronecker-Weber Theorem, since $\mathbb{Q}\left(\sqrt{\Delta_E}\right)$ is abelian over $\mathbb{Q}$. In particular, this is where it is relevant that our elliptic curve be defined over $\mathbb{Q}$ and not over an arbitrary field (even over an arbitrary number field). Note also that this value of d_E minimizes $|d_E|$, subject to (16).

It follows that

$$\varphi_E(G_{\mathbb{Q}}) \leq \left\{g \in \mathrm{GL}_2\left(\hat{\mathbb{Z}}\right) : \epsilon(g) = \chi_E(g)\right\} =: H_E, \tag{17}$$

where the two maps

$$\epsilon : \mathrm{GL}_2\left(\hat{\mathbb{Z}}\right) \to \mathrm{GL}_2(\mathbb{Z}/2\mathbb{Z}) \simeq S_3 \to \{\pm 1\},$$
$$\chi_E : \mathrm{GL}_2\left(\hat{\mathbb{Z}}\right) \to \hat{\mathbb{Z}}^\times \to (\mathbb{Z}/|d_E|\mathbb{Z})^\times \to \{\pm 1\}$$

are defined as follows:

- ϵ is the projection modulo 2, followed by the signature character on the permutation group S_3 (which is also the unique non-trivial multiplicative character on $\mathrm{GL}_2(\mathbb{Z}/2\mathbb{Z})$);
- χ_E is the determinant map, followed by the reduction modulo $|d_E|$, and then followed by the Kronecker symbol $\left(\frac{d_E}{\cdot}\right)$.

In summary, Serre's observation may be rephrased as:

LEMMA 40. *(*[**Se72**, Section 5.5]*)*
For any elliptic curve $E/\mathbb{Q}$, there exists a proper subgroup $H_E < \mathrm{GL}_2\left(\hat{\mathbb{Z}}\right)$ such that

$$\left|\mathrm{GL}_2\left(\hat{\mathbb{Z}}\right) : H_E\right| = 2 \text{ and } \varphi_E(G_{\mathbb{Q}}) \leq H_E.$$

In particular,

$$E/\mathbb{Q} \text{ is a Serre curve } \Leftrightarrow \varphi_E(G_{\mathbb{Q}}) = H_E.$$

One reason for which Serre curves are of interest is that they are handy in computations. For example, the Galois groups of all the division fields of a Serre curve may be determined easily:

PROPOSITION 41. *For any Serre curve $E/\mathbb{Q}$, we have:*

(i) $\mathrm{End}_{\overline{\mathbb{Q}}}(E) \simeq \mathbb{Z}$;
(ii) $E(\mathbb{Q})_{tors}$ *is trivial;*
(iii) *the integer m_E introduced in Theorem 36 satisfies the formula*

$$m_E = \begin{cases} 2\left|\Delta_{sf}\right| & \text{if } \Delta_{sf} \equiv 1 (\operatorname{mod} 4), \\ 4\left|\Delta_{sf}\right| & \text{otherwise,} \end{cases} \tag{18}$$

where Δ_{sf} denotes the squarefree part of the discriminant Δ_E of any Weierstrass model for E;

(iv) *for any integer $m \geq 1$,*

$$[\mathbb{Q}(E[m]) : \mathbb{Q}] = \begin{cases} |\operatorname{GL}_2(\mathbb{Z}/m\mathbb{Z})| & \text{if } m_E \nmid m, \\ \frac{1}{2}|\operatorname{GL}_2(\mathbb{Z}/m\mathbb{Z})| & \text{otherwise.} \end{cases} \tag{19}$$

PROOF. See [**BBCCJMSV**, Proposition 17]. □

Another reason for which Serre curves are of interest is that they dominate the pool of elliptic curves over $\mathbb{Q}$; this result will be discussed in the next section.

7. Elliptic curves over $\mathbb{Q}$: two-parameter families

It is of interest to investigate which properties of elliptic curves are generic, that is, which properties are most likely to hold as we look at all the elliptic curves of an arbitrary set. In Section 12, we will pursue this perspective in relation to our motivating Question 1; in particular, we will regard the given elliptic curve as an arbitrary element of a **two-parameter family of elliptic curves**, which is described as follows.

We consider parameters $A, B > 0$ and denote by

$$\mathcal{F}(A, B)$$

the set of $\mathbb{Q}$-isomorphism classes of elliptic curves $E_{a,b}$ defined by the equation $y^2 = x^3 + ax + b$ with $a, b \in \mathbb{Z}$, $\Delta_{a,b} \neq 0$, and with

$$|a| \leq A, |b| \leq B.$$

Note that

$$|\mathcal{F}(A, B)| \asymp AB. \tag{20}$$

Inside $\mathcal{F}(A, B)$, we distinguish the subsets of: elliptic curves with CM; elliptic curves without CM and which are not Serre curves; elliptic curves which are Serre curves. Specifically, we distinguish the subsets

$$\mathcal{F}_{\mathrm{CM}}(A, B) := \left\{E_{a,b} \in \mathcal{F}(A, B) : \operatorname{End}_{\overline{\mathbb{Q}}}(E_{a,b}) \not\simeq \mathbb{Z}\right\},$$

$$\mathcal{F}_{\mathrm{nonCM,\ nonSerre}}(A, B) := \left\{E_{a,b} \in \mathcal{F}(A, B) : \operatorname{End}_{\overline{\mathbb{Q}}}(E_{a,b}) \simeq \mathbb{Z}, \varphi_{E_{a,b}}(G_{\mathbb{Q}}) < H_E\right\},$$

$$\mathcal{F}_{\mathrm{Serre}}(A, B) := \left\{E_{a,b} \in \mathcal{F}(A, B) : \varphi_{E_{a,b}}(G_{\mathbb{Q}}) = H_E\right\}.$$

It is of interest to know how the size of each of these subsets compares to the size of the whole family.

Using classical number theoretic arguments, it can be proved that the elliptic curves with CM form a zero density subset in the set $\mathcal{F}(A, B)$:

THEOREM 42. *(Upper Bound for CM Curves in Families $\mathcal{F}(A, B)$)* *For any sufficiently large A, B, we have*

$$\frac{|\mathcal{F}_{CM}(A, B)|}{|\mathcal{F}(A, B)|} \ll \frac{1}{A} + \frac{1}{B}.$$

More precisely, upon denoting by

$$\mathcal{F}_j(A,B) := \{(a,b)\in\mathbb{Z}\times\mathbb{Z} : |a|\le A, |b|\le B,$$
$$\Delta_{a,b}\neq 0, \gcd\left(a^3,b^2\right) \textit{ is 12th power free},\ j_{a,b}=j\},$$

we have

$$(21)\quad |\mathcal{F}_0(A,B)| = \#\{b\in\mathbb{Z}\backslash\{0\} : b \textit{ is 6th power free}, |b|\le B\} \sim \frac{2}{\zeta(6)}B,$$

$$(22)\quad |\mathcal{F}_{1728}(A,B)| = \#\{a\in\mathbb{Z}\backslash\{0\} : a \textit{ is 4th power free}, |a|\le A\} \sim \frac{2}{\zeta(4)}A,$$

and, for any $\varepsilon>0$,

$$|\mathcal{F}_j(A,B)| \ll_\varepsilon \min\left\{A^{\frac{1}{2}+\varepsilon}, B^{\frac{1}{3}+\varepsilon}\right\},$$

for each of the j*-invariants of Theorem* 33 *with* $j\neq 0,1728$.

See [**BBCCJMSV**, Lemma 18] for a proof.

The elliptic curves without CM and which are not Serre curves also form a zero density subset in $\mathcal{F}(A,B)$; consequently, Serre curves are the ones which dominate an arbitrary two-parameter family of elliptic curves.

The first proof of this result (stated precisely below) was given by Jones in [**Jo10**], using Gallagher's multidimensional large sieve and arithmetic-geometric arguments, and building on prior work of Duke [**Du97**]:

THEOREM 43. *(Upper Bound for Non-Serre Curves in Families* $\mathcal{F}(A,B)$ [**Jo10**, Thm. 4]*)*

There exists a positive, absolute constant c *such that, for any sufficiently large* A, B, *we have*

$$\frac{|\mathcal{F}_{nonCM,\ nonSerre}(A,B)|}{|\mathcal{F}(A,B)|} \ll \frac{(\log\min\{A,B\})^c}{\sqrt{\min\{A,B\}}}.$$

COROLLARY 44. *(Asymptotic for Serre Curves in Families* $\mathcal{F}(A,B)$ [**Jo10**]*)*
As A, B *approach infinity, we have*

$$\frac{|\mathcal{F}_{Serre}(A,B)|}{|\mathcal{F}(A,B)|}\sim 1.$$

When

$$A = A(x),\ B = B(x)$$

for a given parameter x, Jones [**Jo10**] used Theorem 43 to derive the following upper bound:

THEOREM 45. *(Upper Bound for Non-Serre Curves in Families* $\mathcal{F}(x^2,x^3)$ [**Jo10**]*)*

There exists a positive, absolute, explicit constant c *such that, for any sufficiently large* x, *we have*

$$\frac{1}{|\mathcal{F}(x^2,x^3)|}\#\left\{E\in\mathcal{F}(x^2,x^3) : E \textit{ is not a Serre curve}\right\} \ll \frac{(\log x)^c}{x}.$$

This bound was refined to an asymptotic formula by Radhakrishnan [**Ra**]:

THEOREM 46. *(Asymptotic for Non-Serre Curves in Families $\mathcal{F}(x^2, x^3)$ [**Ra**])*
There exists a positive, absolute, explicit constant C such that, for any $\varepsilon > 0$ and for any sufficiently large $x \gg_\varepsilon 1$, we have

$$\frac{1}{|\mathcal{F}(x^2, x^3)|} \# \left\{E \in \mathcal{F}(x^2, x^3) : E \text{ is not a Serre curve}\right\} = C\frac{1}{x^2} + \mathrm{O}_\varepsilon\left(\frac{1}{x^{3-\varepsilon}}\right).$$

Radhakrishnan's result, based on arithmetic-geometric and group-theoretic arguments, builds on a prior result of Grant [**Grt**]. A corollary to Grant's result is that, on average, the torsion subgroup of the Mordell group $E(\mathbb{Q})$ of an elliptic curve $E/\mathbb{Q}$ is trivial:

THEOREM 47. *(Upper Bound for Curves with Non-Trivial Rational Torsion in Families $\mathcal{F}(x^2, x^3)$ [**Grt**])*
For any sufficiently large x, we have

$$\frac{1}{|\mathcal{F}(x^2, x^3)|} \# \left\{E \in \mathcal{F}(x^2, x^3) : E(\mathbb{Q})_{tors} \neq \{\mathcal{O}\}\right\} \ll \frac{1}{x^2}.$$

To complete the picture regarding the generic behaviour of the Mordell group of an elliptic curve $E/\mathbb{Q}$, we recall the following results about ranks:

THEOREM 48. (i) *(Upper Bound for Average Algebraic Rank in Families $\mathcal{F}(x^2, x^3)$ [**BhSh**])*

$$\limsup_{x\to\infty} \frac{1}{|\mathcal{F}(x^2, x^3)|} \sum_{E\in\mathcal{F}(x^2,x^3)} r_{alg}(E) < \frac{885}{1000}.$$

(ii) *(Upper Bound for Average Analytic Rank in Families $\mathcal{F}(x^2, x^3)$ [**Yo**])*

$$\limsup_{x\to\infty} \frac{1}{|\mathcal{F}(x^2, x^3)|} \sum_{E\in\mathcal{F}(x^2,x^3)} r_{an}(E) \leq \frac{25}{14}.$$

For some related work, see [**CoHa**], [**CoGrJo**], and the references in [**Poo**]. For the purpose of exploring Question 1, Theorems 42 and 43 will be relevant to our Sections 12 – 14.

8. Elliptic curves over $\mathbb{Q}$: reductions modulo primes

We now focus on a pair (E, p), where $E/\mathbb{Q}$ is an elliptic curve defined by (10) and p is a prime with $p \nmid \Delta_E$; in particular, we summarize those arithmetic properties of (E, p) which are relevant to Question 1 and we answer Questions 29 and 30. We start by introducing notation.

We define the integer

$$a_p = a_p(E) := - \sum_{x (\operatorname{mod} p)} \left(\frac{x^3 + ax + b}{p}\right) \tag{23}$$

and observe that

$$\left|\overline{E}(\mathbb{F}_p)\right| = p + 1 - a_p. \tag{24}$$

We define the polynomial

$$P_{E,p}(X) := X^2 - a_p X + p \in \mathbb{Z}[X] \tag{25}$$

and, writing its irreducible factorization as

$$P_{E,p}(X) = (X - \pi_p)(X - \overline{\pi}_p) \in \mathbb{C}[X],$$

we observe that

$$\begin{aligned} \operatorname{disc} P_{E,p} &= a_p^2 - 4p, \\ \pi_p + \overline{\pi}_p &= a_p, \\ \pi_p \cdot \overline{\pi}_p &= p. \end{aligned}$$

THEOREM 49. *(Fundamental Properties of the Frobenius of E mod p)*
For any elliptic curve $E/\mathbb{Q}$ with Weierstrass equation (10) *and for any $p \nmid \Delta_E$, the following properties hold:*

(i) *$\pi_p \in \mathbb{C}\backslash\mathbb{R}$ and $|\pi_p| = \sqrt{p}$;*

(ii) *for any $\ell \neq p$, the representation φ_{E,ℓ^∞} is unramified at p and the well-defined φ_{E,ℓ^∞} (Artin$_p$) has characteristic polynomial $P_{E,p}(X)$;*

(iii) *the p-th power Frobenius map Fr_p on $\overline{\mathbb{F}}_p$ gives rise to an endomorphism $\mathrm{Fr}_p \in \mathrm{End}_{\overline{\mathbb{F}}_p}(\overline{E})$ defined by*

$$\begin{aligned} \mathrm{Fr}_p : \overline{E}\left(\overline{\mathbb{F}}_p\right) &\longrightarrow \overline{E}\left(\overline{\mathbb{F}}_p\right) \\ (x,y) &\mapsto (x^p, y^p), \\ \mathcal{O} &\mapsto \mathcal{O}; \end{aligned}$$

(iv) *the ring $\mathbb{Z}[\mathrm{Fr}_p]$ is either isomorphic to $\mathbb{Z}$, or to the centralizer of Fr_p in $\mathrm{End}_{\mathbb{F}_p}\left(\overline{E}\right)$; in either case, we have the ring embeddings*

$$\mathbb{Z} \subseteq \mathbb{Z}[\mathrm{Fr}_p] \subseteq \mathrm{End}_{\mathbb{F}_p}\left(\overline{E}\right) \subseteq \mathrm{End}_{\overline{\mathbb{F}}_p}\left(\overline{E}\right);$$

(v) *if $\mathbb{Z}[\mathrm{Fr}_p] \not\simeq \mathbb{Z}$, then*

(v1) *$\mathbb{Z}[\mathrm{Fr}_p]$ is an order in an imaginary quadratic field;*

(v2) *$\mathrm{End}_{\mathbb{F}_p}\left(\overline{E}\right)$ is an order in an imaginary quadratic field;*

(v3) *$\mathrm{End}_{\overline{\mathbb{F}}_p}\left(\overline{E}\right)$ is either an order in an imaginary quadratic field (we say that p is* **ordinary***), or an order in a quaternion algebra (we say that p is* **supersingular***); moreover, if $p \geq 5$, then*

$$p \text{ is supersingular } \Leftrightarrow a_p = 0;$$

(vi) *denoting by $\mathbb{Q}_p$ the field of p-adic rational numbers, upon various choices of embeddings $\mathbb{Q} \subseteq \overline{\mathbb{Q}} \subseteq \overline{\mathbb{Q}}_p$, $\mathbb{Q} \subseteq \overline{\mathbb{Q}}_p \subseteq \overline{\mathbb{Q}}_p$ and $\mathrm{End}_{\mathbb{F}_p}\left(\overline{E}\right) \subseteq \mathbb{C}$, we have that one of the roots of $P_{E,p}$, say π_p, is identified with the p-th power Frobenius endomorphism Fr_p from part (iii); this identification gives rise to the ring embeddings*

$$\mathbb{Z} \subseteq \mathbb{Z}[\pi_p] \subseteq \mathrm{End}_{\mathbb{F}_p}(\overline{E}) \subseteq O_{\mathbb{Q}(\pi_p)};$$

moreover, in view of (v1)-(v2), *it gives rise to integers $c_p, c_p' \geq 1$ satisfying*

$$\begin{aligned} \mathbb{Z}[\pi_p] &= \mathbb{Z} + c_p O_{\mathbb{Q}(\pi_p)}, \\ \mathrm{End}_{\mathbb{F}_p}(\overline{E}) &= \mathbb{Z} + c_p' O_{\mathbb{Q}(\pi_p)}, \\ c_p' &\mid c_p. \end{aligned}$$

REMARK 50. *Part (i) of Theorem* 49 *answers Question* 29*: the group $\overline{E}(\mathbb{F}_p)$ has cardinality $p + 1 - a_p$, with a_p growing more slowly than p. Note that this is reminiscent of the group $\mathbb{F}_p^\times$ having cardinality $p - 1$.*

The field $\mathbb{Q}(\pi_p)$, which we shall refer to as the **Frobenius field of E modulo p**, has special properties when the elliptic curve $E/\mathbb{Q}$ is with CM:

THEOREM 51. *(Frobenius Fields for Elliptic Curves with CM)*
For any elliptic curve $E/\mathbb{Q}$, with Weierstrass equation (10) *and such that $O := \operatorname{End}_{\overline{\mathbb{Q}}}(E) \not\simeq \mathbb{Z}$, and for any ordinary prime $p \nmid 2\Delta_E$, we have that*

$$\mathbb{Q}(\pi_p) \simeq O \otimes_{\mathbb{Z}} \mathbb{Q}.$$

PROOF. On one hand, we have the embedding of $\mathbb{Q}$-algebras

$$O \otimes_{\mathbb{Z}} \mathbb{Q} \subseteq \operatorname{End}_{\overline{\mathbb{F}}_p}(\overline{E}) \otimes_{\mathbb{Z}} \mathbb{Q}. \tag{26}$$

On the other hand, since p is an ordinary prime, we have the equalities

$$\operatorname{End}_{\mathbb{F}_p}(\overline{E}) \otimes_{\mathbb{Z}} \mathbb{Q} = \operatorname{End}_{\overline{\mathbb{F}}_p}(\overline{E}) \otimes_{\mathbb{Z}} \mathbb{Q} = \mathbb{Q}(\pi_p). \tag{27}$$

At the same time,

$$[O \otimes_{\mathbb{Z}} \mathbb{Q} : \mathbb{Q}] = [\mathbb{Q}(\pi_p) : \mathbb{Q}] = 2. \tag{28}$$

Putting together (26) - (28), the claim of the theorem follows. □

In connection to Theorem 49, we define

$$\Delta_p := \text{ the discriminant of the order } \operatorname{End}_{\mathbb{F}_p}\left(\overline{E}\right).$$

This is a negative integer satisfying $\Delta_p \equiv 0, 1 (\operatorname{mod} 4)$, for which we define the companion integer

$$\delta_p := \begin{cases} 0 & \text{if } \Delta_p \equiv 0 (\operatorname{mod} 4), \\ 1 & \text{if } \Delta_p \equiv 1 (\operatorname{mod} 4). \end{cases}$$

Additionally, we define the integer

$$b_p := \frac{c_p}{c_p'} \in \mathbb{N}\backslash\{0\},$$

which measures how far the $\mathbb{F}_p$-endomorphism of $\overline{E}$ is from the ring generated by the Frobenius:

$$\operatorname{End}_{\mathbb{F}_p}\left(\overline{E}\right)/\mathbb{Z}[\pi_p] \simeq \mathbb{Z}/b_p\mathbb{Z}.$$

Observing that, from part (vi) of Theorem 49, we have

$$\Delta_p = \left(c_p'\right)^2 \operatorname{disc} O_{\mathbb{Q}(\pi_p)} = \frac{c_p^2}{b_p^2} \operatorname{disc} O_{\mathbb{Q}(\pi_p)} = \frac{1}{b_p^2} \operatorname{disc} \mathbb{Z}[\pi_p],$$

and recalling that

$$\operatorname{disc} \mathbb{Z}[\pi_p] = \operatorname{disc} P_{E,p} = a_p^2 - 4p,$$

we obtain:

LEMMA 52. *For any elliptic curve $E/\mathbb{Q}$ with Weierstrass equation* (10) *and for any $p \nmid \Delta_E$, we have*

$$\Delta_p = \frac{a_p^2 - 4p}{b_p^2}.$$

The integers a_p, b_p, Δ_p encode significant arithmetic information about the curve $E/\mathbb{Q}$, as illustrated by the following result of Duke and Tóth:

THEOREM 53. *(Global Characterization of the Frobenius in Division Fields* [**DuTo**]*)*
For any elliptic curve $E/\mathbb{Q}$ with Weierstrass equation (10) *and for any $p \nmid \Delta_E$, we have that the integral matrix*

$$\begin{pmatrix} \frac{a_p+b_p\delta_p}{2} & b_p \\ \frac{b_p(\Delta_p-\delta_p)}{4} & \frac{a_p-b_p\delta_p}{2} \end{pmatrix},$$

when reduced modulo any integer m coprime to p, represents the class of the Artin symbol $\left(\frac{\mathbb{Q}(E[m])/\mathbb{Q}}{p}\right)$ in $\varphi_{E,m}(G_{\mathbb{Q}})$.

This characterization of the Frobenius is obtained as an application of the following celebrated result of Deuring:

THEOREM 54. *(Deuring's Lifting Lemma for $E/\mathbb{Q}$, 1941)*
For any elliptic curve $E/\mathbb{Q}$ with Weierstrass equation (10) *and for any $p \nmid \Delta_E$, there exist a finite field extension*

$$\mathbb{Q}_p \subseteq K$$

and a pair

$$\left(E^{lift}, \pi^{lift}\right),$$

where E^{lift} is an elliptic curve with a model over the ring of integers O_K of K and where π^{lift} is an endomorphism in $\mathrm{End}_K\left(E^{lift}\right)$, satisfying:

(i) $\mathrm{End}_{\overline{K}}\left(E^{lift}\right) \not\simeq \mathbb{Z}$;
(ii) $\overline{E} \simeq_{\overline{\mathbb{F}}_p} \overline{E}^{lift}$;
(iii) *under the isomorphism of (ii), π_p corresponds to $\pi^{lift} (\operatorname{mod} \mathfrak{p})$ for any prime $\mathfrak{p}$ of K, lying above p.*

Among the multitude of applications to this result let us also mention explicit formulas for the number of points on the reduction of a CM elliptic curve proved by Stark [**St**] and by Rubin & Silverberg [**RuSi09**].

As a corollary to Theorem 53, we obtain the answer to Question 30:

THEOREM 55. *(Group Structure of E mod p)*
For any elliptic curve $E/\mathbb{Q}$ with Weierstrass equation (10) *and for any $p \nmid \Delta_E$, there exist uniquely determined integers $d_{1,p}, d_{2,p} \geq 1$, possibly equal to 1, such that*

$$\begin{aligned} \overline{E}(\mathbb{F}_p) &\simeq \mathbb{Z}/d_{1,p}\mathbb{Z} \times \mathbb{Z}/d_{2,p}\mathbb{Z}, \\ d_{1,p} &\mid d_{2,p}. \end{aligned}$$

Moreover,

$$\begin{aligned} d_{1,p} &= \gcd\left(b_p, \frac{a_p+b_p\delta_p}{2} - 1\right), \\ d_{2,p} &= \frac{p+1-a_p}{d_{1,p}}. \end{aligned}$$

Combining this theorem with part (i) of Theorem 49, we obtain:

COROLLARY 56. *For any elliptic curve $E/\mathbb{Q}$ with Weierstrass equation* (10) *and for any $p \nmid \Delta_E$, we have*

$$1 \leq d_{1,p} \leq \sqrt{p+1-a_p} \leq \sqrt{p}+1$$

and

$$\sqrt{p} - 1 \leq d_{2,p} \leq p + 1 - a_p \leq \left(\sqrt{p} + 1\right)^2.$$

REMARK 57. *Theorem 55 answers Question 30: the abelian group $\overline{E}(\mathbb{F}_p)$ has two elementary divisors, $d_{1,p} \mid d_{2,p}$, and hence is isomorphic to the product of two finite cyclic groups $\mathbb{Z}/d_{1,p}\mathbb{Z} \times \mathbb{Z}/d_{2,p}\mathbb{Z}$. Note that this is reminiscent of the structure of the abelian group $\mathbb{F}_p^\times$, which has one elementary divisor, $p-1$, and is isomorphic to the finite cyclic group $\mathbb{Z}/(p-1)\mathbb{Z}$.*

Henceforth, we shall refer to the integers $d_{1,p}$, $d_{2,p}$ as the **elementary divisors of E modulo p**. Note that $d_{1,p}d_{2,p}$ is the order, while $d_{2,p}$ is the exponent of the group $\overline{E}(\mathbb{F}_p)$.

Observing that the cyclicity of the group $\overline{E}(\mathbb{F}_p)$ implies that the exponent $d_{2,p}$ of the group equals the order $p + 1 - a_p$, it is natural to seek sharper lower bounds for $d_{2,p}$ than those provided by Corollary 56. A first result in this direction was obtained by Schoof, once again as an application to Theorem 54:

THEOREM 58. *(Schoof's Exponent Growth Theorem* [**Scho**]*)*
For any elliptic curve $E/\mathbb{Q}$ defined by (10) and for any $p \nmid \Delta_E$, we have

$$(\sqrt{p} - 1)\,\frac{\sqrt{|\Delta_p|}}{2} \leq d_{2,p}. \tag{29}$$

Furthermore, assuming that $\mathrm{End}_{\overline{\mathbb{Q}}}(E) \simeq \mathbb{Z}$ and that the prime p is ordinary, we have

$$\sqrt{p}\,\frac{\log p}{(\log\log p)^2} \ll_E d_{2,p}. \tag{30}$$

Having answered both Questions 29 and 30, it is time to start our *investigations of the group $\overline{E}(\mathbb{F}_p)$ as a function of p.*

9. Cyclicity question: heuristics and upcoming challenges

In the preceding sections we reviewed results about primes and elliptic curves; in particular, in relation to our targeted exploration of the analogy between the groups $\mathbb{F}_p^\times$ and $\overline{E}(\mathbb{F}_p)$, where $E/\mathbb{Q}$ is a fixed elliptic curve, defined by a Weierstrass equation (10), and where $p \nmid \Delta_E$ is a prime, we highlighted that

$$\mathbb{F}_p^\times \simeq \mathbb{Z}/(p-1)\mathbb{Z}$$

and

$$\left|\mathbb{F}_p^\times\right| = p - 1,$$

while

$$\overline{E}(\mathbb{F}_p) \simeq \mathbb{Z}/d_{1,p}\mathbb{Z} \times \mathbb{Z}/d_{2,p}\mathbb{Z}$$

for uniquely determined integers $d_{1,p}, d_{2,p} \geq 1$ satisfying $d_{1,p} \mid d_{2,p}$, and

$$\left|\overline{E}(\mathbb{F}_p)\right| = p + 1 - a_p$$

for a uniquely determined integer a_p satisfying $|a_p| < 2\sqrt{p}$. We are now ready to explore Question 1 about how often the group $\overline{E}(\mathbb{F}_p)$ is cyclic, i.e. how often $d_{1,p} = 1$.

From Theorems 26 and 28, we know that there are natural restrictions to obtaining a cyclic group $\overline{E}(\mathbb{F}_p)$. Indeed, if $E(\mathbb{Q})_{\mathrm{tors}} \supseteq \mathbb{Z}/2\mathbb{Z} \times \mathbb{Z}/2\mathbb{Z}$, then $\overline{E}(\mathbb{F}_p)$ is not cyclic. In general, how can we characterize the primes for which $\overline{E}(\mathbb{F}_p)$ is cyclic?

Our starting point is the following arithmetic property:

LEMMA 59. *For any elliptic curve $E/\mathbb{Q}$ with Weierstrass equation* (10), *for any prime $p \nmid \Delta_E$, and for any integer m coprime to p, we have that*

$$\overline{E}(\mathbb{F}_p) \supseteq \mathbb{Z}/m\mathbb{Z} \times \mathbb{Z}/m\mathbb{Z} \Leftrightarrow \tfrac{\pi_p - 1}{m} \in \operatorname{End}_{\overline{\mathbb{F}}_p}\left(\overline{E}\right)$$
$$\Leftrightarrow p \text{ splits completely in } \mathbb{Q}(E[m]).$$

PROOF. The first equivalence is derived from the identification between the Frobenius endomorphism $\operatorname{Fr}_p \in \operatorname{End}_{\overline{F}_p}\left(\overline{E}\right)$ and the root π_p (see part (vi) of Theorem 49). For the second equivalence, relying on this identification and using $[m]$ to denote the multiplication by m map on $\overline{E}\left(\overline{\mathbb{F}}_p\right)$, we obtain

$$\begin{aligned}
\frac{\pi_p - 1}{m} \in \operatorname{End}_{\overline{\mathbb{F}}_p}\left(\overline{E}\right) \quad &\Leftrightarrow \quad \exists f \in \operatorname{End}_{\overline{\mathbb{F}}_p}\left(\overline{E}\right) \text{ such that } f \circ [m] = \operatorname{Fr}_p - \operatorname{id}_{\overline{E}(\overline{\mathbb{F}}_p)} \\
&\Leftrightarrow \quad \operatorname{Ker}[m] \leq \operatorname{Ker}\left(\operatorname{Fr}_p - \operatorname{id}_{\overline{E}(\overline{\mathbb{F}}_p)}\right) \\
&\Leftrightarrow \quad \operatorname{Fr}_p |_{\overline{E}[m]} = \operatorname{id}_{\overline{E}[m]} \\
&\Leftrightarrow \quad \left(\frac{\mathbb{Q}(E[m])/\mathbb{Q}}{p}\right) = \left\{\operatorname{id}_{\mathbb{Q}(E[m])}\right\},
\end{aligned}$$

which completes the proof. □

From this lemma, we derive a criterion for the cyclicity of the group $\overline{E}(\mathbb{F}_p)$:

COROLLARY 60. *(Cyclicity Criterion)*
For any elliptic curve $E/\mathbb{Q}$ with Weierstrass equation (10) *and for any prime $p \nmid \Delta_E$, we have that*

$$\overline{E}(\mathbb{F}_p) \text{ is cyclic} \Leftrightarrow p \text{ does not split completely in } \mathbb{Q}(E[\ell]) \text{ for any prime } \ell \neq p.$$

Now let us put this criterion to use. By the Chebotarev Density Theorem 4, the density of primes which split completely in $\mathbb{Q}(E[\ell])$ is $\frac{1}{[\mathbb{Q}(E[\ell]):\mathbb{Q}]}$. Thus, reasoning crudely, we expect that

$$\#\left\{p \leq x : p \nmid \Delta_E, \overline{E}(\mathbb{F}_p) \text{ is cyclic}\right\} \quad \approx \quad \prod_{\ell}\left(1 - \frac{1}{[\mathbb{Q}(E[\ell]) : \mathbb{Q}]}\right)\pi(x).$$

Here, the symbol "$\approx$" is used informally to mean "approximately equal to".

Note that, in order to write down the above infinite product, we rely on two assumptions:

- the product converges – which, recalling Proposition 38 of Section 5, can be proven;
- the events "p splits completely in $\mathbb{Q}(E[\ell])$" are independent as ℓ varies – which, recalling Lemma 40 of Section 6, never holds for an elliptic curve defined over $\mathbb{Q}$.

The latter assumption can be corrected by appealing to the integer m_E introduced in the Open Image Theorems 34 and 36 of Section 5. With this in mind, we propose:

CONJECTURE 61. *(Cyclicity Conjecture)*
For any elliptic curve $E/\mathbb{Q}$ with Weierstrass equation (10), *either we have that $\mathbb{Q}(E[2]) = \mathbb{Q}$, in which case*

$$\#\{p \leq x : p \nmid \Delta_E, \overline{E}(\mathbb{F}_p) \text{ is cyclic}\} \ll_E 1,$$

or we have that $\mathbb{Q}(E(2)) \neq \mathbb{Q}$, *in which case there exists a constant* $C_{cyclic}(E) > 0$ *such that*

$$\#\{p \leq x : p \nmid \Delta_E, \overline{E}(\mathbb{F}_p) \text{ is cyclic}\} \sim C_{cyclic}(E)\ \pi(x).$$

Moreover,

- *if* $\mathrm{End}_{\overline{\mathbb{Q}}}(E) \simeq \mathbb{Z}$, *then*

$$C_{cyclic}(E) := \frac{1 - \frac{\#\{g \in \mathrm{Gal}(\mathbb{Q}(E[m_E])/\mathbb{Q}):\ \gcd(\det g + 1 - \mathrm{tr}\, g,\ m_E) \neq 1\}}{[\mathbb{Q}(E[m_E]):\mathbb{Q}]}}{\prod_{\ell \mid m_E}\left(1 - \frac{1}{\ell}\right)} \times \prod_{\ell \nmid m_E}\left(1 - \frac{1}{(\ell-1)^3(\ell+1)}\right);$$

- *if* $\mathrm{End}_{\overline{\mathbb{Q}}}(E) \not\simeq \mathbb{Z}$, *then*

$$C_{cyclic}(E) := \frac{1 - \frac{\#\{g \in \mathrm{Gal}(\mathbb{Q}(E[\Delta m_E])/\mathbb{Q}):\ \gcd(\det g + 1 - \mathrm{tr}\, g,\ \Delta m_E) \neq 1\}}{[\mathbb{Q}(E[\Delta m_E]):\mathbb{Q}]}}{\prod_{\ell \mid \Delta m_E}\left(1 - \frac{1}{\ell}\right)} \times \prod_{\ell \nmid \Delta m_E}\left(1 - \chi(\ell)\frac{\ell^2 - \ell - 1}{(\ell-1)^2(\ell - \chi(\ell))}\right),$$

where Δ *is the discriminant and* χ *is the Kronecker character of the CM field* $\mathrm{End}_{\overline{\mathbb{Q}}}(E) \otimes_{\mathbb{Z}} \mathbb{Q}$ *of* E.

REMARK 62. *Using Corollaries* 35 *and* 37 *from Section* 5, *it can be proven that, for any elliptic curve* $E/\mathbb{Q}$, *the above constant satisfies*

$$C_{cyclic}(E) = \sum_{m \geq 1} \frac{\mu(m)}{[\mathbb{Q}(E[m]) : \mathbb{Q}]}.$$

We leave the proof to the reader as an exercise.

REMARK 63. *Early computations by Borosh, Moreno, and Porta* [**BoMoPo**] *using 6 elliptic curves* $E/\mathbb{Q}$ *and primes* $p < 5 \times 10^3$ *exhibit cyclic groups* $\overline{E}(\mathbb{F}_p)$. *Recent computations performed by undergraduate students Fitzpatrick, Insley, and Yilmaz under the guidance of the author* [**CoFiInYi**], *using over 350 Serre curves* $E/\mathbb{Q}$ *arising from the work of Daniels* [**Dan**] *and primes* $p < 10^6$, *strongly support the Cyclicity Conjecture; for these curves, more than* 80% *of the primes considered give rise to cyclic groups* $\overline{E}(\mathbb{F}_p)$.

In order to investigate the Cyclicity Conjecture 61, let us point out that the Cyclicity Criterion (Corolary 60) is reminiscent of the criterion upon which the heuristic for Artin's Primitive Root Conjecture 5 is based:

$$\mathbb{F}_p^\times = \langle a (\mathrm{mod}\, p)\rangle \ \Leftrightarrow\ p \text{ does not split completely in } \mathbb{Q}\left(\zeta_\ell, a^{\frac{1}{\ell}}\right) \text{ for any prime } \ell \neq p.$$

Moreover, the conjectural constant in the Cyclicity Conjecture 61 is reminiscent of the conjectural constant in Artin's Primitive Root Conjecture 5. Indeed, in the generic situation of a Serre curve, the infinite product over $\ell \nmid m_E$ occuring in $C_{\mathrm{cyclic}}(E)$ has factors $\left(1 - \frac{1}{[\mathbb{Q}(E[\ell]):\mathbb{Q}]}\right)$, while the infinite product over ℓ in C_{Artin} has factors $\left(1 - \frac{1}{\left[\mathbb{Q}\left(\zeta_\ell, a^{\frac{1}{\ell}}\right):\mathbb{Q}\right]}\right)$.

Given these similarities, we recall that Artin's Primitive Root Conjecture 5 was investigated as a *sieve problem* and proved under GRH by Hooley [**Ho**]. It is then natural to investigate the Cyclicity Conjecture 61 as a sieve problem, as follows:

- We are given
  - an elliptic curve $E/\mathbb{Q}$ defined by a Weierstrass equation (10);
  - a real number $x > 0$ (to be thought of as approaching ∞);
  - a parameter $z = z(x) > 0$ (to be thought of as growing with x);
  - the set $\mathcal{A} := \{p \le x : p \nmid \Delta_E\}$;
  - the set $\mathcal{A}_\ell := \{p \in \mathcal{A} : p \ne \ell, p \text{ splits completely in } \mathbb{Q}(E[\ell])\}$, for each prime $\ell < z$.
- We want to estimate the cardinality

$$\left|\mathcal{A}\backslash \bigcup_{\ell \le z} \mathcal{A}_\ell\right|.$$

- From the Inclusion-Exclusion Principle, we obtain

$$\left|\mathcal{A}\backslash \bigcup_{\ell \le z} \mathcal{A}_\ell\right| = \sum_{m \le m(x)} \mu(m)\,|\mathcal{A}_m|,$$

  where m is a positive, squarefree integer in a suitable range $[1, m(x)]$ defined by $z(x)$, and where $\mathcal{A}_m := \bigcap_{\ell | m} \mathcal{A}_\ell$.
- Rephrasing the above, we obtain

$$\begin{aligned}(31)\quad &\#\{p \le x : p \nmid \Delta_E, \overline{E}(\mathbb{F}_p) \text{ is cyclic}\}\\ &= \sum_{m \le m(x)} \mu(m)\, \#\{p \le x : p \nmid m\Delta_E, p \text{ splits completely in } \mathbb{Q}(E[m])\}.\end{aligned}$$

With this line of thought, observe that, by Lemma 60, a prime p splits completely in $\mathbb{Q}(E[m])$ if and only if $\overline{E}(\mathbb{F}_p)$ contains two copies of $\mathbb{Z}/m\mathbb{Z}$. Consequently, for such a prime p, we have $m^2 \mid p + 1 - a_p$. Recalling that $|a_p| < 2\sqrt{p}$, we deduce that $m < \sqrt{p} + 1$. Hence we may take

$$(32)\qquad m(x) := \sqrt{x} + 1.$$

Moreover, observe that, by the conditional Effective Chebotarev Density Theorem (part (i) of Theorem 16 of Section 1) and the properties of the division fields $\mathbb{Q}(E[m])$ (part (i) of Theorem 32 and Proposition 38 of Section 5), under GRH we obtain

$$\begin{aligned}(33)\qquad &\#\{p \le x : p \nmid m\Delta_E, p \text{ splits completely in } \mathbb{Q}(E[m])\}\\ &= \frac{1}{[\mathbb{Q}(E[m]) : \mathbb{Q}]}\pi(x) + \mathrm{O}_E\left(x^{\frac{1}{2}} \log(mx)\right).\end{aligned}$$

Combining these two observations, the emerging estimate of the accumulated error term is

$$\left|\#\{p \le x : p \nmid \Delta_E, \overline{E}(\mathbb{F}_p) \text{ is cyclic}\} - \sum_{m \le \sqrt{x}+1} \frac{1}{[\mathbb{Q}(E[m]) : \mathbb{Q}]}\pi(x)\right| \ll x \log x,$$

which is even bigger than the trivial bound x. In other words, the Chebotarev Density Theorem, even in its strongest form under GRH, is far from being sufficient to answer Question 1.

Not all hope is lost, however. A similar naive approach towards Artin's Primitive Root Conjecture leads to a similar obstacle. We devote Section 10 to outlining how to refine this approach and make it successful, and we devote Sections 11 and 12 to providing additional theoretical evidence in support of the Cyclicity Conjecture 61. Moreover, we devote Sections 13 and 14 to providing overviews of variations of Question 1, reminiscent of the Hardy-Littlewood Conjectures 6 and 7 of Section 1. Finally, we devote Section 15 to reviewing sample results about function field analogues of the Cyclicity Conjecture 61.

10. Cyclicity question: asymptotic

The heuristical reasoning towards the Cyclicity Conjecture 61, outlined in Section 9, can be morphed into a proof. This was achieved for the first time by Serre [**Se77**], under GRH, via a method inspired by Hooley's conditional proof of Artin's Primitive Root Conjecture 5. After Serre, Cojocaru and Murty obtained several new proofs, conditional or, in some cases, unconditional, and highlighted the growth of the emerging error terms as functions of x and E – see [**Mu83**], [**Co02**], [**Co03**], [**CoMu04**].

One insightful estimate occuring in the above four works, which enabled the authors to overcome the insufficiency of the Chebotarev Density Theorem in the approach discussed in Section 9, may be phrased as follows:

PROPOSITION 64. *For any elliptic curve $E/\mathbb{Q}$ with Weierstrass equation* (10), *and for any $x, y > 0$ with $y = y(x) \leq \sqrt{x} + 1$, growing with x, the following properties hold:*

(i) *under no additional assumptions,*

$$\sum_{m>y} \#\{p \leq x : p \nmid m\Delta_E, p \text{ splits completely in } \mathbb{Q}(E[m])\} \ll \frac{x^{\frac{3}{2}}}{y^2} + \frac{x}{y} + \sqrt{x}\log\frac{x}{y} + \sqrt{x};$$

(ii) *assuming* $\operatorname{End}_{\overline{\mathbb{Q}}}(E) \not\simeq \mathbb{Z}$,

$$\sum_{m>y} \#\{p \leq x : p \nmid m\Delta_E, p \text{ splits completely in } \mathbb{Q}(E[m])\} \ll \frac{x}{y} + \sqrt{x}\log x.$$

PROOF. A proof of part (i) appears in [**Co02**, pp. 343-344]; see also [**CoMu04**, p. 613]. A proof of part (ii) appears in [**Co03**, p. 2569]; see also [**CoMu04**, pp. 616-618]. We outline them below.

(i) Applying part (ii) of Theorem 32, part (ii) of Theorem 49, part (i) of Lemma 60, and (32), followed by elementary estimates, we obtain

$$\begin{aligned}
&\sum_{m>y} \#\{p \leq x : p \nmid m\Delta_E, p \text{ splits completely in } \mathbb{Q}(E[m])\} \\
= &\sum_{y<m\leq \sqrt{x}+1} \#\{p \leq x : p \nmid m\Delta_E, p \text{ splits completely in } \mathbb{Q}(E[m])\} \\
\leq &\sum_{y<m\leq \sqrt{x}+1} \#\{p \leq x : p \nmid m\Delta_E, p \equiv 1 (\operatorname{mod} m) \text{ and } p+1-a_p \equiv 0 (\operatorname{mod} m^2)\} \\
\leq &\sum_{\substack{a\in\mathbb{Z}\setminus\{2\}\\ |a|\leq 2\sqrt{x}}} \sum_{\substack{y<m\leq\sqrt{x}+1\\ m|a-2}} \sum_{\substack{p\leq x, p\nmid\Delta_E\\ a_p=a\\ m^2|p+1-a}} 1 + \sum_{y<m\leq\sqrt{x}+1} \sum_{\substack{p\leq x, p\nmid\Delta_E\\ a_p=2\\ m^2|p+1-a}} 1 \\
\ll &\sum_{y<m\leq\sqrt{x}+1} \left(\frac{x}{m^2}+1\right)\left(\frac{\sqrt{x}}{m}+1\right) + \sum_{y<m\leq\sqrt{x}+1} \left(\frac{x}{m^2}+1\right) \\
\ll &\frac{x^{\frac{3}{2}}}{y^2} + \frac{x}{y} + \sqrt{x}\log\frac{x}{y} + \sqrt{x}.
\end{aligned}$$

(ii) We split the primes p counted in the sum into supersingular and ordinary primes. Since we are seeking upper bounds and since x is large, we may assume that $p \geq 5$. Then, the supersingular primes p which split completely in $\mathbb{Q}(E[m])$ satisfy $a_p = 0$ (see part (v3) of Theorem 49) and, as above, $m^2 \mid p+1-a_p$ and $m \mid p-1$. We infer that, in this case, $m \mid 2$, which is a contradiction with $m > y$. In other words, there is no contribution from supersingular primes to our sum, implying that it suffices to analyze the sum

$$\sum_{m>y} \pi_1^o(x, \mathbb{Q}(E[m])/\mathbb{Q}),$$

where

$$\pi_1^o(x, \mathbb{Q}(E[m])/\mathbb{Q}) := \#\{p \leq x : p \nmid m\Delta_E, p \text{ ordinary}, p \text{ splits completely in } \mathbb{Q}(E[m])\}.$$

Since $\operatorname{End}_{\overline{\mathbb{Q}}}(E) \not\simeq \mathbb{Z}$, recalling Theorem 51 and Lemma 59, any ordinary prime p counted above satisifies

$$\frac{\pi_p - 1}{m} \in \operatorname{End}_{\overline{\mathbb{F}}_p}\left(\overline{E}\right) \leq O_{\mathbb{Q}(\pi_p)} = O_{\mathbb{Q}(\sqrt{-D})},$$

where, by recalling Theorem 33, we have $-D \in \{-1, -2, -3, -7, -11, -19, -43, -67, -163\}$. Upon calculating the norm $N_{\mathbb{Q}(\sqrt{-D})/\mathbb{Q}}(\pi_p)$, we infer that there exist $\alpha, \beta \in \mathbb{Z}$ such that

$$p = (\alpha m + 1)^2 + D\beta^2 m^2 \quad \text{if } -D \equiv 2, 3 (\operatorname{mod} 4),$$

$$p = \left(\frac{\alpha}{2} m + 1\right)^2 + D\frac{\beta^2}{4} m^2 \quad \text{if } -D \equiv 1 (\operatorname{mod} 4).$$

In either case, the number of such integers (not necessarily primes), which are less than or equal to x, is

$$\mathrm{O}\left(\frac{\sqrt{x}}{m\sqrt{D}}\left(\frac{2\sqrt{x}}{m} + 1\right)\right).$$

Consequently,

$$\begin{aligned}\sum_{m>y} \#\{p \leq x : p \nmid m\Delta_E, p \text{ splits completely in } \mathbb{Q}(E[m])\} &= \sum_{m>y} \pi_1^o(x, \mathbb{Q}(E[m])/\mathbb{Q}) \\ &\ll \sum_{m>y} \left(\frac{x}{m^2} + \frac{\sqrt{x}}{m}\right) \\ &\ll \frac{x}{y} + \sqrt{x}\log y \\ &\ll \frac{x}{y} + \sqrt{x}\log x.\end{aligned}$$

This completes the proof. □

Combining this proposition with Chebotarev arguments, we obtain a conditional "Chebotarev on average" theorem for the division fields of an elliptic curve:

THEOREM 65. *For any elliptic curve $E/\mathbb{Q}$ with Weierstrass equation* (10), *assuming GRH we have*

$$\frac{1}{\sqrt{x}} \sum_{m\geq 1} \left(\#\{p \leq x : p \nmid m\Delta_E, p \textit{ splits completely in } \mathbb{Q}(E[m])\} - \frac{1}{[\mathbb{Q}(E[m]):\mathbb{Q}]}\pi(x)\right) = r(E,x), \tag{34}$$

where:

(i) *under no additional assumptions,*

$$r(E,x) = \mathrm{O}_E\left(x^{\frac{1}{3}}(\log x)^{\frac{2}{3}}\right);$$

(ii) *assuming* $\mathrm{End}_{\overline{\mathbb{Q}}}(E) \not\simeq \mathbb{Z}$,

$$r(E,x) = \mathrm{O}_E\left(x^{\frac{1}{4}}(\log x)^{\frac{1}{2}}\right).$$

PROOF. We sketch the proofs, as follows.

(i) The main idea of proof is to make use of the average over m. Recalling (32), the maximal range of m in the sum under consideration is $1 \leq m \leq \sqrt{x}+1$. While not large (compare it with the range $1 \leq m \leq x$ for primes splitting completely in $\mathbb{Q}(\zeta_m)$), it exceeds what can be tackled by a direct application of the Effective Chebotarev Density Theorem, even under GRH (recall (33)). To overcome this obstacle, we choose a parameter $y = y(x) < \sqrt{x}+1$ and split the sum into two, according to whether $1 \leq m \leq y$ or $y < m \leq \sqrt{x}+1$. Over the first range, we apply (33):

$$\sum_{m\leq y} \#\{p \leq x : p \nmid m\Delta_E, p \text{ splits completely in } \mathbb{Q}(E[m])\} = \sum_{m\leq y} \frac{1}{[\mathbb{Q}(E[m]):\mathbb{Q}]}\pi(x) + \mathrm{O}_E\left(yx^{\frac{1}{2}}\log x\right); \tag{35}$$

note that this is the only place in the proof where we will be using GRH. Over the second range, we apply part (i) of Proposition 64:

$$\sum_{y<m\leq\sqrt{x}+1} \#\{p \leq x : p \nmid m\Delta_E, p \text{ splits completely in } \mathbb{Q}(E[m])\} \tag{36}$$
$$\ll \frac{x^{\frac{3}{2}}}{y^2} + \frac{x}{y} + \sqrt{x}\log\frac{x}{y} + \sqrt{x}.$$

By choosing $y \asymp \left(\frac{x}{\log x}\right)^{\frac{1}{3}}$, we obtain

$$\sum_{m\leq\sqrt{x}+1} \#\{p \leq x : p \nmid m\Delta_E, p \text{ splits completely in } \mathbb{Q}(E[m])\}$$
$$= \sum_{m\leq y} \frac{1}{[\mathbb{Q}(E[m]) : \mathbb{Q}]}\pi(x) + \mathrm{O}_E\left(x^{\frac{5}{6}}(\log x)^{\frac{2}{3}}\right).$$

Moreover, by Proposition 38, we obtain

$$\sum_{m>y} \frac{1}{[\mathbb{Q}(E[m]) : \mathbb{Q}]} \ll \sum_{m>y} \frac{\log\log m}{m^{\frac{4}{\gamma}}} \ll \frac{\log\log x}{y^{\frac{4}{\gamma}-1}}, \tag{37}$$

which completes the proof of part (i).

(ii) We proceed as above with the exception of using part (ii), instead of part (i), of Proposition 64, and of making the choice $y \asymp \left(\frac{x}{(\log x)^2}\right)^{\frac{1}{4}}$. □

With more effort, it is possible to relax, or even to remove, the GRH assumption in the above result, and to deduce:

THEOREM 66. *For any elliptic curve $E/\mathbb{Q}$ with Weierstrass equation* (10)*, we have*

$$\frac{1}{\sqrt{x}}\sum_{m\geq 1} \Big(\#\{p \leq x : p \nmid m\Delta_E, p \textit{ splits completely in } \mathbb{Q}(E[m])\} \tag{38}$$
$$- \frac{1}{[\mathbb{Q}(E[m]) : \mathbb{Q}]}\pi(x)\Big) = r(E, x),$$

where:

(i) *assuming a $\frac{3}{4}$-quasi GRH,*

$$r(E, x) = \mathrm{O}_E\left(\frac{x^{\frac{1}{2}}\log\log x}{(\log x)^2}\right);$$

(ii) *assuming $\mathrm{End}_{\overline{\mathbb{Q}}}(E) \not\simeq \mathbb{Z}$, for any $c > 0$,*

$$r(E, x) = \mathrm{O}_{E,c}\left(\frac{x^{\frac{1}{2}}}{(\log x)^c}\right);$$

PROOF. For part (i), proceed as in the proof of Theorem 1.2 of [**Co02**]. For part (ii), proceed as in the unconditional proof of the main theorem given in Section 6 of [**Mu83**] (see also the follow-up [**AkMu**]). □

Note that the proof of the above theorem relies on several results about primes in arithmetic progressions which were recalled in Section 2, such as the Siegel-Walfisz Theorem 9, the Brun-Titchmarsh Theorem 10, and a number field version of the Bombieri-Vinogradov Theorem 12.

REMARK 67. *To prove parts (i) and (ii) of Theorem* 66, *the sum over m is partitioned into two shorter sums, according to whether m is y-smooth or not, as in the classical simple asymptotic sieve. This is the approach used by Serre in* [**Se77**] *and originating in* [**Ho**]. *In contrast, to prove parts (i) and (ii) of Theorem* 65, *the sum over m is partitioned simply according to whether m is less than y or not. This is the approach used by Cojocaru and Murty in* [**CoMu04**]. *While it is based on a simple observation, it has two surprising consequences:*

- *significant improvements in the growth of the error terms;*
- *a departure from the approach on Artin's Primitive Root Conjecture, signaling a contrast between the classical conjecture and the Cyclicity Conjecture* 61.

We are now ready to present theoretical evidence towards the Cyclicity Conjecture 61:

THEOREM 68. *(Cojocaru-Murty-Serre Cyclicity Theorem* [**Co02**], [**CoMu04**], [**Mu83**], [**Se77**]*)*

For any elliptic curve $E/\mathbb{Q}$ with Weierstrass equation (10), *we have*

$$\#\{p \leq x : p \nmid \Delta_E, \overline{E}(\mathbb{F}_p) \text{ is cyclic}\} = \sum_{m \geq 1} \frac{\mu(m)}{[\mathbb{Q}(E[m]) : \mathbb{Q}]} \pi(x) + \mathrm{O}\left(\sqrt{x} \cdot r(E, x)\right),$$

where $r(E, x)$ is as in Theorems 65 *and* 66, *under the assumptions therein.*

PROOF. Starting from (31), we follow the approaches of Theorem 65 and 66, with the only difference that $\mu(m)$ is preserved as such in the sum of the main terms (i.e. over the ranges m y-smooth, $m \leq y$), while being bounded from above by 1 everywhere else. The original proof sources are: [**CoMu04**] under GRH; [**Co02**] under $\frac{3}{4}$-quasi GRH; [**Mu83**], and [**Co03**] for a simpler approach, with a less strong error term, unconditionally for $\mathrm{End}_{\overline{\mathbb{Q}}}(E) \not\simeq \mathbb{Z}$. See also [**Se77**] and [**AkMu**]. □

REMARK 69. *It can be proven that the constant $\sum_{m \geq 1} \frac{\mu(m)}{[\mathbb{Q}(E[m]):\mathbb{Q}]}$ is positive if and only if $\mathbb{Q}(E[2]) \neq \mathbb{Q}$; see* [**CoMu04**]. *Calculations related to this constant appear in* [**CoFiInYi**] *and* [**CoMu04**].

Methods similar to those used in Theorems 65, 66 and 68 can also be used to re-investigate the growth of the exponent $d_{2,p}$ of $\overline{E}(\mathbb{F}_p)$. Recall that, from Corollary 56 and Theorem 58, we already know that, for each prime p, the integer $d_{2,p}$ has a growth similar to that of $\sqrt{p}$. However, it turns out that, for a set of primes p of density 1, the integer $d_{2,p}$ has a growth which is more similar to that of $p + 1 - a_p$:

THEOREM 70. *(Duke's Large Exponent Theorem* [**Du03**]*)*

For any elliptic curve $E/\mathbb{Q}$ with Weierstrass equation (10) *and for any function $f : (0, \infty) \longrightarrow (0, \infty)$ such that $\lim_{x \to \infty} f(x) = \infty$, we have*

$$\#\left\{p \leq x : p \nmid \Delta_E, d_{2,p} \geq \frac{|\overline{E}(\mathbb{F}_p)|}{f(p)}\right\} \sim \pi(x), \tag{39}$$

provided any one of the following holds:

(i) $\operatorname{End}_{\overline{\mathbb{Q}}}(E) \simeq \mathbb{Z}$ *and* $f(x) \asymp x^{\frac{1}{4}}(\log x)^{\frac{1}{2}+\varepsilon} \quad \forall \varepsilon > 0$;
(ii) $\operatorname{End}_{\overline{\mathbb{Q}}}(E) \not\simeq \mathbb{Z}$ *and* $f(x) \asymp (\log x)^{1+\varepsilon} \quad \forall \varepsilon > 0$;
(iii) *GRH and* $f(x) \asymp (\log\log x)^{\frac{1}{3}+\varepsilon} \quad \forall \varepsilon > 0$.

PROOF. We will present a proof which uses Proposition 64 and highlights the intimate relation between the questions regarding the frequency with which $\overline{E}(\mathbb{F}_p)$ is cyclic and the frequency with which $\overline{E}(\mathbb{F}_p)$ has a large exponent. Recalling that $d_{1,p}d_{2,p} = |\overline{E}(\mathbb{F}_p)|$, we deduce that proving (39) is equivalent to proving

$$\#\{p \le x : f(p) < d_{1,p}\} = \mathrm{o}(\pi(x)). \tag{40}$$

To do this, choose a parameter $z = z(x) > 0$, which grows with x and which shall be specified later. Define

$$g(z(x)) := \inf\{f(p) : z < p < x\},$$

which also grows with x, i.e. $\lim_{x\to\infty} g(z(x)) = \infty$. Then

$$\begin{aligned}
\#\{p \le x : f(p) < d_{1,p}\} &= \#\{p \le z : f(p) < d_{1,p}\} + \#\{z < p \le x : f(p) < d_{1,p}\}\\
&\le \pi(z) + \sum_{g(z)\le m} \#\{p \le x : m | d_{1,p}\}\\
&< \frac{2z}{\log z} + \sum_{g(z)\le m\le\sqrt{x}+1} \#\{p \le x : p \text{ splits completely in } \mathbb{Q}(E[m])\}. \qquad (41)
\end{aligned}$$

(i) Using part (i) of Proposition 64 with $y = g(z)$, choosing $z \asymp \frac{x}{\log x}$, and recalling that $f(x) \asymp x^{\frac{1}{4}}(\log x)^{\frac{1}{2}+\varepsilon}$, we obtain (40).

(ii) Using part (ii) of Proposition 64 with $y = g(z)$, choosing $z \asymp \frac{x}{\log x}$, and recalling that $f(x) \asymp (\log x)^{1+\varepsilon}$, we obtain (40).

(iii) We assume GRH. To improve our results, we introduce a new parameter $y = y(x)$, which grows with x, satisfies $g(z) < y < \sqrt{x} + 1$, and shall be specified later. As in parts (i) and (ii), we choose $z \asymp \frac{x}{\log x}$. By part (ii) of the Effective Chebotarev Density Theorem (where GRH is used) and by part (i) of Proposition 64 and (37) (which are unconditional), we obtain

$$\begin{aligned}
&\sum_{g(z)\le m\le\sqrt{x}+1} \#\{p \le x : p \text{ splits completely in } \mathbb{Q}(E[m])\}\\
&= \sum_{g(z)\le m\le y} \frac{1}{[\mathbb{Q}(E[m]) : \mathbb{Q}]}\pi(x) + \mathrm{O}\left(yx^{\frac{1}{2}}\log x\right) + \mathrm{O}\left(\frac{x^{\frac{3}{2}}}{y^2} + \frac{x}{y} + \sqrt{x}\log x\right)\\
&= \mathrm{O}\left(\frac{\log\log x}{f(x)^3}\cdot\pi(x)\right) + \mathrm{O}\left(yx^{\frac{1}{2}}\log x\right) + \mathrm{O}\left(\frac{x^{\frac{3}{2}}}{y^2} + \frac{x}{y} + \sqrt{x}\log x\right).
\end{aligned}$$

Recalling that $f(x) \asymp (\log\log x)^{\frac{1}{3}+\varepsilon}$ and choosing $y \asymp \left(\frac{x}{\log x}\right)^{\frac{1}{2}}$, we obtain (40). □

We refer the reader to [**Du03**] for a formulation of Theorem 70 with fewer conditions on $f(x)$. We conclude this section by remarking that further applications of the methods we discussed have been pursued in several other works, including [**Ak**], [**AkGh**], [**AkFe**], [**FeMu**], [**FrKu**], [**FrPo**], [**Ki**], and [**Wu**].

11. Cyclicity question: lower bound

While Conjecture 61 is known unconditionally only for elliptic curves with CM, in general we have the following unconditional result:

THEOREM 71. *(Gupta-Murty Cyclicity Lower Bound* [**GuMu**]*)*
For any elliptic curve $E/\mathbb{Q}$ *with Weierstrass equation* (10), *assuming that* $\mathbb{Q}(E[2]) \neq \mathbb{Q}$, *we have*

$$\#\{p \leq x : p \nmid \Delta_E, \overline{E}(\mathbb{F}_p) \textit{ is cyclic}\} \gg \frac{x}{(\log x)^2}.$$

PROOF. The main idea of the proof is to capture, among the primes $p \leq x$ with $\overline{E}(\mathbb{F}_p)$ cyclic, a subset of primes in an arithmetic progression which contains at least $\frac{x}{(\log x)^2}$ primes.

To do this, recall from part (ii) of Lemma 60 that a prime p for which $\overline{E}(\mathbb{F}_p)$ is cyclic does not split completely in $\mathbb{Q}(E[2])$. By our hypothesis, the extension $\mathbb{Q} \subseteq \mathbb{Q}(E[2])$ is nontrivial, and since $\mathbb{Q} \subseteq \mathbb{Q}(\sqrt{\Delta_E}) \subseteq \mathbb{Q}(E[2])$, it contains a nontrivial *abelian* extension of $\mathbb{Q}$. Thus there is an arithmetic progression $\alpha(\operatorname{mod} q)$ such that

$$p \equiv \alpha(\operatorname{mod} q) \ \Rightarrow\ p \text{ does not split completely in } \mathbb{Q}(E[2]). \tag{42}$$

With this progression in hand, we remark that a lower bound sieve argument in the style of Fouvry and Iwaniec [**FoIw**] implies the existence of some $\varepsilon > 0$ such that the set

$$\mathcal{S}_\varepsilon(x) := \Big\{p \leq x : p \equiv \alpha(\operatorname{mod} q), \text{ all odd prime factors of } p-1 \text{ are distinct and greater than } x^{\frac{1}{4}+\varepsilon}\Big\}$$

satisfies

$$|\mathcal{S}_\varepsilon(x)| \gg \frac{x}{(\log x)^2}. \tag{43}$$

We now estimate the number of primes $p \in \mathcal{S}_\varepsilon(x)$ for which $\overline{E}(\mathbb{F}_p)$ is cyclic:

(44)

$$\begin{aligned}
&\#\{p \leq x : \overline{E}(\mathbb{F}_p) \text{ is cyclic}\}\\
&\geq \#\{p \leq x : p \text{ does not split completely in } \mathbb{Q}(E[\ell])\ \forall \ell \text{ and}\\
&\qquad \text{all odd prime factors of } p-1 \text{ are distinct and greater than } x^{\frac{1}{4}+\varepsilon}\Big\}\\
&\geq \#\{p \in \mathcal{S}_\varepsilon(x) : p \text{ does not split completely in } \mathbb{Q}(E[\ell])\ \forall \ell \text{ odd}\}\\
&= |\mathcal{S}_\varepsilon(x)| - \#\{p \in \mathcal{S}_\varepsilon(x) : p \text{ splits completely in } \mathbb{Q}(E[\ell]) \text{ for some } \ell \text{ odd}\}.
\end{aligned}$$

To estimate the latter from above, we partition the primes p according to their Frobenius trace a_p. Proceeding similarly to the proof of part (i) of Proposition 64, we obtain that

$$\begin{aligned}
&\#\{p \in \mathcal{S}_\varepsilon(x) : p \text{ splits completely in } \mathbb{Q}(E[\ell]) \text{ for some } \ell \text{ odd}\}\\
&\leq \sum_{\substack{a \in \mathbb{Z}\\ |a| \leq 2\sqrt{x}}} \sum_{3 \leq \ell \leq \sqrt{x}+1} \#\{p \in \mathcal{S}_\varepsilon(x) : a_p = a, p \text{ splits completely in } \mathbb{Q}(E[\ell])\}.
\end{aligned} \tag{45}$$

Note that the primes ℓ under summation satisfy $\ell^2 \mid p+1-a$ and $\ell \mid p-1$, hence $\ell \mid a-2$. Since $p \in \mathcal{S}_\varepsilon(x)$, we must have that $a \neq 2$ and, moreover, that ℓ is determined by a for large x. We denote this prime by ℓ_a.

The double sum in (45) is bounded by

$$\ll \sum_{|a|\le 2\sqrt{x}} \left(\frac{x}{\ell_a^2}+1\right) \ll x^{1-2\varepsilon}.\sqrt{x}.$$

Using this estimate in (44), together with (43), we deduce that

$$\#\{p \le x : \overline{E}(\mathbb{F}_p) \text{ is cyclic}\} \gg \frac{x}{(\log x)^2},$$

which completes the proof. □

12. Cyclicity question: average

To obtain further theoretical evidence for the Cyclicity Conjecture 61, we consider its average versions, that is, we investigate the frequency with which $\overline{E}(\mathbb{F}_p)$ is a cyclic group, when $E/\mathbb{Q}$ is an arbitrary elliptic curve in a family such as the ones introduced in Section 7. Using this perspective, Conjecture 61 is supported by the following result of Banks and Shparlinski:

THEOREM 72. *(Banks-Shparlinski Cyclicity on Average Theorem* [**BaSh**]*)*
For $A, B > 0$, consider the family $\mathcal{F}(A,B)$ of $\mathbb{Q}$-isomorphism classes of elliptic curves $E = E_{a,b}$ defined by (10) *with $a, b \in \mathbb{Z}$ and $|a| \le A$, $|b| \le B$. Then, for any $x > 0$, $\varepsilon > 0$, and $A = A(x), B = B(x)$ such that*

$$x^{\varepsilon} \le A, B \le x^{1+\varepsilon},$$
$$AB \ge x^{1+\varepsilon},$$

we have that, as $x \to \infty$,

$$\frac{1}{|\mathcal{F}(A,B)|} \sum_{E\in\mathcal{F}(A,B)} \#\{p \le x : p \nmid \Delta_E, \overline{E}(\mathbb{F}_p) \textit{ is cyclic}\} \sim C_{cyclic}^{average}\, \pi(x), \tag{46}$$

where

$$C_{cyclic}^{average} := \prod_{\ell}\left(1 - \frac{1}{\ell(\ell-1)(\ell^2-1)}\right) = 0.8137519061068094...$$

It is natural to consider the relationship between the average constant $C_{\text{cyclic}}^{\text{average}}$ and the individual constants $C_{\text{cyclic}}(E)$, especially in light of the similarity between them in the case of an elliptic curve $E/\mathbb{Q}$ without CM (see the definition of $C_{\text{cyclic}}(E)$ in Conjecture 61). In this direction, using arguments based on character sums, Jones ([**Jo09**, Prop. 15 p. 698]) proved that this similarity is even closer in the case of a Serre curve:

THEOREM 73. *(The Cyclicity Constant for a Serre Curve* [**Jo09**]*)*
For any Serre curve $E/\mathbb{Q}$, we have

$$C_{cyclic}(E) = \begin{cases} C_{cyclic}^{average}\left(1 + \frac{\mu(m_E)}{\prod_{\ell\mid m_E}(|\operatorname{GL}_2(\mathbb{Z}/\ell\mathbb{Z})|-1)}\right) & \textit{if } (\Delta_E)_{sf} \equiv 1 (\operatorname{mod} 4), \\ C_{cyclic}^{average} & \textit{otherwise}, \end{cases}$$

where $(\Delta_E)_{sf}$ denotes the squarefree part of the discriminant Δ_E of any Weierstrass model for E.

Moreover, Jones [**Jo09**] proved that the average constant $C_{\text{cyclic}}^{\text{average}}$ is indeed the average of all individual constants $C_{\text{cyclic}}(E)$:

THEOREM 74. *(Jones' Cyclicity Constant on Average Theorem* [**Jo09**]*)*
For $A, B > 0$, consider the family $\mathcal{F}(A, B)$ of $\mathbb{Q}$-isomorphism classes of elliptic curves $E = E_{a,b}$ defined by (10) with $a, b \in \mathbb{Z}$ and $|a| \leq A, |b| \leq B$. Then, for any $x > 0$ and $A = A(x) > 0$, $B = B(x) > 0$ such that

$$\lim_{x\to\infty} \frac{(\log A(x))^7 \cdot \log B(x)}{B(x)} = 0, \tag{47}$$

we have that, as $x \to \infty$,

$$\frac{1}{|\mathcal{F}(A, B)|} \sum_{E\in\mathcal{F}(A,B)} C_{cyclic}(E) \sim C_{cyclic}^{average}. \tag{48}$$

REMARK 75. *In view of Corollary 44 and Theorem 73, we see that, while the group $\overline{E}(\mathbb{F}_p)$ is not always cyclic, it is expected to be so for a majority of primes and for a majority of elliptic curves. Specifically, for the density 1 subset of Serre curves $E/\mathbb{Q}$ of an arbitrary two-parameter family $\mathcal{F}(A(x), B(x))$ satisfying (47), about 80% of the primes p lead to cyclic groups $\overline{E}(\mathbb{F}_p)$. This result indicates a stronger similarity between the groups $\overline{E}(\mathbb{F}_p)$ and $\mathbb{F}_p^\times$ than seen in Section 8.*

We devote the next two subsections to summaries of the proofs of Theorems 72 and 74.

12.1. Cyclicity: averaging the prime counting function. We will outline the proof of Theorem 72 using the ideas of [**BaSh**] and drawing inspiration from the presentations in [**BaCoDa**] and [**CoIwJo**].

For a prime p and a pair of integers (a, b), we define

$$w_p(a,b) := \begin{cases} 1 & \text{if } p \nmid \Delta_{a,b} \text{ and } \overline{E}_{a,b}(\mathbb{F}_p) \text{ is cyclic,} \\ 0 & \text{otherwise.} \end{cases}$$

With this, we define the bilinear forms

$$\mathcal{S}(A, B; x) := \sum_{|a|\leq A} \sum_{\substack{|b|\leq B \\ \Delta_{a,b}\neq 0}} \sum_{p\leq x} w_p(a,b), \tag{49}$$

$$\mathcal{S}^*(A, B; x) := \sum_{|a|\leq A} \sum_{\substack{|b|\leq B \\ \Delta_{a,b}\neq 0}} \sum_{\substack{p\leq x \\ p\nmid ab}} w_p(a,b), \tag{50}$$

which are related to each other via the inequality

$$|\mathcal{S}(A, B; x) - \mathcal{S}^*(A, B; x)| \leq (2A+1)(2B+1)\log(AB). \tag{51}$$

Our goal is to obtain an asymptotic formula for $\mathcal{S}^*(A, B; x)$.

We partition $\mathcal{F}(A, B)$ into subsets of curves according to their Weierstrass models modulo p. Note that, without any relevant loss, we may restrict the sum over $p \leq x$ to primes $5 \leq p \leq x$. The symbol "$*$" next to the sigma sums below signifies that we are only summing over invertible residue classes modulo p. We obtain

$$\begin{aligned}
\mathcal{S}^*(A,B;x) &= \sum_{5\le p\le x} \sideset{}{^*}\sum_{s(\operatorname{mod} p)} \sideset{}{^*}\sum_{t(\operatorname{mod} p)} \sideset{}{^*}\sum_{\substack{|a|\le A\\ a\equiv s(\operatorname{mod} p)}} \sideset{}{^*}\sum_{\substack{|b|\le B\\ b\equiv t(\operatorname{mod} p)\\ p\nmid\Delta_{a,b}}} w_p(a,b)\\
&= \sum_{5\le p\le x} \sideset{}{^*}\sum_{s(\operatorname{mod} p)} \sideset{}{^*}\sum_{t(\operatorname{mod} p)} w_p(s,t) \sideset{}{^*}\sum_{\substack{|a|\le A\\ a\equiv s(\operatorname{mod} p)}} \sideset{}{^*}\sum_{\substack{|b|\le B\\ b\equiv t(\operatorname{mod} p)\\ p\nmid\Delta_{a,b}}} 1\\
&=: \sum_{5\le p\le x} \sideset{}{^*}\sum_{s(\operatorname{mod} p)} \sideset{}{^*}\sum_{t(\operatorname{mod} p)} w_p(s,t)\ \gamma(s,t),
\end{aligned}$$

where the notation $\gamma(s,t)$ for the double sum over s and t was introduced for simplifying the exposition in the next step.

For each $p \ge 5$, we partition the set of Weierstrass models modulo p into $\mathbb{F}_p$-isomorphism classes. For this, recall (13) of Proposition 19 that, given pairs of residue classes $(s,t)(\operatorname{mod} p)$, $(s',t')(\operatorname{mod} p)$, the elliptic curves $E_{s,t}$, $E_{s',t'}$ are $\mathbb{F}_p$-isomorphic if and only if there exists $u(\operatorname{mod} p)$ invertible satisfying $s' \equiv su^4(\operatorname{mod} p)$ and $t' \equiv tu^6(\operatorname{mod} p)$. For ease of notation, we shall use $\widehat{(s,t)}$ for the coset of $(s,t)(\operatorname{mod} p)$ modulo this $\mathbb{F}_p$-isomorphism, and $\hat{u}$ for the coset of $u(\operatorname{mod} p)$ modulo multiplication by ± 1. By Theorem 24, for a fixed $p \ge 5$ we obtain:

$$\begin{aligned}
&\sideset{}{^*}\sum_{s(\operatorname{mod} p)} \sideset{}{^*}\sum_{t(\operatorname{mod} p)} w_p(s,t)\ \gamma(s,t)\\
&= \sum_{\substack{\widehat{(s,t)}\\ p\nmid\Delta(s,t)}} \sum_{\hat{u}} w_p(su^4,tu^6)\ \gamma(su^4,tu^6)\\
&= \sum_{\widehat{(s,t)}} w_p(s,t) \sum_{\hat{u}} \gamma(su^4,tu^6)\\
&= \sideset{}{^*}\sum_{s(\operatorname{mod} p)} \sideset{}{^*}\sum_{t(\operatorname{mod} p)} w_p(s,t)\ \frac{|\operatorname{Aut}(E_{s,t})|}{p-1} \sum_{\hat{u}} \gamma(su^4,tu^6)\\
&= \frac{1}{p-1} \sideset{}{^*}\sum_{s(\operatorname{mod} p)} \sideset{}{^*}\sum_{t(\operatorname{mod} p)} w_p(s,t) \sideset{}{^*}\sum_{u(\operatorname{mod} p)} \gamma(su^4,tu^6)\\
&= \frac{1}{p-1} \sideset{}{^*}\sum_{s(\operatorname{mod} p)} \sideset{}{^*}\sum_{t(\operatorname{mod} p)} w_p(s,t) \sideset{}{^*}\sum_{u(\operatorname{mod} p)} \left(\sideset{}{^*}\sum_{\substack{|a|\le A\\ a\equiv su^4(\operatorname{mod} p)}} 1 \right)\left(\sideset{}{^*}\sum_{\substack{|b|\le B\\ b\equiv tu^6(\operatorname{mod} p)}} 1 \right).
\end{aligned}$$

We use χ_1 and χ_2 to denote arbitrary Dirichlet characters modulo p, and χ_0 to denote the trivial character modulo p. Applying the orthogonality relations, we obtain:

$$\frac{1}{p-1}\sum_{s(\text{mod}\,p)}{}^{*}\sum_{t(\text{mod}\,p)}{}^{*}w_p(s,t)\sum_{u(\text{mod}\,p)}{}^{*}\left(\sum_{\substack{|a|\le A\\ a\equiv su^4(\text{mod}\,p)}}{}^{*}1\right)\left(\sum_{\substack{|b|\le B\\ b\equiv tu^6(\text{mod}\,p)}}{}^{*}1\right)$$

$$=\frac{1}{(p-1)^3}\sum_{s(\text{mod}\,p)}{}^{*}\sum_{t(\text{mod}\,p)}{}^{*}w_p(s,t)\sum_{u(\text{mod}\,p)}{}^{*}\left(\sum_{\chi_1}\overline{\chi}_1(su^4)\sum_{|a|\le A}\chi_1(a)\right)$$
$$\times\left(\sum_{\chi_2}\overline{\chi}_2(tu^6)\sum_{|b|\le B}\chi_2(b)\right)$$

$$=\frac{1}{(p-1)^3}\sum_{s(\text{mod}\,p)}{}^{*}\sum_{t(\text{mod}\,p)}{}^{*}w_p(s,t)\sum_{\chi_1}\overline{\chi}_1(s)\sum_{\chi_2}\overline{\chi}_2(t)\left(\sum_{u(\text{mod}\,p)}{}^{*}\overline{\chi}_1^4\overline{\chi}_2^6(u)\right)$$
$$\times\left(\sum_{|a|\le A}\chi_1(a)\right)\left(\sum_{|b|\le B}\chi_2(b)\right)$$

$$=\frac{1}{(p-1)^2}\sum_{s(\text{mod}\,p)}{}^{*}\sum_{t(\text{mod}\,p)}{}^{*}w_p(s,t)\sum_{\chi_1}\overline{\chi}_1(s)\sum_{\substack{\chi_2\\ \chi_1^4\chi_2^6=\chi_0}}\overline{\chi}_2(t)$$
$$\times\left(\sum_{|a|\le A}\chi_1(a)\right)\left(\sum_{|b|\le B}\chi_2(b)\right)$$

$$=\frac{1}{(p-1)^2}\sum_{\chi_1}\sum_{\substack{\chi_2\\ \chi_1^4\chi_2^6=\chi_0}}\left(\sum_{s(\text{mod}\,p)}{}^{*}\sum_{t(\text{mod}\,p)}{}^{*}w_p(s,t)\,\overline{\chi}_1(s)\overline{\chi}_2(t)\right)\left(\sum_{|a|\le A}\chi_1(a)\right)\left(\sum_{|b|\le B}\chi_2(b)\right).$$

We partition the character sum above into smaller character sums, according to whether: $\chi_1=\chi_2=\chi_0$; $\chi_1\neq\chi_0$, $\chi_2=\chi_0$; $\chi_1=\chi_0$, $\chi_2\neq\chi_0$; $\chi_1\neq\chi_0$, $\chi_2\neq\chi_0$. More precisely, we write

$$\mathcal{S}^*(A,B;x)=$$

$$\sum_{5\le p\le x}\frac{1}{(p-1)^2}\left(\sum_{s(\text{mod}\,p)}{}^{*}\sum_{t(\text{mod}\,p)}{}^{*}w_p(s,t)\right)\left(\sum_{\substack{|a|\le A\\ p\nmid a}}1\right)\left(\sum_{\substack{|b|\le B\\ p\nmid b}}1\right)$$

$$+\sum_{5\le p\le x}\frac{1}{(p-1)^2}\sum_{\substack{\chi\neq\chi_0\\ \chi^4=\chi_0}}\left(\sum_{s(\text{mod}\,p)}{}^{*}\sum_{t(\text{mod}\,p)}{}^{*}w_p(s,t)\,\overline{\chi}(s)\right)\left(\sum_{|a|\le A}\chi(a)\right)\left(\sum_{\substack{|b|\le B\\ p\nmid b}}1\right)$$

$$+\sum_{5\le p\le x}\frac{1}{(p-1)^2}\sum_{\substack{\chi\neq\chi_o\\ \chi^6=\chi_0}}\left(\sum_{s(\text{mod}\,p)}{}^{*}\sum_{t(\text{mod}\,p)}{}^{*}w_p(s,t)\,\overline{\chi}(t)\right)\left(\sum_{\substack{|a|\le A\\ p\nmid a}}1\right)\left(\sum_{|b|\le B}\chi(b)\right)$$

$$+\sum_{5\le p\le x}\frac{1}{(p-1)^2}\sum_{\chi_1\neq\chi_0}\sum_{\substack{\chi_2\neq\chi_0\\ \chi_1^4\chi_2^6=\chi_0}}\left(\sum_{s(\text{mod}\,p)}{}^{*}\sum_{t(\text{mod}\,p)}{}^{*}w_p(s,t)\,\overline{\chi}_1(s)\overline{\chi}_2(t)\right)$$

$$\times \left(\sum_{|a|\le A} \chi_1(a)\right)\left(\sum_{|b|\le B} \chi_2(b)\right)$$

and we denote each of these sums by

$$\mathcal{S}_0(A,B;x),\ \mathcal{S}_4(A,B;x),\ \mathcal{S}_6(A,B;x), \text{ and } \mathcal{S}_\infty(A,B;x),$$

respectively. The main term in the asymptotic growth of $\mathcal{S}^*(A,B;x)$ is encoded in $\mathcal{S}_0(A,B;x)$.

Let us first focus on $\mathcal{S}_4(A,B;x)$ and $\mathcal{S}_6(A,B;x)$. By trivially estimating $|w_p(s,t)|$ and $|\chi(s)|$, $|\chi(t)|$, we obtain

$$\begin{aligned}
\mathcal{S}_4(A,B;x) &\le \sum_{5\le p\le x}\sum_{\substack{\chi\ne\chi_0\\ \chi^4=\chi_0}}\left|\sum_{|a|\le A}\chi(a)\right|\left(\sum_{\substack{|b|\le B\\ p\nmid b}} 1\right),\\
\mathcal{S}_6(A,B;x) &\le \sum_{5\le p\le x}\sum_{\substack{\chi\ne\chi_0\\ \chi^6=\chi_0}}\left(\sum_{\substack{|a|\le A\\ p\nmid a}} 1\right)\left|\sum_{|b|\le B}\chi(b)\right|.
\end{aligned}$$

This leads to estimating sums of the form

$$\mathcal{S}(A,x) := \sum_{p\le x}\sum_{\chi\ne\chi_0}\left|\sum_{|a|\le A}\chi(a)\right| \quad \text{and} \quad \mathcal{S}^{(m)}(A,x) := \sum_{p\le x}\sum_{\substack{\chi\ne\chi_0\\ \operatorname{ord}\chi=m}}\left|\sum_{|a|\le A}\chi(a)\right|$$

for $m\in\{4,6\}$.

We proceed as in the proof of [**BaCoDa**, Lemma 6], namely, we first note that there are at most 4 (respectively at most 6) characters satisfying $\chi^4=\chi_0$ (respectively $\chi^6=\chi_0$) and we then use Hölder's Inequality for an arbitrary positive integer k:

$$\begin{aligned}
\mathcal{S}^{(m)}(A,x) &\le 2\sum_{p\le x}\sum_{\substack{\chi\ne\chi_0\\ \chi^m=\chi_0}}\left|\sum_{a\le A}\chi(a)\right|\\
&\le 2\left(\sum_{p\le x}\sum_{\substack{\chi\ne\chi_0\\ \chi^m=\chi_0}} 1\right)^{1-\frac{1}{2k}}\left(\sum_{p\le x}\sum_{\chi\ne\chi_0}\left|\sum_{a\le A}\chi(a)\right|^{2k}\right)^{\frac{1}{2k}}\\
&\le 2\,\pi(x)^{1-\frac{1}{2k}}\left(\sum_{p\le x}\sum_{\chi\ne\chi_0}\left|\sum_{a\le A}\chi(a)\right|^{2k}\right)^{\frac{1}{2k}}\\
&= 2\,\pi(x)^{1-\frac{1}{2k}}\left(\sum_{p\le x}\sum_{\chi\ne\chi_0}\left|\sum_{a\le A^k}\tau_k(a)\chi(a)\right|^{2}\right)^{\frac{1}{2k}}\\
&\ll_k \left(\frac{x}{\log x}\right)^{1-\frac{1}{2k}}\left(\left(x^2+A^k\right)A^k\left(\log\left(A^k\right)\right)^{k^2-1}\right)^{\frac{1}{2k}}
\end{aligned}$$

$$\ll_k \sqrt{A}\,\frac{x^{1-\frac{1}{2k}}}{(\log x)^{1-\frac{k}{2}}}\left(x^{\frac{1}{k}}+\sqrt{A}\right),$$

where in the fourth line above, $\tau_k(a)$ is the number of ways of writing a as the product of k positive integers at most A, and where, in the fifth line above, we used Chebysheff's Theorem and the Large Sieve Inequality (see Section 1).

In summary, we proved that

$$\mathcal{S}_4(A,B;x)+\mathcal{S}_6(A,B;x)\ll_k \frac{x^{1-\frac{1}{2k}}}{(\log x)^{1-\frac{k}{2}}}\left(x^{\frac{1}{k}}\left(\sqrt{A}+\sqrt{B}\right)+(A+B)\right)\quad \forall k\geq 1.$$

Next, let us focus on $\mathcal{S}_\infty(A,B;x)$. Again, we proceed as in the proof of [**BaCoDa**, Lemma 6], as follows. A double application of the Cauchy-Schwarz Inequality gives

(52)

$$\begin{aligned}
\mathcal{S}_\infty(A,B;x) \leq & \left(\sum_{5\leq p\leq x}\sum_{\chi_1\neq\chi_0}\sum_{\substack{\chi_2\neq\chi_0\\ \chi_1^4\chi_2^6=\chi_0}}\frac{1}{p(p-1)^3}\left|\sideset{}{^*}\sum_{s(\operatorname{mod} p)}\sideset{}{^*}\sum_{t(\operatorname{mod} p)} w_p(s,t)\,\overline{\chi}_1(s)\overline{\chi}_2(t)\right|^2\right)^{\frac{1}{2}}\\
&\times\left(\sum_{5\leq p\leq x}\sum_{\chi_1\neq\chi_0}\sum_{\substack{\chi_2\neq\chi_0\\ \chi_1^4\chi_2^6=\chi_0}}\frac{p}{\phi(p)}\left|\sum_{\substack{|a|\leq A\\ p\nmid a}}\chi_1(a)\right|^2\left|\sum_{\substack{|b|\leq B\\ p\nmid b}}\chi_2(b)\right|^2\right)^{\frac{1}{2}}\\
\ll & \left(\sum_{5\leq p\leq x}\sum_{\chi_1\neq\chi_0}\sum_{\substack{\chi_2\neq\chi_0\\ \chi_1^4\chi_2^6=\chi_0}}\frac{1}{p(p-1)^3}\left|\sideset{}{^*}\sum_{s(\operatorname{mod} p)}\sideset{}{^*}\sum_{t(\operatorname{mod} p)} w_p(s,t)\,\overline{\chi}_1(s)\overline{\chi}_2(t)\right|^2\right)^{\frac{1}{2}}\\
&\times\left(\sum_{5\leq p\leq x}\sum_{\chi\neq\chi_0}\frac{p}{\phi(p)}\left|\sum_{\substack{|a|\leq A\\ p\nmid a}}\chi(a)\right|^4\right)^{\frac{1}{4}}\\
&\times\left(\sum_{5\leq p\leq x}\sum_{\chi\neq\chi_0}\frac{p}{\phi(p)}\left|\sum_{\substack{|b|\leq B\\ p\nmid b}}\chi(b)\right|^4\right)^{\frac{1}{4}},
\end{aligned}$$

where, when writing the second and third factors above, we used, as earlier, that, for a fixed p and for a Dirichlet character χ_1 (respectively χ_2) modulo p, there exist at most 4 characters χ_1 and at most 6 characters χ_2 such that $\chi_1^4\chi_2^6=\chi_0$.

To estimate the first factor, we complete the sums over χ_1, χ_2 to sums over all characters mod p and we use the orthogonality relations:

$$\sum_{\chi_1\neq\chi_0} \sum_{\substack{\chi_2\neq\chi_0 \\ \chi_1^4\chi_2^6=\chi_0}} \left| \sum_{s (\operatorname{mod} p)}{}^{*} \sum_{t (\operatorname{mod} p)}{}^{*} w_p(s,t)\, \overline{\chi}_1(s)\overline{\chi}_2(t) \right|^2$$
$$\leq \sum_{\chi_1}\sum_{\chi_2} \sum_{s (\operatorname{mod} p)}{}^{*} \sum_{t (\operatorname{mod} p)}{}^{*} w_p(s,t)\, \overline{\chi}_1(s)\overline{\chi}_2(t) \sum_{s' (\operatorname{mod} p)}{}^{*} \sum_{t' (\operatorname{mod} p)}{}^{*} w_p(s',t')\, \chi_1(s')\chi_2(t')$$
$$= \sum_{s (\operatorname{mod} p)}{}^{*} \sum_{t (\operatorname{mod} p)}{}^{*} \sum_{s' (\operatorname{mod} p)}{}^{*} \sum_{t' (\operatorname{mod} p)}{}^{*} w_p(s,t)\, w_p(s',t') \sum_{\chi_1} \chi_1(s^{-1}s') \sum_{\chi_2} \chi_2(t^{-1}t')$$
$$= (p-1)^2 \sum_{s (\operatorname{mod} p)}{}^{*} \sum_{t (\operatorname{mod} p)}{}^{*} |w_p(s,t)|^2$$
$$\leq (p-1)^4.$$

Then, summing over p and using Chebysheff's Theorem, we deduce that

$$\left(\sum_{5\leq p\leq x} \sum_{\chi_1\neq\chi_0} \sum_{\substack{\chi_2\neq\chi_0 \\ \chi_1^4\chi_2^6=\chi_0}} \frac{1}{p(p-1)^3} \left| \sum_{s (\operatorname{mod} p)}{}^{*} \sum_{t (\operatorname{mod} p)}{}^{*} w_p(s,t)\, \overline{\chi}_1(s)\overline{\chi}_2(t) \right|^2 \right)^{\frac{1}{2}} \ll \frac{x^{\frac{1}{2}}}{(\log x)^{\frac{1}{2}}}. \tag{53}$$

To estimate the second and third factors in (52), we expand out the squares, use the Large Sieve Inequality, and obtain

$$\sum_{5\leq p\leq x} \sum_{\chi_1\neq\chi_0} \sum_{\substack{\chi_2\neq\chi_0 \\ \chi_1^4\chi_2^6=\chi_0}} \frac{p}{\phi(p)} \left(\sum_{\substack{|a|\leq A \\ p\nmid b}} \chi_1(a)^4 \right)^4 \ll \left(x^2+A^2\right) A^2 \log A, \tag{54}$$

$$\sum_{5\leq p\leq x} \sum_{\chi_1\neq\chi_0} \sum_{\substack{\chi_2\neq\chi_0 \\ \chi_1^4\chi_2^6=\chi_0}} \frac{p}{\phi(p)} \left(\sum_{\substack{1\leq b\leq B \\ p\nmid b}} \chi_2(b)^4 \right)^4 \ll \left(x^2+B^2\right) B^2 \log B. \tag{55}$$

By putting together (52) - (55), we now obtain

$$\mathcal{S}_\infty(A,B;x) \ll x^{\frac{1}{2}} \left(x^2+A^2\right)^{\frac{1}{4}} \left(x^2+B^2\right)^{\frac{1}{4}} \sqrt{AB}. \tag{56}$$

Finally, let us analyze $\mathcal{S}_0(A,B;x)$. This requires an estimate of the number of elliptic curves over a fixed finite field $\mathbb{F}_p$ which have a cyclic group of $\mathbb{F}_p$-rational points. Such an estimate was obtained by Vlăduţ [**Vl**] as an application of geometric results of Howe [**Ho**, p. 245] (see also [**Ge06**] and [**Ge08**] for closely related results) and may be stated as follows:

THEOREM 76. ([**Vl**, Lemma 6.1, pp. 22–23])
For any prime $p \geq 5$, we have

$$\left| \sum_{s(\mathrm{mod}\, p)}{}^{*} \sum_{t(\mathrm{mod}\, p)}{}^{*} w_p(s,t) - \prod_{\ell | p-1} \left(1 - \frac{1}{\ell(\ell^2-1)}\right) \cdot (p-1)^2 \right| \leq p^{\frac{3}{2}+\mathrm{o}(1)}. \tag{57}$$

Using this result, together with Proposition 14 from Section 1, we obtain

$$\begin{aligned} \mathcal{S}_0(A,B;x) = |\mathcal{F}(A,B)| \prod_{\ell} \left(1 - \frac{1}{\ell(\ell-1)^2(\ell+1)}\right) \cdot \pi(x) \\ + \mathrm{O}\left(\frac{x}{(\log x)^k} AB\right) \quad \forall k \geq 1. \end{aligned}$$

It is now time to put everything together:

$$\begin{aligned} \mathcal{S}(A,B;x) &= |\mathcal{F}(A,B)| \prod_{\ell} \left(1 - \frac{1}{\ell(\ell-1)^2(\ell+1)}\right) \cdot \pi(x) \\ &+ \mathrm{O}_k\left(\frac{x}{(\log x)^k} AB\right) \\ &+ \mathrm{O}_k\left(\frac{x^{1-\frac{1}{2k}}}{(\log x)^{1-\frac{k}{2}}} \left(x^{\frac{1}{k}}\left(\sqrt{A}+\sqrt{B}\right) + (A+B)\right)\right) \\ &+ \mathrm{O}\left(x^{\frac{1}{2}}\left(x^2+A^2\right)^{\frac{1}{4}}\left(x^2+B^2\right)^{\frac{1}{4}}\sqrt{AB}\right), \end{aligned}$$

where $k \geq 1$ is arbitrary. Recalling that $|\mathcal{F}(A,B)| \asymp AB$ (see (20) in Section 7) and recalling that A, B are chosen such that $x^{\varepsilon} \leq A, B \leq x^{1+\varepsilon}$, $AB \geq x^{1+\varepsilon}$, we see that the above implies the asymptotic formula (46) claimed in Theorem 72.

12.2. Cyclicity: averaging the individual constants. We outline the proof of a more general version of Theorem 74, following [**Jo09**]. Precisely, for an arbitrary integer $k \geq 1$, we estimate, from above, the k-th moment

$$\frac{1}{|\mathcal{F}(A,B)|} \sum_{E \in \mathcal{F}(A,B)} \left| C_{\mathrm{cyclic}}(E) - C_{\mathrm{cyclic}}^{\mathrm{average}} \right|^k, \tag{58}$$

by distinguishing between elliptic curves with CM, elliptic curves without CM and which are not Serre curves, and elliptic curves which are Serre curves in $\mathcal{F}(A,B)$; the contribution coming from Serre curves is shown to dominate.

Let us observe that, for any elliptic curve $E/\mathbb{Q}$, we have

$$C_{\mathrm{cyclic}}(E) = \sum_{m \geq 1} \frac{\mu(m)}{[\mathbb{Q}(E[m]) : \mathbb{Q}]} \leq 1.$$

Thus, for any subset $\mathcal{F} \subseteq \mathcal{F}(A,B)$, we have

$$\frac{1}{|\mathcal{F}(A,B)|} \sum_{E \in \mathcal{F}} \left| C_{\mathrm{cyclic}}(E) - C_{\mathrm{cyclic}}^{\mathrm{average}} \right|^k \ll \frac{|\mathcal{F}|}{|\mathcal{F}(A,B)|}.$$

In particular, recalling Theorems 42 and 43 of Section 7, we obtain that

$$\frac{1}{|\mathcal{F}(A,B)|} \sum_{E \in \mathcal{F}_{\mathrm{CM}}(A,B)} \left| C_{\mathrm{cyclic}}(E) - C_{\mathrm{cyclic}}^{\mathrm{average}} \right|^k \ll \frac{1}{A} + \frac{1}{B}, \tag{59}$$

(60)

$$\frac{1}{|\mathcal{F}(A,B)|}\sum_{E\in\mathcal{F}_{\text{nonCM, nonSerre}}(A,B)}\left|C_{\text{cyclic}}(E)-C_{\text{cyclic}}^{\text{average}}\right|^k \ll \frac{(\log\min\{A,B\})^c}{\sqrt{\min\{A,B\}}}.$$

We will now focus on estimating the difference of the constants as we average over the Serre curves in the family. In this case, upon applying Theorems 73, followed by Proposition 41, we obtain

(61)

$$\begin{aligned}
&\frac{1}{|\mathcal{F}(A,B)|}\sum_{E\in\mathcal{F}_{\text{Serre}}(A,B)}\left|C_{\text{cyclic}}(E)-C_{\text{cyclic}}^{\text{average}}\right|^k\\
\leq\ &\frac{1}{|\mathcal{F}(A,B)|}\sum_{\substack{E\in\mathcal{F}_{\text{Serre}}(A,B)\\ (\Delta_E)_{\text{sf}}\equiv 1 (\text{mod } 4)}}\frac{\left|C_{\text{cyclic}}^{\text{average}}\right|^k}{\prod_{\ell|2(\Delta_E)_{\text{sf}}}\left(|\mathrm{GL}_2(\mathbb{Z}/\ell\mathbb{Z})|-1\right)^k}\\
\ll_k\ &\frac{1}{AB}\sum_{\substack{|a|\leq A,|b|\leq B\\ 4a^3+27b^2\neq 0}}\frac{1}{\prod_{\ell|2(4a^3+27b^2)_{\text{sf}}}\left(|\mathrm{GL}_2(\mathbb{Z}/\ell\mathbb{Z})|-1\right)^k}\\
\ll_k\ &\frac{1}{AB}\sum_{\substack{|a|\leq A,|b|\leq B\\ 4a^3+27b^2\neq 0}}\frac{1}{|(4a^3+27b^2)_{\text{sf}}|^k},
\end{aligned}$$

where $(4a^3+27b^2)_{\text{sf}}$ denotes the squarefree part of $4a^3+27b^2$.

By counting ideals of bounded norm in various quadratic fields, we can prove:

LEMMA 77. [**Jo09**, Lemma 22, pp. 705–708]
For any sufficiently large $A, B, z > 0$,

$$\begin{aligned}
&\#\left\{(a,b)\in\mathbb{Z}\times\mathbb{Z} : |a|\leq A, |b|\leq B, 4a^3+27b^2\neq 0, \left|(4a^3+27b^2)_{sf}\right|\leq z\right\}\\
&\qquad\ll z\,A(\log A)^7(\log B)+B.
\end{aligned}$$

Then, upon fixing a parameter $z=z(x)$, to be defined later, and upon partitioning the curves $E_{a,b}\in\mathcal{F}_{\text{Serre}}(A,B)$ according to whether $\left|(4a^3+27b^2)_{\text{sf}}\right|$ is less, or greater, than z, we obtain:

$$\begin{aligned}
\frac{1}{AB}\sum_{|a|\leq A}\sum_{\substack{|b|\leq B\\ \Delta_{a,b}\neq 0}}\frac{1}{|(4a^3+27b^2)_{\text{sf}}|^k}\leq\ &\frac{1}{AB}\sum_{|a|\leq A}\sum_{\substack{|b|\leq B\\ \Delta_{a,b}\neq 0\\ (\Delta_{a,b})_{\text{sf}}\leq z}}\frac{1}{|(4a^3+27b^2)_{\text{sf}}|^k}\\
&+\frac{1}{AB}\sum_{|a|\leq A}\sum_{\substack{|b|\leq B\\ (\Delta_{a,b})_{\text{sf}}>z}}\frac{1}{z^k}\\
&\ll z\,A(\log A)^7(\log B)+B+\frac{1}{z^k}.
\end{aligned}$$

Choosing

$$z\asymp\left(\frac{B}{(\log A)^7\,(\log B)}\right)^{\frac{1}{k+1}},$$

we deduce that

(62) $$\frac{1}{|\mathcal{F}(A,B)|} \sum_{E \in \mathcal{F}_{\text{Serre}}(A,B)} \left| C_{\text{cyclic}}(E) - C_{\text{cyclic}}^{\text{average}} \right|^k \ll_k \left(\frac{(\log A)^7 (\log B)}{B} \right)^{\frac{k}{k+1}}.$$

The bounds (59), (60), and (62), put together, lead to an upper bound for the k-th moment (58), and then to a complete proof of Theorem 74.

13. Primality of $p+1-a_p$

We will now consider further facets of Question 1. For example, observing that any finite group of prime order is cyclic, we ask:

QUESTION 78. Given an elliptic curve $E/\mathbb{Q}$, how often is the order of the group $\overline{E}(\mathbb{F}_p)$ a prime?

To investigate this question, we note that, by (24) and part (i) of Theorem 49, we are asking for the frequency with which both p and $p+1-a_p$ are prime, the integer a_p growing more slowly than p itself. This is reminiscent of the Twin Prime Conjecture 6, whose heuristic is based on the Cramér model that a random positive, non-unit integer n is a prime with probability $\frac{1}{\log n}$, and on the simple observation that an integer is a prime if and only if it is not divisible by any smaller prime (see [**So**] for lectures on this heuristic).

Drawing inspiration from classical approaches towards the Twin Prime Conjecture (see [**HaRi**]), we tackle Question 78 similarly, for example as the sieve problem

$$\left| \mathcal{A} \backslash \bigcup_{\ell \leq z} \mathcal{A}_\ell \right|,$$

where

$$\mathcal{A} := \{p \leq x : p \nmid \Delta_E\}$$

and

$$\mathcal{A}_\ell := \{p \in \mathcal{A} : p \neq \ell, p+1-a_p \equiv 0 (\operatorname{mod} \ell)\},$$

for each prime $\ell < z$ and some suitable parameter $z = z(x)$.

To proceed with this approach, we need the cardinality of the set $\mathcal{A}_\ell$, or, more generally, of the set $\mathcal{A}_m := \bigcap_{\ell | m} \mathcal{A}_\ell$. This can be derived from conditional and unconditional versions of the Chebotarev Density Theorem 16, combined with Theorem 53:

(63)
$$|\mathcal{A}_m| = \# \left\{ p \leq x : p \nmid m\Delta_E, \left(\frac{\mathbb{Q}(E[m])/\mathbb{Q}}{p} \right) \subseteq \mathcal{C}_{m,E}^{\text{prime}} \right\} \sim \frac{|\mathcal{C}_{m,E}^{\text{prime}}|}{[\mathbb{Q}(E[m]) : \mathbb{Q}]} \pi(x),$$

where

$$\mathcal{C}_{m,E}^{\text{prime}} := \{g \in \operatorname{Gal}(\mathbb{Q}(E[m])/\mathbb{Q}) : \det g + 1 - \operatorname{tr} g \equiv 0 (\operatorname{mod} m)\}.$$

By reasoning crudely based on this approach, we derive the expectation

$$\begin{aligned}\#\left\{p \leq x : p \nmid \Delta_E, \left|\overline{E}(\mathbb{F}_p)\right| \text{ is prime}\right\} &\approx \prod_{\ell}\left(1 - \frac{|\mathcal{C}_{\ell,E}^{\text{prime}}|}{[\mathbb{Q}(E[\ell]) : \mathbb{Q}]}\right)\left(1 - \frac{1}{\ell}\right)^{-1} \\ &\quad \times \sum_{p \leq x} \frac{1}{\log(p + 1 - a_p)} \\ &\approx \prod_{\ell}\left(1 - \frac{|\mathcal{C}_{\ell,E}^{\text{prime}}|}{[\mathbb{Q}(E[\ell]) : \mathbb{Q}]}\right)\left(1 - \frac{1}{\ell}\right)^{-1} \\ &\quad \times \int_2^x \frac{1}{\log(t+1)} \cdot \frac{dt}{\log t},\end{aligned}$$

where $\left(1 - \frac{1}{\ell}\right)^{-1}$ is a correction factor introduced to remedy the initial assumption that $p + 1 - a_p$ behaves like a random integer, and where the infinite product over ℓ arises by making the assumption that the events $p + 1 - a_p \equiv 0 (\operatorname{mod} \ell)$ are independent.

Note that, similarly to our crude heuristic from Section 9 for the Cyclicity Question 1, while this latter assumption never holds for an elliptic curve $E/\mathbb{Q}$, the obstruction to the independence of the mod ℓ events may be accounted for by using the integer m_E. With this in mind, the above expectation may be refined to:

CONJECTURE 79. *(Primality Conjecture)*
For any elliptic curve $E/\mathbb{Q}$ with Weierstrass equation (10), *either we have that there exists an elliptic curve $E'/\mathbb{Q}$ with $E'(\mathbb{Q})_{tors} \neq \{\mathcal{O}\}$ and $E' \sim_{\mathbb{Q}} E$, in which case*

$$\#\{p \leq x : p \nmid \Delta_E, \left|\overline{E}(\mathbb{F}_p)\right| \textit{ is prime}\} \ll_E 1,$$

or we have that E is not $\mathbb{Q}$-isogenous with any elliptic curve defined over $\mathbb{Q}$ with a non-trivial torsion subgroup, in which case there exists a constant $C_{prime}(E) \geq 0$ such that

$$\#\{p \leq x : p \nmid \Delta_E, \left|\overline{E}(\mathbb{F}_p)\right| \textit{ is prime}\} \sim C_{prime}(E) \int_2^x \frac{1}{\log(t+1)} \cdot \frac{dt}{\log t}.$$

Moreover,

- *if $\operatorname{End}_{\overline{\mathbb{Q}}}(E) \simeq \mathbb{Z}$, then*

$$C_{prime}(E) := \frac{1 - \frac{\#\{g \in \operatorname{Gal}(\mathbb{Q}(E[m_E])/\mathbb{Q}) :\, \gcd(\det g + 1 - \operatorname{tr} g,\ m_E) \neq 1\}}{[\mathbb{Q}(E[m_E]):\mathbb{Q}]}}{\prod_{\ell \mid m_E}\left(1 - \frac{1}{\ell}\right)} \times \prod_{\ell \nmid m_E}\left(1 - \frac{\ell^2 - \ell - 1}{(\ell - 1)^3(\ell + 1)}\right);$$

- *if $\operatorname{End}_{\overline{\mathbb{Q}}}(E) \not\simeq \mathbb{Z}$, then*

$$C_{prime}(E) := \frac{1 - \frac{\#\{g \in \operatorname{Gal}(\mathbb{Q}(E[\Delta m_E])/\mathbb{Q}) :\, \gcd(\det g + 1 - \operatorname{tr} g,\ \Delta m_E) \neq 1\}}{[\mathbb{Q}(E[\Delta m_E]):\mathbb{Q}]}}{\prod_{\ell \mid \Delta m_E}\left(1 - \frac{1}{\ell}\right)} \times \prod_{\ell \nmid \Delta m_E}\left(1 - \chi(\ell)\frac{\ell^2 - \ell - 1}{(\ell - 1)^2(\ell - \chi(\ell))}\right),$$

  where Δ is the discriminant and χ is the Kronecker character of the CM field of E.

This conjecture originates in [**Kob**], where the heuristic assumed the independence of the division fields of the elliptic curve. In [**Zy11**], Zywina proposed a modified version of Koblitz's Conjecture, similar to the conjecture stated above. The positivity of the constant $C_{\text{prime}}(E)$ is an open question, whose study has unravelled subtle arithmetic properties of elliptic curves. While we will not address it here, we refer the reader to discussions on this topic made in [**Jo10bis**] and [**Zy11**].

Pursuing the analogy between the sieve problem set earlier and the Twin Prime Conjecture 6 viewed as the sieve problem $\#\{p \leq x : p+2 \not\equiv 0 (\bmod \ell) \ \forall \ell < z(x)\}$, several results towards Question 78 can be proven, including the following upper and lower bounds:

THEOREM 80. *(Upper Bounds related to the Primality Conjecture* [**Co05**], [**DaWu**], [**Zy11**]*)*

For any elliptic curve $E/\mathbb{Q}$ with Weierstrass equation (10), *we have:*

(i) *under no additional assumptions,*

$$\#\{p \leq x : p \nmid \Delta_E,\ |\overline{E}(\mathbb{F}_p)| \text{ is prime}\} \ll_E \frac{x}{(\log x)(\log\log x)};$$

(ii) *assuming a δ-quasi-GRH for any fixed $\frac{1}{2} \leq \delta < 1$,*

$$\#\{p \leq x : p \nmid \Delta_E,\ |\overline{E}(\mathbb{F}_p)| \text{ is prime}\} \ll_E \frac{x}{(\log x)^2};$$

(iii) *assuming* $\mathrm{End}_{\overline{\mathbb{Q}}}(E) \not\simeq \mathbb{Z}$,

$$\#\{p \leq x : p \nmid \Delta_E,\ |\overline{E}(\mathbb{F}_p)| \text{ prime}\} \ll \frac{x}{(\log x)^2}.$$

THEOREM 81. *(Lower Bounds related to the Primality Conjecture* [**Co05**], [**DaWu**], [**IwJU**], [**JU08**], [**MiMu**], [**StWe**]*)*

For any elliptic curve $E/\mathbb{Q}$ with Weierstrass equation (10), *we have:*

(i) *assuming GRH,*

$$\#\{p \leq x : p \nmid \Delta_E,\ |\overline{E}(\mathbb{F}_p)| = P_8\} \gg_E \frac{x}{(\log x)^2};$$

(ii) *assuming* $\mathrm{End}_{\overline{\mathbb{Q}}}(E) \not\simeq \mathbb{Z}$,

$$\#\{p \leq x : p \nmid \Delta_E,\ |\overline{E}(\mathbb{F}_p)| = P_2\} \gg \frac{x}{(\log x)^2}.$$

Here, P_k denotes an integer which has at most k distinct prime factors.

Theorem 80 is due to Cojocaru [**Co05**], with an improvement in part (i) by Zywina [**Zy11**] (see also the follow-up [**DaWu**]). Part (i) of Theorem 81 is due to David & Wu [**DaWu**], who built on work of Miri & Murty [**MiMu**] and of Steuding & Weng [**StWe**]; part (ii) of Theorem 81 is due to Jimenez-Urroz [**JU08**], who built on work of Iwaniec & Jimenez-Urroz [**IwJU**] and of Cojocaru [**Co05**].

As for the Cyclicity Conjecture, to obtain further theoretical evidence towards the Primality Conjecture we may investigate the latter on average over a two-parameter family of elliptic curves. In this direction, using techniques similar to those presented in Section 12, the following results have been proven:

THEOREM 82. *(Balog-Cojocaru-David Primality on Average Theorem* [**BaCoDa**]*)* *For $A, B > 0$, consider the family $\mathcal{F}(A,B)$ of $\mathbb{Q}$-isomorphism classes of elliptic curves $E = E_{a,b}$ defined by (10) with $a, b \in \mathbb{Z}$ and $|a| \le A, |b| \le B$. Then, for any $x > 0$, $\varepsilon > 0$, and $A = A(x), B = B(x)$ such that*

$$A, B > x^{\varepsilon},$$

$$AB > x \log^{10} x,$$

we have

(64)

$$\frac{1}{|\mathcal{F}(A,B)|} \sum_{E \in \mathcal{F}(A,B)} \#\{p \le x : p \nmid \Delta_E, |\overline{E}(\mathbb{F}_p)| \text{ is prime}\} \sim C_{prime}^{average} \int_2^x \frac{1}{(\log(t+1)} \cdot \frac{dt}{\log t},$$

where

$$C_{prime}^{average} := \frac{2}{3} \prod_{\ell \neq 2} \frac{\ell^4 - 2\ell^3 - \ell^2 + 3\ell}{(\ell-1)^3(\ell+1)} = \prod_{\ell} \left(1 - \frac{\ell^2 - \ell - 1}{(\ell-1)^3(\ell+1)}\right)$$
$$= 0.505166168239435774...$$

We remark that, while Theorem 82 is a result in the realm of elliptic curves, one of the key results used in its proof is Theorem 15 about twin-prime pairs, recorded in Section 1. The passage from elliptic curves to primes is achieved using a formula of Deuring, proven in detail in [**Bi**] (see also [**Ge03**] for closely related work), for the number of elliptic curves over $\mathbb{F}_p$ with a fixed Frobenius trace.

As a companion to Theorem 82, Jones [**Jo09**] proved that the average constant $C_{\text{prime}}^{\text{average}}$ is closely related to $C_{\text{prime}}(E)$ in the case of a Serre curve and that, in general, $C_{\text{prime}}^{\text{average}}$ is the average of all individual constants $C_{\text{prime}}(E)$:

PROPOSITION 83. *(The Primality Constant for a Serre Curve* [**Jo09**]*)* *For any Serre curve $E/\mathbb{Q}$, we have*

$$C_{prime}(E) = \begin{cases} C_{prime}^{average} \left(1 + \prod_{\ell | (\Delta_E)_{sf}} \frac{1}{\ell^3 - 2\ell^2 - \ell + 3}\right) & \text{if } (\Delta_E)_{sf} \equiv 1 (\text{mod } 4), \\ C_{prime}^{average} & \text{otherwise}, \end{cases}$$

where $(\Delta_E)sf$ denotes the squarefree part of the discriminant Δ_E of any Weierstrass model for E.

THEOREM 84 (Jones' Primality Constant on Average Theorem [**Jo09**]). *For $A, B > 0$, consider the family $\mathcal{F}(A,B)$ of $\mathbb{Q}$-isomorphism classes of elliptic curves $E = E_{a,b}$ defined by (10) with $a, b \in \mathbb{Z}$ and $|a| \le A, |b| \le B$. Then, for any $x > 0$ and $A = A(x), B = B(x)$ such that*

$$\lim_{x \to \infty} \frac{(\log A(x))^7 \cdot \log B(x)}{B(x)} = 0,$$

we have

$$\frac{1}{|\mathcal{F}(A,B)|} \sum_{E \in \mathcal{F}(A,B)} C_{prime}(E) \sim C_{prime}^{average}. \tag{65}$$

14. Anomalous primes

Observing that, when $a_p = 1$, the order of the group $\overline{E}(\mathbb{F}_p)$ is a prime, we ask:

QUESTION 85. Given an elliptic curve $E/\mathbb{Q}$, how often do we have $a_p = 1$?

Primes p for which $a_p = 1$ have been studied by several mathematicians for over four decades, including Mazur [**Ma77**], who called them **anomalous primes** (N.B. Rather, an anomalous prime is defined to be a prime of good reduction satisfying $a_p \equiv 1 (\operatorname{mod} p)$; when $p \geq 7$, this is equivalent to $a_p = 1$). In [**LaTr76**], Lang & Trotter proposed a precise conjecture about these primes and, in general, about primes p for which $a_p = a$, for some fixed integer a.

To investigate anomalous primes, recall that, by part (i) of Theorem 49, we have $|a_p| < 2\sqrt{p}$. Thus, if a_p were to behave like a random integer, we would expect

$$\#\{p \leq x : p \nmid \Delta_E, a_p = 1\} \approx \sum_{p \leq x} \frac{1}{4\sqrt{p}}.$$

However, there is more about the integers a_p to consider. By part (ii) of Theorem 49, the integers a_p carry important arithmetic information about the curve E. In particular, for any prime $\ell \neq p$, we know that $a_p (\operatorname{mod} \ell)$ is the trace of $\varphi_{E,\ell}\left(\frac{\mathbb{Q}(E[\ell])/\mathbb{Q}}{p}\right)$, viewed as an element of $\mathrm{GL}_2(\mathbb{Z}/\ell\mathbb{Z})$. Combining this property with the simple observation that

$$a_p = 1 \;\Leftrightarrow\; a_p \equiv 1 (\operatorname{mod} \ell) \text{ for any prime } \ell < |a_p - 1|, \tag{66}$$

we can investigate Question 85 via the Chebotarev Density Theorem, i.e. via variations of the asymptotic formula

$$\#\left\{p \leq x : p \nmid \ell\Delta_E, a_p \equiv 1 (\operatorname{mod} \ell)\right\} \sim \frac{\left|\mathcal{C}_{\ell,E}^{\text{anomalous}}\right|}{[\mathbb{Q}(E[\ell]) : \mathbb{Q}]}\pi(x), \tag{67}$$

where

$$\mathcal{C}_{\ell,E}^{\text{anomalous}} := \left\{g \in \mathrm{Gal}(\mathbb{Q}(E[\ell])/\mathbb{Q}) : \operatorname{tr} \varphi_{E,\ell}(g) \equiv 1 (\operatorname{mod} \ell)\right\}.$$

In addition to this arithmetic information, we know that the angles $\theta_p \in [0, \pi]$ defined by $\frac{a_p}{2\sqrt{p}} = \cos\theta_p$ obey the Sato-Tate equidistribution law if $E/\mathbb{Q}$ is without CM and obey the Hecke equidistribution law if $E/\mathbb{Q}$ is with CM. Briefly, this means that, for any interval $I := [\alpha, \beta] \subseteq [0, \pi]$, we have

$$\lim_{x \to \infty} \frac{\#\left\{p \leq x : p \nmid \Delta_E, \theta_p \in I\right\}}{\pi(x)} = \begin{cases} \int_I \frac{2}{\pi} \sin^2\theta \, d\theta & \text{if } E/\mathbb{Q} \text{ is without CM}, \\ \frac{\delta_I}{2} + \frac{\beta - \alpha}{2\pi} & \text{if } E/\mathbb{Q} \text{ is with CM}, \end{cases}$$

where

$$\delta_I := \begin{cases} 1 & \text{if } \frac{\pi}{2} \in I, \\ 0 & \text{otherwise}; \end{cases}$$

equivalently, this means that, for any interval $J \subseteq [-1, 1]$, we have

$$\lim_{x \to \infty} \frac{\#\left\{p \leq x : p \nmid \Delta_E, \frac{a_p}{2\sqrt{p}} \in J\right\}}{\pi(x)} = \int_J \Phi(t) \, dt,$$

where

$$\Phi(t) := \begin{cases} \frac{2}{\pi}\sqrt{1-t^2} & \text{if } E/\mathbb{Q} \text{ is without CM,} \\ \\ \frac{1}{\sqrt{1-t^2}} & \text{if } E/\mathbb{Q} \text{ is with CM.} \end{cases}$$

We refer the reader to [**Ca**] and [**Cl**] for an overview of this topic.

A crude heuristic based on the above remarks leads to the rough expectation

$$\#\left\{p \leq x : p \nmid \Delta_E, a_p = 1\right\} \quad \approx \quad \prod_{\ell} \frac{\ell \left|C_{E,\ell}^{\text{anomalous}}\right|}{[\mathbb{Q}(E[\ell]) : \mathbb{Q}]} \sum_{p \leq x} \Phi\left(\frac{1}{2\sqrt{p}}\right) \frac{1}{2\sqrt{p}};$$

a more refined heuristic, as detailed in [**LaTr76**], leads to:

Conjecture 86. *(Anomalous Primes Conjecture)*
For any elliptic curve $E/\mathbb{Q}$ with Weierstrass equation (10), *either we have that $E(\mathbb{Q})_{tors} \neq \{\mathcal{O}\}$, in which case*

$$\#\{p \leq x : p \nmid \Delta_E, a_p = 1\} \ll_E 1,$$

or we have that $E(\mathbb{Q})_{tors} = \{\mathcal{O}\}$, in which case there exists a constant $C_{anomalous}(E) > 0$ such that

$$\#\{p \leq x : p \nmid \Delta_E, a_p = 1\} \sim C_{anomalous}(E) \int_2^x \frac{1}{2\sqrt{t}} \cdot \frac{dt}{\log t}.$$

Moreover,

- *if* $\operatorname{End}_{\overline{\mathbb{Q}}}(E) \simeq \mathbb{Z}$, *then*

$$C_{anomalous}(E) := \frac{2}{\pi} \times \frac{m_E \ \#\left\{g \in \operatorname{Gal}(\mathbb{Q}(E[m_E])/\mathbb{Q}) : \operatorname{tr} g \equiv 1 (\operatorname{mod} m_E)\right\}}{[\mathbb{Q}(E[m_E]) : \mathbb{Q}]} \times \prod_{\ell \nmid m_E} \frac{\ell\left(\ell^2 - \ell - 1\right)}{(\ell-1)^2(\ell+1)};$$

- *if* $\operatorname{End}_{\overline{\mathbb{Q}}}(E) \not\simeq \mathbb{Z}$, *then*

$$C_{anomalous}(E) := \frac{1}{2\pi} \times \frac{m_E \ \#\left\{g \in \operatorname{Gal}(\mathbb{Q}(E[m_E])/\mathbb{Q}) : \operatorname{tr} g \equiv 1 (\operatorname{mod} m_E)\right\}}{[\mathbb{Q}(E[m_E]) : \mathbb{Q}]} \times \prod_{\ell \nmid m_E} \frac{\ell^2 - (1 + \chi(\ell))\,\ell}{(\ell-1)(\ell-\chi(\ell))},$$

where χ is the Kronecker character of the CM field of E.

For a refinement of this conjecture, see [**BaJo**]. Regarding the constant, note that it is not difficult to justify the torsion assumption in the conjecture: if $E(\mathbb{Q})_{\text{tors}}$ is non-trivial, then for some prime ℓ we have that $\operatorname{Im} \varphi_{E,\ell}$ is contained in the group of invertible matrices of the form $\begin{pmatrix} 1 & * \\ 0 & * \end{pmatrix}$; in turn, this group does not contain elements of trace 1. For a detailed discussion about the constant, see [**Ka**] and [**Jo09**].

While Conjecture 86 remains open, several related partial results have been proven; we provide a brief summary of them below. From the start, we point out that no lower bounds are known.

Upper bounds related to the conjecture have been established via observation (66), as applications of the effective version of the Chebotarev Density Theorem. In particular, in the generic case of an elliptic curve without CM, by building on work of Serre [**Se72**], Murty & Murty & Saradha [**MuMuSa**] proved the best known such bound (a slight improvement of the exponent of the $\log x$ factor was obtained in [**Zy15**]); the bound is conditional upon GRH. Unconditionally, Serre [**Se72**], followed by Wan [**Wan**] and Murty [**KMu**], proved a zero density result, derived from an upper bound which has no savings in the exponent of x.

THEOREM 87. *(Upper Bounds related to the Anomalous Prime Conjecture* [**MuMuSa**], [**KMu**], [**Se72**], [**Wan**], [**Zy15**]*)*

For any elliptic curve $E/\mathbb{Q}$ with Weierstrass equation (10) *and such that* $\operatorname{End}_{\overline{\mathbb{Q}}}(E) \simeq \mathbb{Z}$, *we have:*

(i) *unconditionally, for any $\varepsilon > 0$,*

$$\#\{p \leq x : p \nmid \Delta_E, a_p(E) = 1\} \ll_{E,\varepsilon} \frac{x}{(\log x)^{\frac{4}{3}-\varepsilon}};$$

(ii) *under GRH,*

$$\#\{p \leq x : p \nmid \Delta_E, a_p(E) = 1\} \ll_E \frac{x^{\frac{4}{5}}}{(\log x)^{\frac{3}{5}}}.$$

In the CM case, these upper bounds can be improved substantially by establishing connections between anomalous primes and classical arithmetic questions. For instance, by recalling Lemma 52 of Section 8, we have

$$4p - a_p^2 = c_p^2 \left|\operatorname{disc} O_{\mathbb{Q}(\pi_p)}\right|$$

and so p is represented by an integral quadratic polynomial, whose coefficients depend on both E and p. If $E/\mathbb{Q}$ is with CM, then, by imposing the condition $a_p = 1$ on the above equation, we deduce that p is represented by an integral quadratic polynomial which depends solely on E, and not on p. Indeed, under these assumptions, we obtain that:[1]

(a) $p = 3n^2 + 3n + 1$ for some $n \in \mathbb{Z}$, if $K = \mathbb{Q}\left(\sqrt{-3}\right)$;
(b) $p = 7n^2 + 7n + 2$ for some $n \in \mathbb{Z}$, if $K = \mathbb{Q}\left(\sqrt{-7}\right)$;
(c) $p = 11n^2 + 11n + 3$ for some $n \in \mathbb{Z}$, if $K = \mathbb{Q}\left(\sqrt{-11}\right)$;
(d) $p = 19n^2 + 19n + 5$ for some $n \in \mathbb{Z}$, if $K = \mathbb{Q}\left(\sqrt{-19}\right)$;
(e) $p = 43n^2 + 43n + 11$ for some $n \in \mathbb{Z}$, if $K = \mathbb{Q}\left(\sqrt{-43}\right)$;
(f) $p = 67n^2 + 67n + 17$ for some $n \in \mathbb{Z}$, if $K = \mathbb{Q}\left(\sqrt{-67}\right)$;
(g) $p = 163n^2 + 163n + 41$ for some $n \in \mathbb{Z}$, if $K = \mathbb{Q}\left(\sqrt{-163}\right)$.

Question 85 is thus reminiscent of the classical Hardy-Littlewood Conjecture 7 from Section 2.

Using the above connection to primes represented by quadratic polynomials, we can apply sieve methods in the classical setting and deduce:

THEOREM 88. *For any elliptic curve $E/\mathbb{Q}$ with Weierstrass equation* (10) *and such that* $\operatorname{End}_{\overline{\mathbb{Q}}}(E) \not\simeq \mathbb{Z}$, *we have, unconditionally,*

$$\#\{p \leq x : p \nmid \Delta_E, a_p = 1\} \ll \frac{\sqrt{x}}{\log x}.$$

[1]Note that, if $\operatorname{disc} K \equiv 2, 3 \pmod 4$, that is, if K is one of the fields $\mathbb{Q}\left(\sqrt{-1}\right)$ or $\mathbb{Q}\left(\sqrt{-2}\right)$, then there are no primes p for which $a_p = 1$.

For some elliptic curves with CM, an even stronger connection between anomalous primes and the Hardy-Littlewood Conjecture 7 holds:

THEOREM 89. *(Equivalence between the Anomalous Prime Conjecture and the Hardy-Littlewood Conjecture* [**Qi**]*)*

For $D \in \mathbb{Z}$, not a square or a cube in $\mathbb{Q}\left(\sqrt{-3}\right)$, denote by E_D the elliptic curve defined by $y^2 = x^3 + D$.

(i) *If Conjecture 7 is true, then, for any D as above and not of the form $80d^6$ for some $0 \neq d \in \mathbb{Z}\left[\frac{1+\sqrt{-3}}{2}\right]$ with $d^6 \in \mathbb{Z}$, there exists a positive constant $C(D)$ such that*

$$\#\left\{p \leq x : p \nmid \Delta_{E_D}, a_p = 1\right\} \sim C(D)\frac{\sqrt{x}}{\log x}. \tag{68}$$

(ii) *If there exists some $D \in \mathbb{Z}$, not a square or a cube in $\mathbb{Q}\left(\sqrt{-3}\right)$, for which* (68) *holds, then Conjecture 7 holds for the polynomial $12X^2 + 18X + 7$.*

Further evidence for Conjecture 86 has been provided by results on average. Indeed, by building on the work of Fouvry & Murty [**FoMu**] on supersingular primes, David & Pappalardi [**DaPa**], Baier [**Ba**], and Banks & Shparlinski [**BaSh**] proved asymptotic formulae for the number of anomalous primes of an elliptic curve, averaged over a two-parameter family as discussed in Section 7. For instance:

THEOREM 90. *(Anomalous Primes on Average Theorem* [**Ba**]*)*

For $A, B > 0$, consider the family $\mathcal{F}(A, B)$ of $\mathbb{Q}$-isomorphism classes of elliptic curves $E = E_{a,b}$ defined by (10) *with $a, b \in \mathbb{Z}$ and $|a| \leq A, |b| \leq B$. Then, for any $x > 0$, $\varepsilon > 0$, and $A = A(x), B = B(x)$ such that*

$$A, B > x^{\varepsilon},$$

$$AB > x^{\frac{3}{2}+\varepsilon},$$

we have

$$\frac{1}{|\mathcal{F}(A, B)|} \sum_{E \in \mathcal{F}(A,B)} \#\{p \leq x : p \nmid \Delta_E, a_p = 1\} \sim C_{anomalous}^{average} \int_2^x \frac{1}{2\sqrt{t}} \cdot \frac{dt}{\log t}, \tag{69}$$

where

$$C_{anomalous}^{average} := \frac{2}{\pi} \prod_{\ell} \frac{\ell(\ell^2 - \ell - 1)}{(\ell - 1)(\ell^2 - 1)} \approx 0.3916056127252347549356 2...$$

For similar results for other families of elliptic curves, see the work of K. James, such as [**Ja**].

As was the case with Theorem 82, while Theorem 90 is a result in the realm of elliptic curves, one of the key results used in its proof is a theorem about primes in arithmetic progressions, namely the Barban-Davenport-Halberstam Theorem 11 stated in Section 2. The passage from elliptic curves to primes is achieved, once again, using Deuring's formula for the number of elliptic curves over $\mathbb{F}_p$ with a fixed Frobenius trace.

As a companion to Theorem 90, Jones [**Jo09**] proved that the average constant $C_{\text{anomalous}}^{\text{average}}$ is closely related to $C_{\text{anomalous}}(E)$ in the case of a Serre curve, and that, in general, $C_{\text{anomalous}}^{\text{average}}$ is the average of all individual constants $C_{\text{anomalous}}(E)$:

PROPOSITION 91. *(The Anomalous Primes Constant for a Serre Curve* [**Jo09**]*)* *For any Serre curve $E/\mathbb{Q}$, define*

$$W := \frac{(\Delta_E)_{sf}}{\gcd((\Delta_E)_{sf}, 2)},$$

$$k := \begin{cases} 1 & \text{if } (\Delta_E)_{sf} \equiv 1 (\operatorname{mod} 4), \\ 2 & \text{if } (\Delta_E)_{sf} \equiv 3 (\operatorname{mod} 4), \\ 3 & \text{if } (\Delta_E)_{sf} \equiv 2 (\operatorname{mod} 4), \end{cases}$$

$$\chi_4 : \mathbb{Q} \longrightarrow \{\pm 1\}, \quad \chi_4(x) := \begin{cases} -1 & \text{if } x \in \mathbb{Z} \text{ and } x \equiv -1 (\operatorname{mod} 4), \\ 1 & \text{otherwise}, \end{cases}$$

and

$$\delta := (-1)^{\omega(W) + \frac{W+1}{2} + 1} \chi_4 \left(-\frac{(\Delta_E)_{sf}}{2} \right).$$

Then

$$C_{anomalous}(E) = \begin{cases} C_{anomalous}^{average} \left(1 + \frac{\delta \ W}{\#\{g \in \mathrm{GL}_2(\mathbb{Z}/2W\mathbb{Z}) : \mathrm{tr}\, g \equiv 1 (\operatorname{mod} 2W)\}} \right) & \text{if } k = 1, \\ C_{anomalous}^{average} & \text{otherwise.} \end{cases}$$

THEOREM 92 (Anomalous Primes Constant on Average Theorem [**Jo09**]). *For $A, B > 0$, consider the family $\mathcal{F}(A, B)$ of $\mathbb{Q}$-isomorphism classes of elliptic curves $E = E_{a,b}$ defined by* (10) *with $a, b \in \mathbb{Z}$ and $|a| \le A, |b| \le B$. Then, for any $x > 0$ and $A = A(x), B = B(x)$ such that*

$$\lim_{x \to \infty} \frac{(\log A(x))^7 \cdot \log B(x)}{B(x)} = 0,$$

we have

$$\frac{1}{|\mathcal{F}(A, B)|} \sum_{E \in \mathcal{F}(A,B)} C_{anomalous}(E) \sim C_{anomalous}^{average}. \tag{70}$$

15. Global perspectives

Question 1 may also be formulated in function field settings, as we briefly discuss below.

15.1. Cyclicity: elliptic curves over function fields. Let K be a global field of characteristic $p \ge 5$ and constant field $\mathbb{F}_q$. Let E/K be an elliptic curve over K with j-invariant $j_E \notin \mathbb{F}_q$. All but finitely many primes $\wp$ of K are of good reduction for E/K. We denote by $\mathcal{P}_E$ the collection of these primes, and for each $\wp \in \mathcal{P}_E$, we consider the residue field $\mathbb{F}_\wp$ at $\wp$ and the abelian group $\overline{E}(\mathbb{F}_\wp)$ defined by the reduction of E modulo $\wp$. From the theory of torsion points for elliptic curves, there exist uniquely determined integers $d_{1,\wp}, d_{2,\wp} \ge 1$, possibly equal to 1, such that

$$\begin{aligned} \overline{E}(\mathbb{F}_\wp) &\simeq \mathbb{Z}/d_{1,\wp}\mathbb{Z} \times \mathbb{Z}/d_{2,\wp}\mathbb{Z}, \\ d_{1,\wp} &\mid d_{2,\wp}. \end{aligned}$$

In analogy with Theorems 68 and 70, in this setting we have:

THEOREM 93. *(Cojocaru-Tóth Cyclicity and Large Exponent Theorem* [**CoTo**]*)*

(i) *The Dirichlet density of the set*

$$\{\wp \in \mathcal{P}_E : \overline{E}(\mathbb{F}_\wp) \text{ is cyclic}\}$$

exists and equals

$$\sum_{\substack{m \geq 1 \\ (m,p)=1}} \frac{\mu(m)}{[K(E[m]) : K]},$$

where $\mu(m)$ *is the Möbius function of* m *and* $K(E[m])$ *is the* m*-th division field of* E/K.

(ii) *Let* $f : (0, \infty) \longrightarrow (0, \infty)$ *be such that* $\lim_{x\to\infty} f(x) = \infty$. *The Dirichlet density of the set*

$$\left\{\wp \in \mathcal{P}_E : d_{2,\wp} > \frac{|\overline{E}(\mathbb{F}_\wp)|}{f(\deg \wp)}\right\}$$

exists and equals 1.

15.2. Cyclicity: Drinfeld modules. Another function field analogue of Question 1 can be formulated in the setting of Drinfeld modules. For this, let q be a prime power, $A := \mathbb{F}_q[T]$, $K := \mathbb{F}_q(T)$, and Ψ a generic Drinfeld A-module over K, of rank $r \geq 2$. All but finitely many primes $\wp$ of K are of good reduction for Ψ. We denote by $\mathcal{P}_\Psi$ the collection of these primes, and for each $\wp \in \mathcal{P}_\Psi$, we consider the residue field $\mathbb{F}_\wp$ at $\wp$ and the A-module structure on $\mathbb{F}_\wp$, denoted $\overline{\Psi}(\mathbb{F}_\wp)$, defined by the reduction of Ψ modulo $\wp$. We denote by $\left|\chi\left(\overline{\Psi}\left(\mathbb{F}_\wp\right)\right)\right|_\infty$ the norm (defined by the prime at infinity $\frac{1}{T}$ of K) of the Euler-Poincaré characteristic of the A-module $\overline{\Psi}(\mathbb{F}_\wp)$. From the theory of torsion points for Drinfeld modules and that of finitely generated modules over a PID, there exist uniquely determined monic polynomials $d_{1,\wp}, \ldots, d_{r,\wp} \in A$, possibly equal to 1, such that

$$\overline{\Psi}(\mathbb{F}_\wp) \simeq_A A/d_{1,\wp}A \times \ldots \times A/d_{r,\wp}A \tag{71}$$

and

$$d_{1,\wp} \mid \ldots \mid d_{r,\wp}.$$

The polynomials $d_{1,\wp}, \ldots, d_{r,\wp}$ are the elementary divisors of the A-module $\overline{\Psi}(\mathbb{F}_\wp)$, with the rth one, the exponent, having the property that $d_{r,\wp}\lambda = 0$ for all $\lambda \in \overline{\Psi}(\mathbb{F}_\wp)$; here, $d_{r,\wp}\lambda := \overline{\Psi}(d_{r,\wp})(\lambda)$.

In analogy with Theorems 68 and 70, we have:

THEOREM 94. *(Cojocaru-Shulman Cyclicity and Large Exponent Theorem* [**CoSh15**]*)*

(i) *The Dirichlet density of the set*

$$\{\wp \in \mathcal{P}_\Psi : d_{1,\wp} = 1\}$$

exists and equals

$$\sum_{\substack{m \in A \\ m \text{ monic}}} \frac{\mu_A(m)}{[K(\Psi[m]) : K]}, \tag{72}$$

where $\mu_A(m)$ *is the Möbius function of* m *and* $K(\Psi[m])$ *is the* m*-th division field of* Ψ.

(ii) *Assume that* $r = 2$ *and let* $f : (0, \infty) \longrightarrow (0, \infty)$ *be such that* $\lim_{x \to \infty} f(x) = \infty$. *The Dirichlet density of the set*

$$\left\{ \wp \in \mathcal{P}_\Psi : |d_{2,\wp}|_\infty > \frac{\left|\chi\left(\overline{\Psi}\left(\mathbb{F}_\wp\right)\right)\right|_\infty}{q^{f(\deg \wp)}} \right\}$$

exists and equals 1.

For additional results, see [**CoSh13**] and [**KuLi**].

16. Final remarks

While we have explored several facets of Question 1, several others arise when we expand the question in both depth and breadth.

Indeed, remaining in the context of elliptic curves $E/\mathbb{Q}$, it can be observed that the cyclicity of $\overline{E}(\mathbb{F}_p)$ relates to other questions about the integers a_p and b_p introduced in Section 8. For example, by noting that $b_p = 1$ implies that $\overline{E}(\mathbb{F}_p)$ is cyclic, we are led to exploring the asymptotic behaviour of

$$\#\{p \leq x : p \nmid \Delta_E, b_p = 1\},$$

and by noting that $|\overline{E}(\mathbb{F}_p)|$ is squarefree implies that $\overline{E}(\mathbb{F}_p)$ is cyclic, we are led to exploring the asymptotic behaviour of

$$\#\{p \leq x : p \nmid \Delta_E, p + 1 - a_p \text{ is squarefree}\}.$$

Similarly to the Cyclicity Conjecture 61, these questions have conjectural answers based on Chebotarev heuristics and have been answered, partially, in [**Co08**] and [**CoDu**].

Other facets of Question 1 arise when pursuing related explorations in the more general contexts of an elliptic curve defined over a global field K, of a higher dimensional abelian variety over a global field K, and of a generic Drinfeld module.

We hope that more techniques will be developed to overcome the intrinsic obstacles connecting all these explorations and that more results will follow.

APPENDIX: REDUCTIONS MODULO PRIMES OF SERRE CURVES: COMPUTATIONAL DATA

Alina Carmen Cojocaru, Matthew Fitzpatrick, Thomas Insley, and Hakan Yilmaz

Department of Mathematics,
Statistics and Computer Science
University of Illinois at Chicago
851 S Morgan St, 322 SEO
Chicago, 60607, IL, USA

The purpose of this appendix is to provide numerical data supporting the Cyclicity Conjecture 61 and its variants. This is the richest amount of such data in the literature and can certainly be developed further.

The data has been collected by performing computations in SAGE on over 350 Serre curves chosen from the one-parameter family

$$y^2 + xy = x^3 + t \quad \text{(to be referred to as "Serre curve } t\text{")},$$

where

$$t \text{ varies over rational primes in the range } 2 \le t < 3000.$$

Additional data has been collected from computations performed on a few Serre curves not from this family, including

$$\begin{aligned} y^2 + y &= x^3 - x^2 + x - 1 \quad \text{(to be referred to as "Serre curve 123b1")},\\ y^2 + y &= x^3 - x \quad \text{(to be referred to as "Serre curve 37a1")},\\ y^2 &= x^3 - x + 1 \quad \text{(to be referred to as "Serre curve 92b1")},\\ y^2 + xy &= x^3 + x^2 - 182317x + 29887645 \quad \text{(to be referred to as "Serre curve 222e1")},\\ y^2 + xy &= x^3 - x^2 - 10x - 10 \quad \text{(to be referred to as "Serre curve 170e1")}. \end{aligned}$$

All these curves arise from work of H. B. Daniels. The primes p with respect to which the curves are reduced and studied vary in the range $5 \le p \le 1299720$, which comprises about 10^5 primes.

The reason our computations are performed on Serre curves is that, as reviewed in Sections 6 - 7 of the paper, Serre curves are effective for calculations which involve degrees of division fields (see Proposition 41). Moreover, Serre curves are generic among elliptic curves over $\mathbb{Q}$ (see Theorem 43 and Corollary 44), hence are the most natural candidates to consider when checking conjectures about arbitrary elliptic curves over $\mathbb{Q}$.

While the usefulness in computations of Serre curves has been known for several decades – indeed, in their monograph [**LaTr76**], Lang and Trotter collected numerical data supporting their conjectures by using four Serre curves – it is only recent that an infinite set of explicit Serre curves has been described. Precisely, building on his doctoral thesis, Daniels [**Dan**] proved that for all prime values of t, the curve $y^2 + xy = x^3 + t$ is a Serre curve. Daniels also proved that the additional five elliptic curves mentioned above (Serre curves 123b1, 37a1, 92b1, 222e1, and 170e1) are Serre curves, which was communicated to the authors privately.

The properties investigated in our computations for each given curve, to be called E, include:

(i) for which primes p in the given range we have $\overline{E}(\mathbb{F}_p)$ is cyclic;
(ii) for which primes p in the given range we have $d_{1,p} = d$ for some given $d \in \mathbb{N}\backslash\{0\}$;
(iii) for which primes p in the given range we have $\left|\overline{E}(\mathbb{F}_p)\right|$ is squarefree;
(iv) for which primes p in the given range we have $\left|\overline{E}(\mathbb{F}_p)\right|$ is prime;
(v) for which primes p in the given range we have $\left|\overline{E}(\mathbb{F}_p)\right| = p$, i.e. $a_p(E) = 1$.

We counted these primes and compared the results with the main terms predicted by the corresponding conjectures on the reductions $\overline{E}/\mathbb{F}_p$ such as the Cyclicity Conjecture 61, the Primality Conjecture 79, and the Anomalous Prime Conjecture 86. In particular, we created pie charts for the counts and we calculated and created growth graphs for the functions:

$$\frac{\#\left\{p \le x : p \nmid \Delta_E, \overline{E}(\mathbb{F}_p) \text{ is cyclic}\right\}}{C_{\text{cyclic}}(E)\ \int_2^x \frac{dt}{\log t}}, \tag{73}$$

$$\frac{\#\left\{p \le x : p \nmid \Delta_E, \overline{E}(\mathbb{F}_p) \text{ is cyclic}\right\} - C_{\text{cyclic}}(E)\ \int_2^x \frac{dt}{\log t}}{\left(C_{\text{cyclic}}(E)\ \int_2^x \frac{dt}{\log t}\right)^{\frac{1}{2}}}, \tag{74}$$

$$\frac{\#\left\{p \le x : p \nmid \Delta_E, \left|\overline{E}(\mathbb{F}_p)\right| \text{ is prime}\right\}}{C_{\text{prime}}(E)\ \int_2^x \frac{1}{\log(t+1)} \cdot \frac{dt}{\log t}}, \tag{75}$$

$$\frac{\#\left\{p \le x : p \nmid \Delta_E, \left|\overline{E}(\mathbb{F}_p)\right| \text{ is prime}\right\} - C_{\text{prime}}(E)\ \int_2^x \frac{1}{\log(t+1)} \cdot \frac{dt}{\log t}}{\left(C_{\text{prime}}(E)\ \int_2^x \frac{1}{\log(t+1)} \cdot \frac{dt}{\log t}\right)^{\frac{1}{2}}}, \tag{76}$$

$$\frac{\#\left\{p \le x : p \nmid \Delta_E, a_p = 1\right\}}{C_{\text{anomalous}}(E)\ \int_2^x \frac{1}{2\sqrt{t}} \cdot \frac{dt}{\log t}}, \tag{77}$$

$$\frac{\#\left\{p \le x : p \nmid \Delta_E, a_p = 1\right\} - C_{\text{anomalous}}(E)\ \int_2^x \frac{1}{2\sqrt{t}} \cdot \frac{dt}{\log t}}{\left(C_{\text{anomalous}}(E)\ \int_2^x \frac{1}{2\sqrt{t}} \cdot \frac{dt}{\log t}\right)^{\frac{1}{2}}}. \tag{78}$$

Thanks to the computations tackling (ii) above and thanks to the recent results in [**BBCCJMSV**], we were also able to create growth graphs for functions relevant to elliptic curve analogues of the Titchmarsh Divisor Problem, such as:

$$\frac{\sum_{\substack{p \le x \\ p \nmid \Delta_E}} d_{1,p}}{C_{d_1, nonCM}(E)\ \int_2^x \frac{dt}{\log t}}, \tag{79}$$

$$\frac{\sum_{\substack{p \le x \\ p \nmid \Delta_E}} d_{1,p} - C_{d_1, nonCM}(E)\ \int_2^x \frac{dt}{\log t}}{\left(C_{d_1, nonCM}(E)\ \int_2^x \frac{dt}{\log t}\right)^{\frac{1}{2}}}, \tag{80}$$

$$\frac{\sum_{\substack{p \le x \\ p \nmid \Delta_E}} \tau(d_{1,p})}{C_{\tau(d_1)}(E)\ \int_2^x \frac{dt}{\log t}}, \tag{81}$$

$$\frac{\sum_{\substack{p\leq x\\ p\nmid\Delta_E}} \tau(d_{1,p}) - C_{\tau(d_1)}(E)\ \int_2^x \frac{dt}{\log t}}{\left(C_{\tau(d_1)}(E)\ \int_2^x \frac{dt}{\log t}\right)^{\frac{1}{2}}}, \tag{82}$$

$$\frac{\sum_{\substack{p\leq x\\ p\nmid\Delta_E}} d_{2,p}}{\frac{1}{2}C_{d_2}(E)\ \int_2^{x^2} \frac{dt}{\log t}}, \tag{83}$$

$$\frac{\sum_{\substack{p\leq x\\ p\nmid\Delta_E}} d_{2,p} - \frac{1}{2}C_{d_2}(E)\ \int_2^{x^2} \frac{dt}{\log t}}{\left(\frac{1}{2}C_{d_2}(E)\ \int_2^{x^2} \frac{dt}{\log t}\right)^{\frac{1}{2}}}, \tag{84}$$

where

$$C_{d_1,\mathrm{nonCM}}(E) := \sum_{m\geq 1} \frac{\phi(m)}{[\mathbb{Q}(E[m]) : \mathbb{Q}]},$$

$$C_{\tau(d_1)}(E) := \sum_{m\geq 1} \frac{1}{[\mathbb{Q}(E[m]) : \mathbb{Q}]},$$

$$C_{d_2}(E) := \sum_{m\geq 1} \frac{(-1)^{\omega(m)}\phi(\mathrm{rad}\, m)}{m[\mathbb{Q}(E[m]) : \mathbb{Q}]},$$

with $\omega(m) := \sum_{\ell|m} 1$ denoting the number of distinct prime factors of m and $\mathrm{rad}(m) := \prod_{\ell|m} \ell$ denoting the product of distinct prime factors of m.

Our data provides strong evidence for all the conjectures investigated. Below is a representative sample of the pie charts and graphs obtained, chosen at random.

Pie charts for primes p with given $d_{1,p}$

The labels represent:

d, the number of primes p with $d_{1,p} = d$, the percentage of primes p in our range with $d_{1,p} = d$

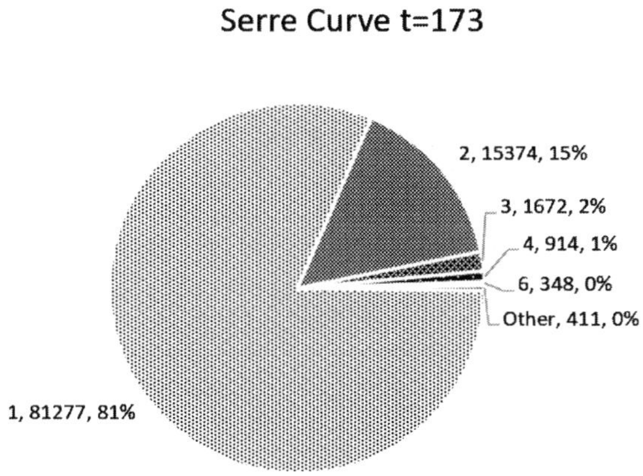

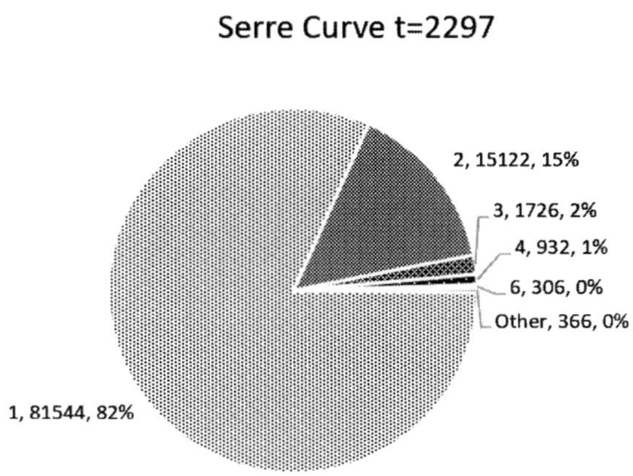

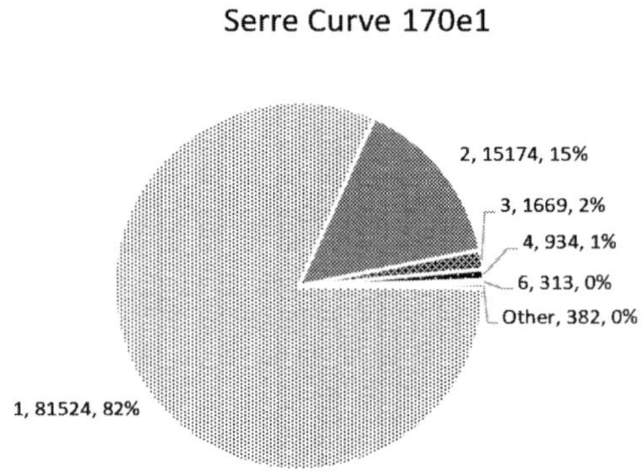
Serre Curve 170e1
2, 15174, 15%
3, 1669, 2%
4, 934, 1%
6, 313, 0%
Other, 382, 0%
1, 81524, 82%

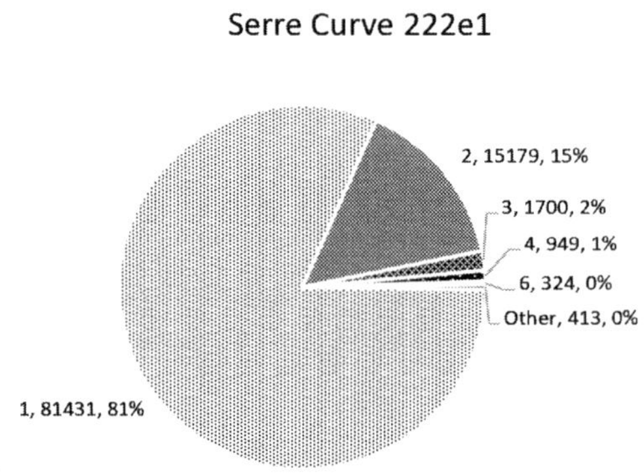
Serre Curve 222e1
2, 15179, 15%
3, 1700, 2%
4, 949, 1%
6, 324, 0%
Other, 413, 0%
1, 81431, 81%

Pie charts for primes p with $d_{1,p} = 1$

The labels represent:

the property of p (e.g., $a_p = 1$; $p + 1 - a_p$ is prime, but $a_p \neq 1$; $p + 1 - a_p$ is squarefree, but not prime), the number of primes p with the given property, the percentage of primes p in our range with the given property

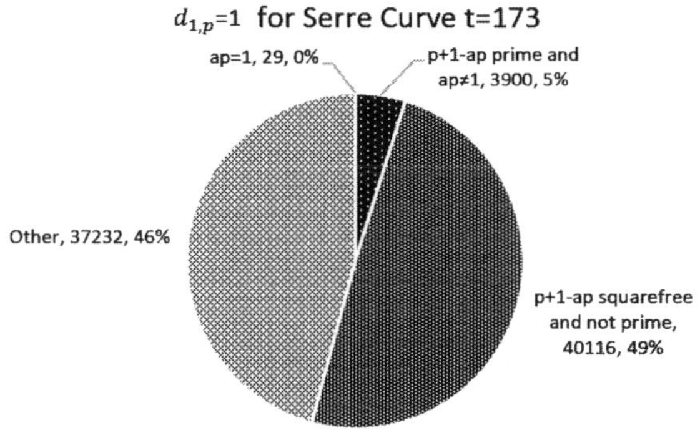

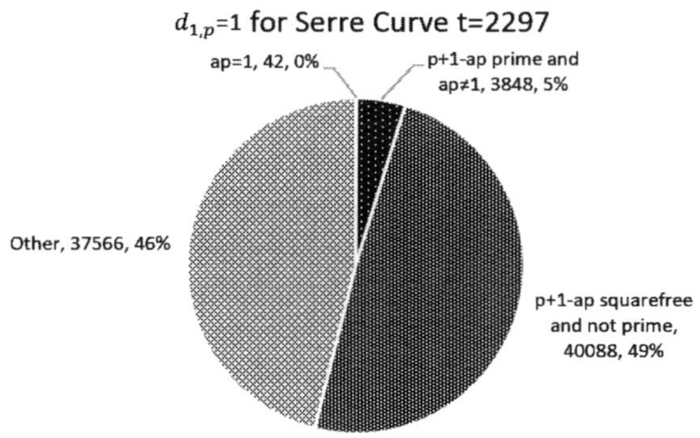

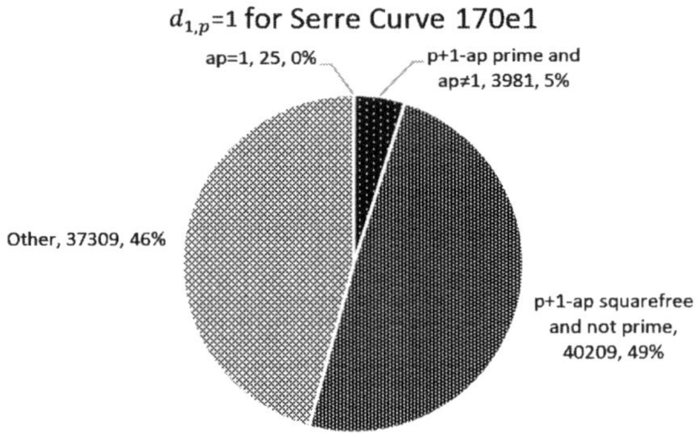
$d_{1,p}$=1 for Serre Curve 170e1
ap=1, 25, 0%
p+1-ap prime and ap≠1, 3981, 5%
Other, 37309, 46%
p+1-ap squarefree and not prime, 40209, 49%

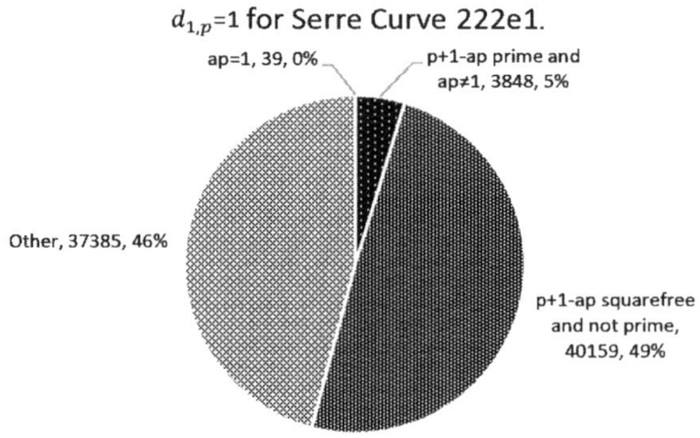
$d_{1,p}$=1 for Serre Curve 222e1.
ap=1, 39, 0%
p+1-ap prime and ap≠1, 3848, 5%
Other, 37385, 46%
p+1-ap squarefree and not prime, 40159, 49%

.

Graph for (73) for the Serre curve $t = 173$

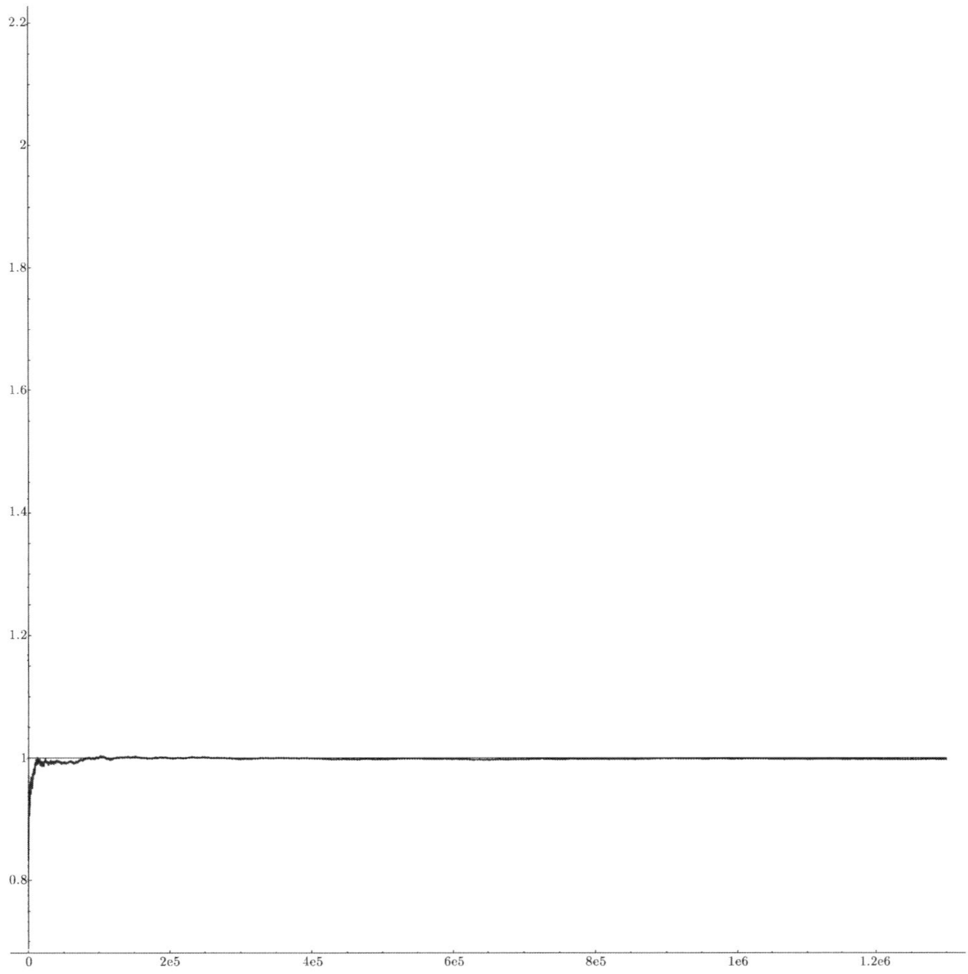

Graph for (75) **for the Serre curve** $t = 2297$

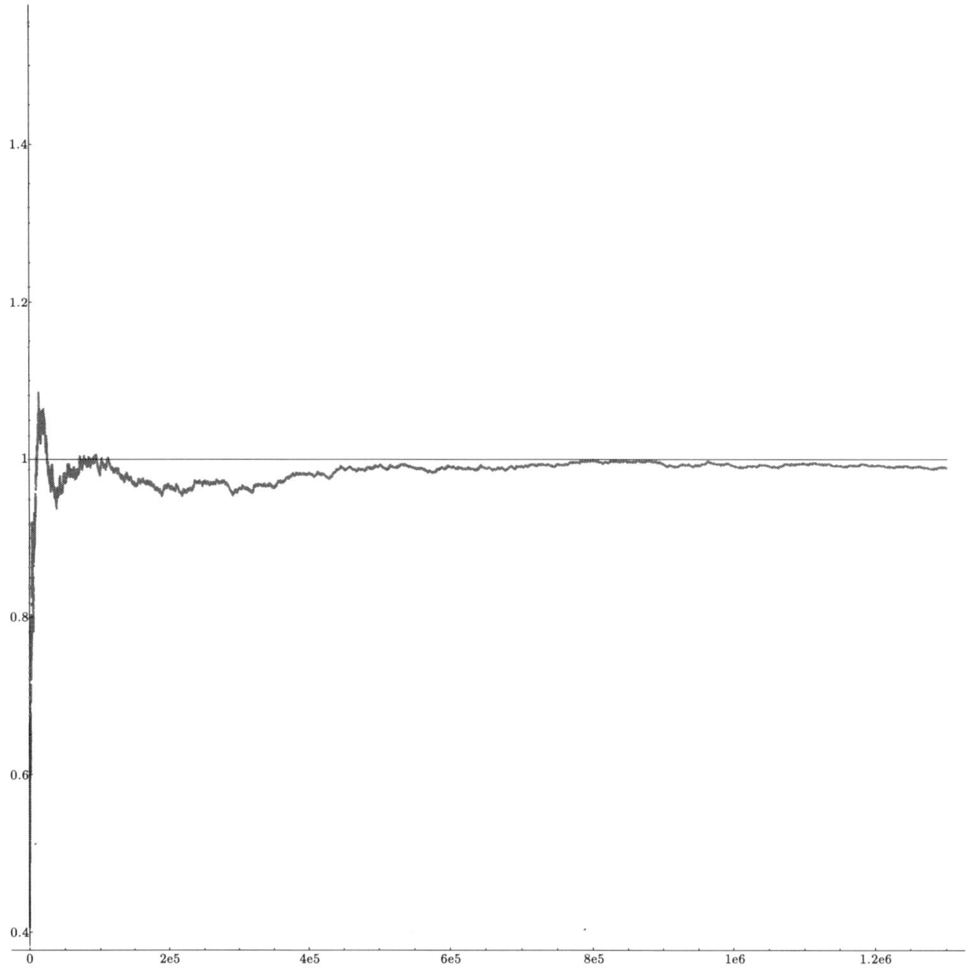

Graph for (79) for the Serre curve 170e1

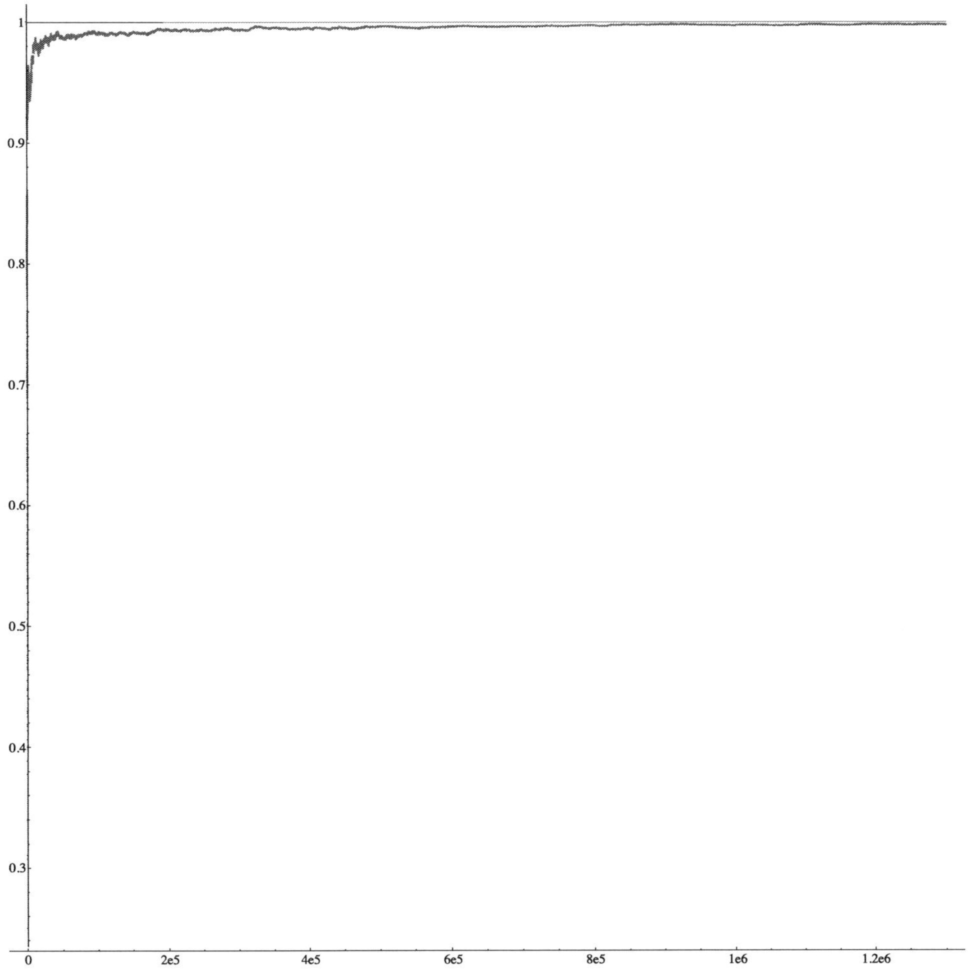

Graph for (81) for the Serre curve 222e1

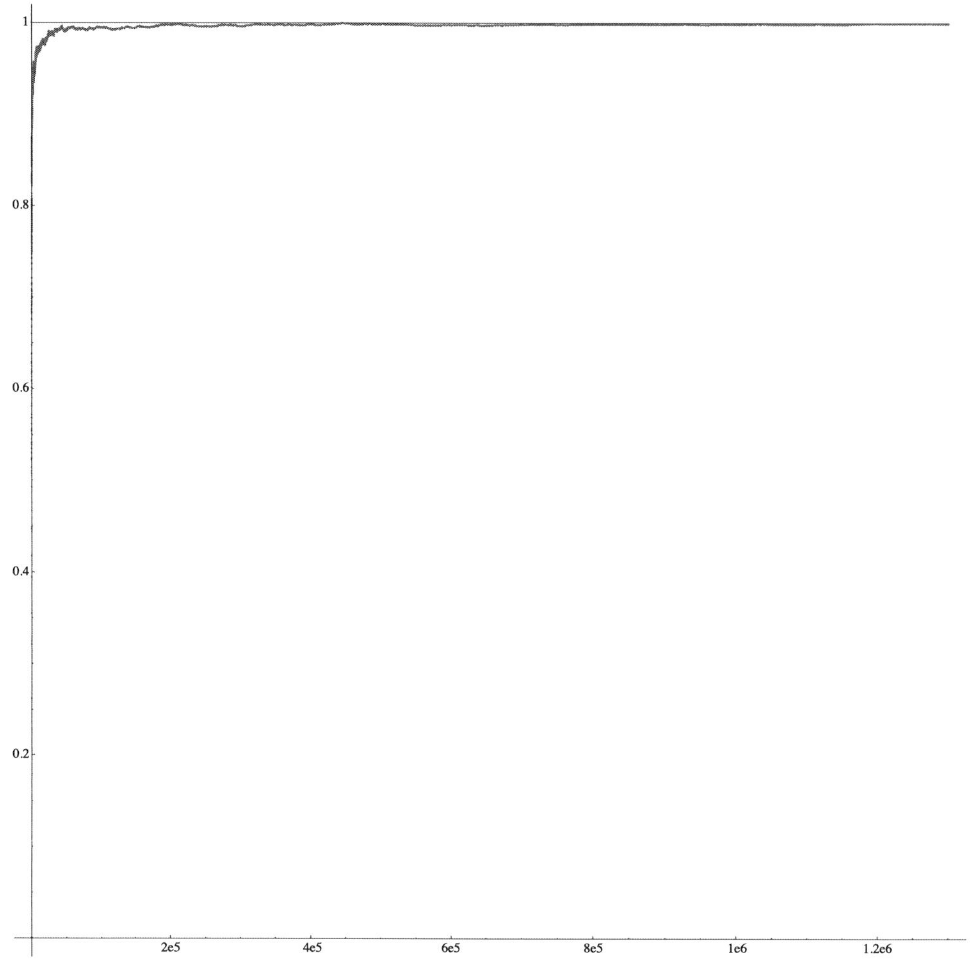

Graph for (83) **for the Serre curve** $t = 197$

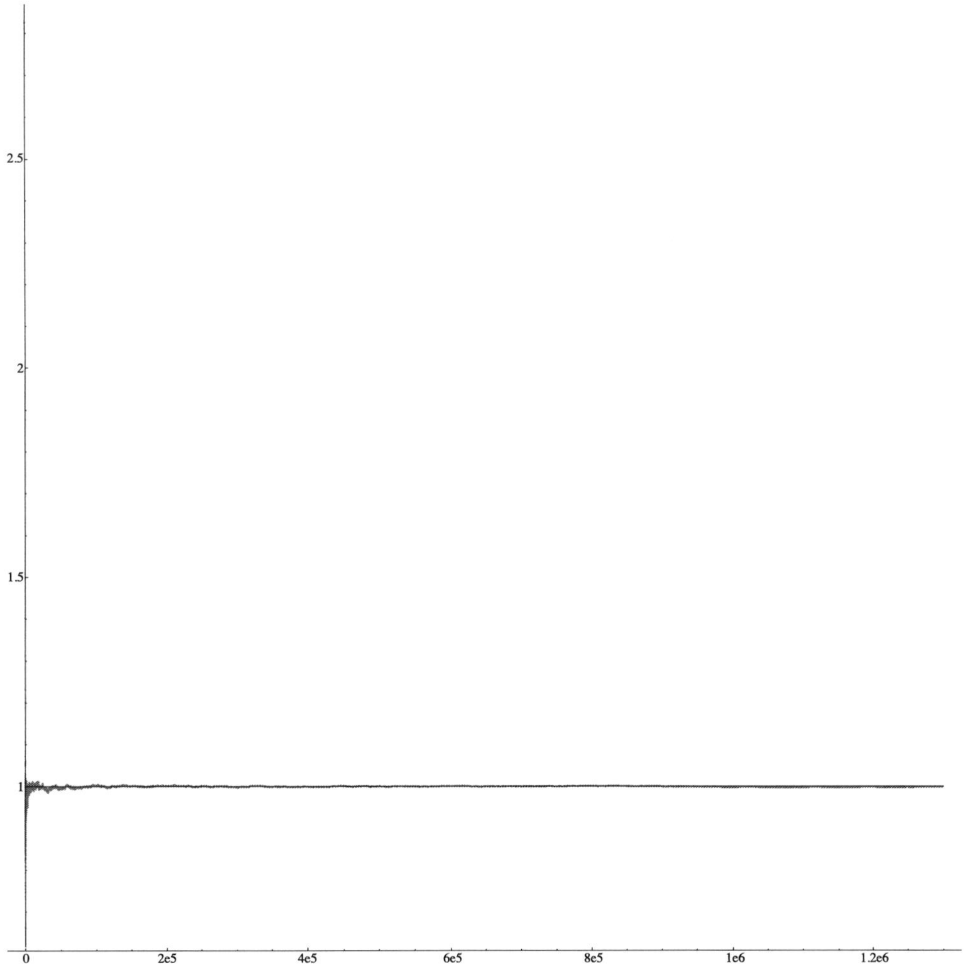

Acknowledgments

The paper is based on lectures given by the author in multiple venues, including: the CIMPA-ICTP research school "Algebraic curves over finite fields" held at the University of the Phillipines Dillman, Manila, in 2013; a mini-course given as a Shapiro Visiting Professor at Penn State University, University Park, Pennsylvania, in 2015; the Arizona Winter School "Analytic methods in arithmetic geometry" held at the University of Arizona, Tucson, USA, in 2016. The author is deeply grateful to the organizers of these events: Francesco Pappalardi, Valerio Talamanca, and Michel Waldschmidt (CIMPA-ICTP); Mihran Papikian (Penn State); Alina Bucur, Bryden Cais, Mirela Ciperiani, Romyar Sharifi, and David Zureick-Brown (Arizona Winter School). The author is also deeply grateful to the student and faculty participants in the lectures for their keen interest and stimulating feedback. Moreover, the author is thankful to Alina Bucur and David Zurick-Brown for their patience and support. A shorter version of these lectures appeared in [**Co17**].

References

[Ak] A. Akbary, *On the greatest prime divisor of N_p*, J. Ramanujan Math. Soc. **23** (2008), no. 3, 259–282. MR2446601

[AkGh] A. Akbary and D. Ghioca, *A geometric variant of Titchmarsh divisor problem*, Int. J. Number Theory **8** (2012), no. 1, 53–69, DOI 10.1142/S1793042112500030. MR2887882

[AkFe] A. Akbary and A. T. Felix, *On invariants of elliptic curves on average*, Acta Arith. **168** (2015), no. 1, 31–70, DOI 10.4064/aa168-1-3. MR3337211

[AkMu] A. Akbary and V. K. Murty, *An analogue of the Siegel-Walfisz theorem for the cyclicity of CM elliptic curves mod p*, Indian J. Pure Appl. Math. **41** (2010), no. 1, 25–37, DOI 10.1007/s13226-010-0002-4. MR2650098

[Ap] T. M. Apostol, *Introduction to analytic number theory*, Springer-Verlag, New York-Heidelberg, 1976. Undergraduate Texts in Mathematics. MR0434929

[Ba] S. Baier, *The Lang-Trotter conjecture on average*, J. Ramanujan Math. Soc. **22** (2007), no. 4, 299–314. MR2376806

[BaJo] S. Baier and N. Jones, *A refined version of the Lang-Trotter conjecture*, Int. Math. Res. Not. IMRN **3** (2009), 433–461, DOI 10.1093/imrn/rnn136. MR2482121

[BaCoDa] A. Balog, A. C. Cojocaru, and C. David, *Average twin prime conjecture for elliptic curves*, Amer. J. Math. **133** (2011), no. 5, 1179–1229, DOI 10.1353/ajm.2011.0033. MR2843097

[BaSh] W. D. Banks and I. E. Shparlinski, *Sato-Tate, cyclicity, and divisibility statistics on average for elliptic curves of small height*, Israel J. Math. **173** (2009), 253–277, DOI 10.1007/s11856-009-0091-0. MR2570668

[BBCCJMSV] R. Bell, C. Blakestad, A. C. Cojocaru, A. Cowan, N. Jones, V. Matei, G. Smith and I. Vogt, *Constants in Titchmarsh divisor problems for elliptic curves,* to appear in Research in Number Theory.

[Bi] B. J. Birch, *How the number of points of an elliptic curve over a fixed prime field varies*, J. London Math. Soc. **43** (1968), 57–60, DOI 10.1112/jlms/s1-43.1.57. MR0230682

[BiSD] B. J. Birch and H. P. F. Swinnerton-Dyer, *Notes on elliptic curves. II*, J. Reine Angew. Math. **218** (1965), 79–108, DOI 10.1515/crll.1965.218.79. MR0179168

[BhSh] M. Bhargava and A. Shankar, *Binary quartic forms having bounded invariants, and the boundedness of the average rank of elliptic curves*, Ann. of Math. (2) **181** (2015), no. 1, 191–242, DOI 10.4007/annals.2015.181.1.3. MR3272925

[BoMoPo] I. Borosh, C. J. Moreno, and H. Porta, *Elliptic curves over finite fields. II*, Math. Comput. **29** (1975), 951–964. MR0404264

[Br] F. Breuer, *Torsion bounds for elliptic curves and Drinfeld modules*, J. Number Theory **130** (2010), no. 5, 1241–1250, DOI 10.1016/j.jnt.2009.11.009. MR2607312

[Ca] H. Carayol, *La conjecture de Sato-Tate (d'après Clozel, Harris, Shepherd-Barron, Taylor)* (French, with French summary), Astérisque **317** (2008), Exp. No. 977, ix, 345–391. Séminaire Bourbaki. Vol. 2006/2007. MR2487739

[Cl] L. Clozel, *The Sato-Tate conjecture*, Current developments in mathematics, 2006, Int. Press, Somerville, MA, 2008, pp. 1–34. MR2459304

[CoPhD] A. C. Cojocaru, *Cyclicity of elliptic curves modulo p*, ProQuest LLC, Ann Arbor, MI, 2002. Thesis (Ph.D.) Queen's University (Canada). MR2703535

[Co02] A. C. Cojocaru, *On the cyclicity of the group of $\mathbb{F}_p$-rational points of non-CM elliptic curves*, J. Number Theory **96** (2002), no. 2, 335–350. MR1932460

[Co03] A. C. Cojocaru, *Cyclicity of CM elliptic curves modulo p*, Trans. Amer. Math. Soc. **355** (2003), no. 7, 2651–2662, DOI 10.1090/S0002-9947-03-03283-5. MR1975393

[Co04] A. C. Cojocaru, *Questions about the reductions modulo primes of an elliptic curve*, Number theory, CRM Proc. Lecture Notes, vol. 36, Amer. Math. Soc., Providence, RI, 2004, pp. 61–79. MR2076566

[Co05] A. C. Cojocaru, *Reductions of an elliptic curve with almost prime orders*, Acta Arith. **119** (2005), no. 3, 265–289, DOI 10.4064/aa119-3-3. MR2167436

[Co17] A. C. Cojocaru, *Primes, elliptic curves and cyclic groups: a synopsis*, Rev. Roumaine Math. Pures Appl. **62** (2017), no. 1, 3–40. MR3626431

[Co05bis] A. C. Cojocaru, *Reductions of an elliptic curve with almost prime orders*, Acta Arith. **119** (2005), no. 3, 265–289, DOI 10.4064/aa119-3-3. MR2167436

[Co08] A. C. Cojocaru, *Square-free orders for CM elliptic curves modulo p*, Math. Ann. **342** (2008), no. 3, 587–615, DOI 10.1007/s00208-008-0249-9. MR2430992

[CoDu] A. C. Cojocaru and W. Duke, *Reductions of an elliptic curve and their Tate-Shafarevich groups*, Math. Ann. **329** (2004), no. 3, 513–534, DOI 10.1007/s00208-004-0517-2. MR2127988

[CoGrJo] A. C. Cojocaru, D. Grant, and N. Jones, *One-parameter families of elliptic curves over $\mathbb{Q}$ with maximal Galois representations*, Proc. Lond. Math. Soc. (3) **103** (2011), no. 4, 654–675, DOI 10.1112/plms/pdr001. MR2837018

[CoFiInYi] A. C. Cojocaru, M. Fitzpatrick, T. Insley, and H. Yilmaz, *Reductions modulo primes of Serre curves: computational data,* appendix to present paper.

[CoHa] A. C. Cojocaru and C. Hall, *Uniform results for Serre's theorem for elliptic curves*, Int. Math. Res. Not. **50** (2005), 3065–3080, DOI 10.1155/IMRN.2005.3065. MR2189500

[CoIwJo] A. C. Cojocaru, H. Iwaniec and N. Jones, *The average asymptotic behaviour of the Frobenius fields of an elliptic curve*, in preparation.

[CoMu04] A. C. Cojocaru and M. R. Murty, *Cyclicity of elliptic curves modulo p and elliptic curve analogues of Linnik's problem*, Math. Ann. **330** (2004), no. 3, 601–625, DOI 10.1007/s00208-004-0562-x. MR2099195

[CoMu05] A. C. Cojocaru and M. R. Murty, *An introduction to sieve methods and their applications*, London Mathematical Society Student Texts, vol. 66, Cambridge University Press, Cambridge, 2006. MR2200366

[CoSh13] A. C. Cojocaru and A. M. Shulman, *An average Chebotarev density theorem for generic rank 2 Drinfeld modules with complex multiplication*, J. Number Theory **133** (2013), no. 3, 897–914, DOI 10.1016/j.jnt.2012.07.001. MR2997774

[CoSh15] A. C. Cojocaru and A. M. Shulman, *The distribution of the first elementary divisor of the reductions of a generic Drinfeld module of arbitrary rank*, Canad. J. Math. **67** (2015), no. 6, 1326–1357, DOI 10.4153/CJM-2015-006-9. MR3415655

[CoTo] A. C. Cojocaru and Á. Tóth, *The distribution and growth of the elementary divisors of the reductions of an elliptic curve over a function field*, J. Number Theory **132** (2012), no. 5, 953–965, DOI 10.1016/j.jnt.2011.08.007. MR2890521

[Dan] H. B. Daniels, *An infinite family of Serre curves*, J. Number Theory **155** (2015), 226–247, DOI 10.1016/j.jnt.2015.03.016. MR3349445

[Dav] H. Davenport, *Multiplicative number theory*, 3rd ed., Graduate Texts in Mathematics, vol. 74, Springer-Verlag, New York, 2000. Revised and with a preface by Hugh L. Montgomery. MR1790423

[DaPa] C. David and F. Pappalardi, *Average Frobenius distributions of elliptic curves*, Internat. Math. Res. Notices **4** (1999), 165–183, DOI 10.1155/S1073792899000082. MR1677267

[DaWu] C. David and J. Wu, *Almost prime values of the order of elliptic curves over finite fields*, Forum Math. **24** (2012), no. 1, 99–119, DOI 10.1515/form.2011.051. MR2879973

[De] M. Deuring, *Die Typen der Multiplikatorenringe elliptischer Funktionenkörper* (German), Abh. Math. Sem. Hansischen Univ. **14** (1941), 197–272, DOI 10.1007/BF02940746. MR0005125

[Du97] W. Duke, *Elliptic curves with no exceptional primes* (English, with English and French summaries), C. R. Acad. Sci. Paris Sér. I Math. **325** (1997), no. 8, 813–818, DOI 10.1016/S0764-4442(97)80118-8. MR1485897

[Du03] W. Duke, *Almost all reductions modulo p of an elliptic curve have a large exponent* (English, with English and French summaries), C. R. Math. Acad. Sci. Paris **337** (2003), no. 11, 689–692, DOI 10.1016/j.crma.2003.10.006. MR2030403

[DuTo] W. Duke and Á. Tóth, *The splitting of primes in division fields of elliptic curves*, Experiment. Math. **11** (2002), no. 4, 555–565 (2003). MR1969646

[FeMu] A. T. Felix and M. R. Murty, *On the asymptotics for invariants of elliptic curves modulo p*, J. Ramanujan Math. Soc. **28** (2013), no. 3, 271–298. MR3113386

[FoIw] É. Fouvry and H. Iwaniec, *Primes in arithmetic progressions*, Acta Arith. **42** (1983), no. 2, 197–218, DOI 10.4064/aa-42-2-197-218. MR719249

[FoMu] É. Fouvry and M. R. Murty, *On the distribution of supersingular primes*, Canad. J. Math. **48** (1996), no. 1, 81–104, DOI 10.4153/CJM-1996-004-7. MR1382477

[FrIw] J. Friedlander and H. Iwaniec, *Opera de cribro*, American Mathematical Society Colloquium Publications, vol. 57, American Mathematical Society, Providence, RI, 2010. MR2647984

[FrKu] T. Freiberg and P. Kurlberg, *On the average exponent of elliptic curves modulo p*, Int. Math. Res. Not. IMRN **8** (2014), 2265–2293, DOI 10.1093/imrn/rns280. MR3194018

[FrPo] T. Freiberg and P. Pollack, *The average of the first invariant factor for reductions of CM elliptic curves* mod p, Int. Math. Res. Not. IMRN **21** (2015), 11333–11350, DOI 10.1093/imrn/rnv027. MR3456045

[Ge03] E.-U. Gekeler, *Frobenius distributions of elliptic curves over finite prime fields*, Int. Math. Res. Not. **37** (2003), 1999–2018, DOI 10.1155/S1073792803211272. MR1995144

[Ge06] E.-U. Gekeler, *The distribution of group structures on elliptic curves over finite prime fields*, Doc. Math. **11** (2006), 119–142. MR2226271

[Ge08] E.-U. Gekeler, *Statistics about elliptic curves over finite prime fields*, Manuscripta Math. **127** (2008), no. 1, 55–67, DOI 10.1007/s00229-008-0192-9. MR2429913

[Grt] D. Grant, *A formula for the number of elliptic curves with exceptional primes*, Compositio Math. **122** (2000), no. 2, 151–164, DOI 10.1023/A:1001874400583. MR1775416

[Grv] A. Granville, *Primes in intervals of bounded length*, Bull. Amer. Math. Soc. (N.S.) **52** (2015), no. 2, 171–222, DOI 10.1090/S0273-0979-2015-01480-1. MR3312631

[GuMu] R. Gupta and M. R. Murty, *Cyclicity and generation of points mod p on elliptic curves*, Invent. Math. **101** (1990), no. 1, 225–235, DOI 10.1007/BF01231502. MR1055716

[HaRi] H. Halberstam and H.-E. Richert, *Sieve methods*, Academic Press [A subsidiary of Harcourt Brace Jovanovich, Publishers], London-New York, 1974. London Mathematical Society Monographs, No. 4. MR0424730

[HaWr] G. H. Hardy and E. M. Wright, *An introduction to the theory of numbers*, 6th ed., Oxford University Press, Oxford, 2008. Revised by D. R. Heath-Brown and J. H. Silverman; With a foreword by Andrew Wiles. MR2445243

[Ho] E. W. Howe, *On the group orders of elliptic curves over finite fields*, Compositio Math. **85** (1993), no. 2, 229–247. MR1204781

[Hu] M. N. Huxley, *The large sieve inequality for algebraic number fields. III. Zero-density results*, J. London Math. Soc. (2) **3** (1971), 233–240, DOI 10.1112/jlms/s2-3.2.233. MR0276196

[InWeLu] K.-H. Indlekofer, S. Wehmeier, and L. G. Lucht, *Mean behaviour and distribution properties of multiplicative functions*, Comput. Math. Appl. **48** (2004), no. 12, 1947–1971, DOI 10.1016/j.camwa.2004.01.015. MR2116969

[IwJU] H. Iwaniec and J. Jiménez Urroz, *Orders of CM elliptic curves modulo p with at most two primes*, Ann. Sc. Norm. Super. Pisa Cl. Sci. (5) **9** (2010), no. 4, 815–832. MR2789476

[IwKo] H. Iwaniec and E. Kowalski, *Analytic number theory*, American Mathematical Society Colloquium Publications, vol. 53, American Mathematical Society, Providence, RI, 2004. MR2061214

[Ja] K. James, *Variants of the Sato-Tate and Lang-Trotter conjectures*, Frobenius distributions: Lang-Trotter and Sato-Tate conjectures, Contemp. Math., vol. 663, Amer. Math. Soc., Providence, RI, 2016, pp. 175–184, DOI 10.1090/conm/663/13354. MR3502943

[JU08] J. Jiménez Urroz, *Almost prime orders of CM elliptic curves modulo p*, Algorithmic number theory, Lecture Notes in Comput. Sci., vol. 5011, Springer, Berlin, 2008, pp. 74–87, DOI 10.1007/978-3-540-79456-1_4. MR2467838

[JU16] J. Jiménez-Urroz, *Some problems of elliptic curves in number theory*, Highly composite: papers in number theory, Ramanujan Math. Soc. Lect. Notes Ser., vol. 23, Ramanujan Math. Soc., Mysore, 2016, pp. 123–135. MR3692731

[JiQi] Q. Ji and H. Qin, *CM elliptic curves and primes captured by quadratic polynomials*, Asian J. Math. **18** (2014), no. 4, 707–726, DOI 10.4310/AJM.2014.v18.n4.a7. MR3275725

[Jo09] N. Jones, *Averages of elliptic curve constants*, Math. Ann. **345** (2009), no. 3, 685–710, DOI 10.1007/s00208-009-0373-1. MR2534114

[Jo10] N. Jones, *Almost all elliptic curves are Serre curves*, Trans. Amer. Math. Soc. **362** (2010), no. 3, 1547–1570, DOI 10.1090/S0002-9947-09-04804-1. MR2563740

[Jo10bis] T. Bandman, F. Grunewald, and B. Kunyavskiĭ, *Geometry and arithmetic of verbal dynamical systems on simple groups*, Groups Geom. Dyn. **4** (2010), no. 4, 607–655, DOI 10.4171/GGD/98. With an appendix by Nathan Jones. MR2727656

[Ka] N. M. Katz, *Lang-Trotter revisited*, Bull. Amer. Math. Soc. (N.S.) **46** (2009), no. 3, 413–457, DOI 10.1090/S0273-0979-09-01257-9. MR2507277

[Ki] S. Kim, *Average behaviors of invariant factors in Mordell-Weil groups of CM elliptic curves modulo p*, Finite Fields Appl. **30** (2014), 178–190, DOI 10.1016/j.ffa.2014.07.003. MR3249828

[Kob] N. Koblitz, *Primality of the number of points on an elliptic curve over a finite field*, Pacific J. Math. **131** (1988), no. 1, 157–165. MR917870

[Kow] E. Kowalski, *Analytic problems for elliptic curves*, J. Ramanujan Math. Soc. **21** (2006), no. 1, 19–114. MR2226355

[KuLi] W. Kuo and Y.-R. Liu, *Cyclicity of finite Drinfeld modules*, Journal of the London Mathematical Society 80, 2009, pp. 567–584.

[LaOd] J. C. Lagarias and A. M. Odlyzko, *Effective versions of the Chebotarev density theorem*, Algebraic number fields: L-functions and Galois properties (Proc. Sympos., Univ. Durham, Durham, 1975), Academic Press, London, 1977, pp. 409–464. MR0447191

[LaTr76] S. Lang and H. Trotter, *Frobenius distributions in* GL_2*-extensions: Distribution of Frobenius automorphisms in* GL_2*-extensions of the rational numbers*, Lecture Notes in Mathematics, Vol. 504, Springer-Verlag, Berlin-New York, 1976. MR0568299

[LaTr77] S. Lang and H. Trotter, *Primitive points on elliptic curves*, Bull. Amer. Math. Soc. **83** (1977), no. 2, 289–292, DOI 10.1090/S0002-9904-1977-14310-3. MR0427273

[Ma77] B. Mazur, *Modular curves and the Eisenstein ideal*, Inst. Hautes Études Sci. Publ. Math. **47** (1977), 33–186 (1978). MR488287

[Ma78] B. Mazur, *Rational isogenies of prime degree (with an appendix by D. Goldfeld)*, Invent. Math. **44** (1978), no. 2, 129–162, DOI 10.1007/BF01390348. MR482230

[MiMu] S. A. Miri and V. K. Murty, *An application of sieve methods to elliptic curves*, Progress in cryptology—INDOCRYPT 2001 (Chennai), Lecture Notes in Comput. Sci., vol. 2247, Springer, Berlin, 2001, pp. 91–98, DOI 10.1007/3-540-45311-3_9. MR1934487

[MoVa] H. L. Montgomery and R. C. Vaughan, *Multiplicative number theory. I. Classical theory*, Cambridge Studies in Advanced Mathematics, vol. 97, Cambridge University Press, Cambridge, 2007. MR2378655

[Mo] P. Moree (with contributions by A. C. Cojocaru, W. Gajda and H. Graves), *Artin's primitive root conjecture—a survey*, Integers 12A, 2012, John Selfridge Memorial Issue, #A13.

[Mu83] M. R. Murty, *On Artin's conjecture*, J. Number Theory **16** (1983), no. 2, 147–168, DOI 10.1016/0022-314X(83)90039-2. MR698163

[Mu87] M. R. Murty, *On the supersingular reduction of elliptic curves*, Proc. Indian Acad. Sci. Math. Sci. **97** (1987), no. 1-3, 247–250 (1988), DOI 10.1007/BF02837827. MR983618

[MuMuSa] M. R. Murty, V. K. Murty, and N. Saradha, *Modular forms and the Chebotarev density theorem*, Amer. J. Math. **110** (1988), no. 2, 253–281, DOI 10.2307/2374502. MR935007

[KMu] V. K. Murty, *Modular forms and the Chebotarev density theorem. II*, Analytic number theory (Kyoto, 1996), London Math. Soc. Lecture Note Ser., vol. 247, Cambridge Univ. Press, Cambridge, 1997, pp. 287–308, DOI 10.1017/CBO9780511666179.019. MR1694997

[Ol] L. D. Olson, *Points of finite order on elliptic curves with complex multiplication*, Manuscripta Math. **14** (1974), 195–205, DOI 10.1007/BF01171442. MR0352104

[Pol] P. Pollack, *A Titchmarsh divisor problem for elliptic curves*, Math. Proc. Cambridge Philos. Soc. **160** (2016), no. 1, 167–189, DOI 10.1017/S0305004115000614. MR3432335

[Poo] B. Poonen, *Average rank of elliptic curves [after Manjul Bhargava and Arul Shankar]*, Astérisque **352** (2013), Exp. No. 1049, viii, 187–204. Séminaire Bourbaki. Vol. 2011/2012. Exposés 1043–1058. MR3087347

[Qi] H. Qin, *Anomalous primes of the elliptic curve* $E_D\colon y^2 = x^3 + D$, Proc. Lond. Math. Soc. (3) **112** (2016), no. 2, 415–453, DOI 10.1112/plms/pdv072. MR3471254

[Ra] V. Radhakrishnan, *Asymptotic formula for the number of non-Serre curves in a two-parameter family,* PhD Thesis, University of Colorado at Boulder, 2008.

[RuSi02] K. Rubin and A. Silverberg, *Ranks of elliptic curves*, Bull. Amer. Math. Soc. (N.S.) **39** (2002), no. 4, 455–474, DOI 10.1090/S0273-0979-02-00952-7. MR1920278

[RuSi09] K. Rubin and A. Silverberg, *Point counting on reductions of CM elliptic curves*, J. Number Theory **129** (2009), no. 12, 2903–2923, DOI 10.1016/j.jnt.2009.01.020. MR2560842

[Sa] P. Sarnak, *Equidistribution and primes: Géométrie différentielle, physique mathématique, mathématiques et société. II*, Astérisque **322** (2008), 225–240. MR2521658

[Scha] W. Schaal, *On the large sieve method in algebraic number fields*, J. Number Theory **2** (1970), 249–270, DOI 10.1016/0022-314X(70)90052-1. MR0272745

[Scho] R. Schoof, *The exponents of the groups of points on the reductions of an elliptic curve*, Arithmetic algebraic geometry (Texel, 1989), Progr. Math., vol. 89, Birkhäuser Boston, Boston, MA, 1991, pp. 325–335. MR1085266

[Se72] J.-P. Serre, *Propriétés galoisiennes des points d'ordre fini des courbes elliptiques* (French), Invent. Math. **15** (1972), no. 4, 259–331, DOI 10.1007/BF01405086. MR0387283

[Se77] J-P. Serre, *Résumé des cours de 1977-1978,* Annuaire du Collège de France 1978, pp. 67–70.

[Se81] J.-P. Serre, *Quelques applications du théorème de densité de Chebotarev* (French), Inst. Hautes Études Sci. Publ. Math. **54** (1981), 323–401. MR644559

[Si] J. H. Silverman, *The arithmetic of elliptic curves*, Graduate Texts in Mathematics, vol. 106, Springer-Verlag, New York, 1986. MR817210

[So] K. Soundararajan, *The distribution of prime numbers*, Equidistribution in number theory, an introduction, NATO Sci. Ser. II Math. Phys. Chem., vol. 237, Springer, Dordrecht, 2007, pp. 59–83, DOI 10.1007/978-1-4020-5404-4_4. MR2290494

[St] H. M. Stark, *Counting points on CM elliptic curves*, Rocky Mountain J. Math. **26** (1996), no. 3, 1115–1138, DOI 10.1216/rmjm/1181072041. MR1428490

[StWe] J. Steuding and A. Weng, *On the number of prime divisors of the order of elliptic curves modulo p*, Acta Arith. **117** (2005), no. 4, 341–352, DOI 10.4064/aa117-4-2. MR2140162

[Te] G. Tenenbaum, *Introduction to analytic and probabilistic number theory,* 3rd edition, Graduate Texts in Mathematics Vol. 163, American Mathematical Society, 2015.

[Vl] S. G. Vlăduţ, *Cyclicity statistics for elliptic curves over finite fields*, Finite Fields Appl. **5** (1999), no. 1, 13–25, DOI 10.1006/ffta.1998.0225. MR1667099

[Wan] D. Q. Wan, *On the Lang-Trotter conjecture*, J. Number Theory **35** (1990), no. 3, 247–268, DOI 10.1016/0022-314X(90)90117-A. MR1062334

[Was] L. C. Washington, *Elliptic curves: Number theory and cryptography*, Discrete Mathematics and its Applications (Boca Raton), Chapman & Hall/CRC, Boca Raton, FL, 2003. MR1989729

[We55] A. Weil, *On a certain type of characters of the idèle-class group of an algebraic number-field*, Science Council of Japan, Tokyo, 1956, pp. 1–7. MR0083523

[We55bis] A. Weil, *On the theory of complex multiplication*, Proceedings of the International Symposium on Algebraic Number Theory, Tokyo & Nikko, 1955, Science Council of Japan, Tokyo, 1956, pp. 9–22. MR0083177

[Wi] A. Wiles, *The Birch and Swinnerton-Dyer conjecture*, The millennium prize problems, Clay Math. Inst., Cambridge, MA, 2006, pp. 31–41. MR2238272

[Wu] J. Wu, *The average exponent of elliptic curves modulo p*, J. Number Theory **135** (2014), 28–35, DOI 10.1016/j.jnt.2013.08.009. MR3128449

[Yo] M. P. Young, *Low-lying zeros of families of elliptic curves*, J. Amer. Math. Soc. **19** (2006), no. 1, 205–250, DOI 10.1090/S0894-0347-05-00503-5. MR2169047

[Zy11] D. Zywina, *A refinement of Koblitz's conjecture*, Int. J. Number Theory **7** (2011), no. 3, 739–769, DOI 10.1142/S1793042111004411. MR2805578

[Zy15] D. Zywina, *Bounds for the Lang-Trotter conjectures*, SCHOLAR—a scientific celebration highlighting open lines of arithmetic research, Contemp. Math., vol. 655, Amer. Math. Soc., Providence, RI, 2015, pp. 235–256, DOI 10.1090/conm/655/13206. MR3453123

Department of Mathematics, Statistics and Computer Science, University of Illinois at Chicago, 851 S Morgan St, 322 SEO, Chicago, 60607, Illinois –and– Institute of Mathematics "Simion Stoilow" of the Romanian Academy, 21 Calea Grivitei St, Bucharest, 010702, Sector 1, Romania

Email address: cojocaru@uic.edu

Contemporary Mathematics
Volume **740**, 2019
https://doi.org/10.1090/conm/740/14902

Growth and expansion in algebraic groups over finite fields

Harald Andrés Helfgott

CONTENTS

1. Introduction

This text is meant to serve as a brief introduction to the study of growth in groups of Lie type, with $\mathrm{SL}_2(\mathbb{F}_q)$ and some of its subgroups as the key examples. They are an edited version of the notes I distributed at the Arizona Winter School in 2016. Those notes were, in turn, based in part on the survey [**Hel15**] and in part on the notes for courses I gave on the subject in Cusco [**Hela**] and Göttingen.

Given the format of the Arizona Winter School, the emphasis here is on reaching the frontiers of current research as soon as possible, and not so much on giving a comprehensive overview of the field. For that the reader is referred to [**Hel15**] and its bibliography, or to [**Kow13**] and [**Tao15**]. At the same time – again motivated by the school's demands – we will take a brief look at several applications at the end.

It will be necessary to be minimally conversant with some of the basic classical vocabulary of algebraic geometry (as in the first chapter of Mumford's Red Book [**Mum99**]), and with some notions on algebraic groups (such as SL_2) and Lie algebras (such as $\mathfrak{sl}_2$). A very brief compendium of what will be needed can be found in §4.1. It is often helpful (and only rarely misleading) to be willing to believe that matters work out in much the same way over finite field as they do over the reals.

The purpose of these notes is expository, not historical, though I have tried to give key references. The origins of several ideas are traced in greater detail in

2010 *Mathematics Subject Classification.* Primary 20F69; 20D60; 11B30; 05C25.

[**Hel15**]. In §1.2, we will give a summary of the results we later prove and also of results and open questions of the same kind. We will go over some important related questions and applications later, in §6.

1.1. Basic questions and concepts: diameter, growth, expansion. Let A be a finite subset of a group G. Consider the sets

$$\begin{aligned} &A, \\ A\cdot A &= \{x\cdot y : x,y\in A\}, \\ A\cdot A\cdot A &= \{x\cdot y\cdot z : x,y,z\in A\}, \\ &\dots \\ A^k &= \{x_1x_2\dots x_k : x_i\in A\}. \end{aligned}$$

Write $|S|$ for the *size* of a finite set S, meaning simply the number of elements of S. A question arises naturally: how does $|A^k|$ grow as k grows?

This kind of question has been studied from the perspective of additive combinatorics (for G abelian) and geometric group theory (G infinite, $k\to\infty$). There are also some crucial related concepts coming from other fields: *diameters* and *expanders*, to start with.

Diameters. Let A be a set of generators of G. When G is infinite, a central question is how $|A^k|$ behaves as a function of k as $k\to\infty$. When G is finite, that question does not make much sense, as $|A^k|$ obviously stays constant as soon as $A^k=G$. Instead, let us ask ourselves what is the least value of k such that $A^k=G$. This value of k is called the *diameter*. It is finite because, for A generating G, $A^j\neq G$ implies $|A^{j+1}|>|A^j|$. (Why is this last statement true?)

The term *diameter* comes from geometry. What we have is not just an analogy – we can actually put our basic terms in a geometrical framework, as geometric group theory does. A *Cayley graph* $\Gamma(G,A)$ is the graph having $V=G$ as its set of vertices and $E=\{(g,ag): g\in G, a\in A\}$ as its set of edges. Define the length of a path in the graph as the number of edges in it, and the distance $d(v,w)$ between two vertices v, w in the graph as the length of the shortest path between them. The *diameter* of a graph is the maximum of the distance $d(v,w)$ over all vertices v, w. It is easy to see that the diameter of G with respect to A, as we defined it above, equals the diameter of the graph $\Gamma(G,A)$.

Product theorems. A central question of additive combinatorics is as follows: for finite subsets A of an abelian group $(G,+)$, when exactly is it that $A+A$ is much larger than A? In non-abelian groups $(G,\cdot)$, the right form of the question turns out to be: given a set of generators A of G, when is A^3 much larger than A? (We will see later why it is better to ask about $A^3=A\cdot A\cdot A$ rather than $A^2=A\cdot A$ here.)

It is clear that, if we show that, for any generating set A of G,

$$\text{either}\quad |A^3| \text{ is much larger than } A \quad\text{or}\quad A^3=G, \tag{1.1}$$

then A^k grows rapidly until roughly the point where $A^k=G$: simply apply (1.1) to A, A^3, A^9, etc., in place of A. In particular, (1.1) yields an upper bound on the diameter of G with respect to A. We call a result of the form (1.1) a *product theorem*.

Expansion. We say that a graph is an *vertex expander* with parameter $\delta>0$ (or *δ-vertex expander*) if, for every subset S of the set of vertices V satisfying (say)

$|S| \leq |V|/2$, the number of vertices $v \in V$ not in S such that at least one edge connects v to some element of S is at least $\delta|S|$. (We may think of S as being a set of infected individuals; then we are saying that the number of the newly infected will always be at least $\delta|S|$, unless the disease has reached a near-saturation point.)

Two closely connected notions are that of *edge expansion* and *spectral expansion*. First, some basic terms. A graph is *regular* if, for any vertex v, the number of vertices w such that (v, w) is an edge equals a constant d, and the number of vertices w such that (w, v) is an edge also equals a constant (which must also be d, by a simple counting argument). We call d the *degree* or *valency* of the graph. A Cayley graph $\Gamma(G, A)$ is always regular of degree $d = |A|$.

A regular graph $\Gamma = (V, E)$ of degree d is a *δ-edge expander* if, for every $S \subset V$ satisfying $|S| \leq |V|/2$, the number of edges having one vertex in S and one outside S is at least $\delta d|S|$. It is clear that, if Γ is a δ-vertex expander, then it is a (δ/d)-edge expander, and, if it is a δ-edge expander, then it is a δ-vertex expander.

We say that a graph Γ is *symmetric* to mean that (v, w) is an edge if and only if (w, v) is an edge. If Γ is a Cayley graph $\Gamma(G, A)$, then Γ is symmetric provided that $A^{-1} = \{g^{-1} : g \in A\}$ equals A. We will generally assume that $A^{-1} = A$ without much loss of generality. (Replace A by $A \cup A^{-1}$ otherwise.)

Given a regular graph Γ with a set of vertices V, the *adjacency operator* $\mathscr{A}$ is the linear operator taking any given function $f : V \to \mathbb{C}$ to the function $\mathscr{A}f : V \to \mathbb{C}$ defined by

$$\mathscr{A}f(v) = \frac{1}{d} \sum_{w:(v,w) \text{ is an edge}} f(w). \tag{1.2}$$

Assume that the graph Γ is symmetric. Then $\mathscr{A}$ is a symmetric operator, and thus has full real spectrum. Its largest eigenvalue is 1; it corresponds to constant eigenfunctions. If every eigenvalue λ of $\mathscr{A}$ corresponding to non-constant eigenfunctions satisfies $\lambda \leq 1 - \delta$ for some $\delta > 0$, we say that Γ is a δ-spectral expander, or a *δ-expander* for short.

If a regular, symmetric graph is a δ-spectral expander, then it is a $(\delta/2)$-edge expander, and, if it is a δ-edge expander, then it is a $(\delta^2/2)$-spectral expander. This fact is non-trivial; it is called the *Cheeger-Alon-Milman* inequality [**AM85**], by analogy with the *Cheeger inequality* on manifolds [**Che70**].

The notion of spectral expansion is natural, not just because of the analogy with surfaces and their Laplacians, but, among other reasons, because of random walks: a drunken mathematician left to wander in a spectral expander Γ will be anywhere with about the same probability after only a short while. To put matters more formally – as we shall see in §6.1, spectral expansion implies small *mixing time*.

Since the diameter of a graph is bounded by its (ℓ_∞-)mixing time, it follows immediately that spectral expansion implies small diameter. We can also prove this implication going through edge and vertex expansion: if a graph is a δ-vertex expander, it is very easy to see that its diameter is $\ll (\log|G|)/\delta$; apply, then, the Cheeger-Alon-Milman inequality.

1.2. A brief overview of results on growth and diameter. Let us first review some basic terms from group theory. A group G is *simple* if it has no normal subgroups other than itself and the identity. A *subnormal series* of a group G is a sequence of subgroups

$$\{e\} = H_0 \lhd H_1 \lhd H_2 \lhd \cdots \lhd H_k = G, \tag{1.3}$$

i.e., H_i is normal in H_{i+1} for every $0 \leq i < k$. A *decomposition series* is a subnormal series in which every quotient H_{i+1}/H_i is simple. It is clear that every finite group has a decomposition series.

In some limited sense, questions on growth behave well under taking quotients, and thus reduce to the case of simple groups, at least if our decomposition series of bounded length. (To be precise: for how product theorems behave under taking quotients, see exercises 2.8 and 2.9. For the behavior of diameters under quotients, look up *Schreier generators*.) It thus makes sense to focus on simple groups.

1.2.1. *Simple groups: what to expect?* Some special cases of the following conjecture are arguably older "folklore".

CONJECTURE 1. (*Babai*, [**BS92**, Conj. 1.7]) *Let G be finite, simple and non-abelian. Let A be any set of generators of G. Then*

$$\operatorname{diam}(\Gamma(G, A)) \ll (\log |G|)^C,$$

where C and the implied constant are absolute constants.

(See §1.3 for definitions of asymptotic notation.)

What about finite, simple, *abelian* groups G? They are the groups $G = \mathbb{Z}/p\mathbb{Z}$. In that case, diameters can be very large: for instance, $\operatorname{diam}\Gamma(\mathbb{Z}/p\mathbb{Z}, \{1\}) = p-1$. In general, when G is abelian, the question of which subsets $A \subset \mathbb{Z}/p\mathbb{Z}$ satisfy $|A + A| > K|A|$ for given K is classical, and difficult; for K a constant, it is answered by a suitable generalization of Freiman's theorem [**GR07**]. (Freiman had done the case $G = \mathbb{Z}$; see [**Fre73**], or the exposition [**Bil99**].) The strongest result on the abelian case to date is that of Sanders ([**San12**]; based in part on [**CS10**]).

The Classification of Finite Simple Groups[1] tells us that all finite, simple, non-abelian groups G fall into three classes:

(a) simple groups of Lie type, that is, matrix groups over finite fields (such as $\mathrm{PSL}_n(\mathbb{F}_q)$ or $\mathrm{PSp}_{2n}(\mathbb{F}_q)$), including some generalizations (twisted groups);
(b) alternating groups $\mathrm{Alt}(n)$. The simple group $\mathrm{Alt}(n)$ is the unique subgroup of index 2 of the group $\mathrm{Sym}(n)$ of all permutations of n elements;
(c) a finite list of exceptions, including, for example, the "monster group".

We can put (c) out of our minds, since it has a finite number of elements, and we are aiming for asymptotic statements.

1.2.2. *Simple groups of Lie type (and bounded rank).* Our main goal in these notes will be to prove the following theorem.

THEOREM 1.1. *Let $G = \mathrm{SL}_2(K)$ or $G = \mathrm{PSL}_2(K)$, K a field. Let $A \subset G$ be a set of generators of G. Then either*

$$|A^3| \geq |A|^{1+\delta} \tag{1.4}$$

or

$$A^3 = G, \tag{1.5}$$

where $\delta > 0$ is an absolute constant.

[1] Famed in mathematical lore as the theorem whose proof would be of the size of a large encyclopedia, were it all in one place.

Here $\mathrm{PSL}_2(K) = \mathrm{SL}_2(K)/\{I, -I\}$, where $\mathrm{SL}_2(K)$ is, of course, the group of 2-by-2 matrices with entries in a field K and determinant 1. The group $\mathrm{PSL}_2(K)$ is simple for $K = \mathbb{F}_q$ finite. It is a group of Lie type; indeed, it will be our white mouse, in that it is convenient to work with, but sufficiently complex to be a good example of a large class.

Theorem 1.1 was first proved in [**Hel08**] for $K = \mathbb{F}_p$, with $A^k = G$ (k a constant) instead of $A^3 = G$. It then underwent a series of generalizations ([**BG08a**], [**Din11**], [**Hel11**], [**GH11**], [**BGT11**] and [**PS16**], among others). By now, we know it for every simple group of Lie type of bounded rank ([**BGT11**], [**PS16**]). The "bounded rank" condition means simply that the constant δ in the inequality $|A^3| \geq |A|^{1+\delta}$ depends on the rank of the group. (The rank of SL_n is $n-1$, that of SO_n is $\lfloor n/2 \rfloor$, etc.) In fact, there are examples (due to Pyber) that show that δ has to depend on the rank.

We will give a proof of Thm. 1.1 that descends from, but is not the same as, the proof in [**Hel08**]; it has strong influences from [**Hel11**], [**BGT11**] and [**PS16**]. In particular, the proof we shall give generalizes readily to SL_n and other higher-rank groups; many of our intermediate results will be stated for SL_n, and the ideas carry over to other group families.

EXERCISE 1.2. *Let K be a finite field. Let $G = \mathrm{PSL}_2(K)$ or $G = \mathrm{SL}_2(K)$. Let $S \subset G$ generate G. Using Thm.* 1.1, *prove that the diameter of $\Gamma(G, S)$ is $\ll (\log |G|)^C$, where C and the implied constant are absolute. Indeed, $C = O(1/\delta)$, where δ is the absolute constant in* (1.4)*. Hint: apply Thm.* 1.1 *repeatedly, with S equal to A, A^3, A^9,...*

In other words, Babai's conjecture holds for $G = \mathrm{PSL}_2(\mathbb{F}_q)$. The bound $\operatorname{diam}\Gamma(G, A) \ll (\log |G|)^C$ also holds for all other simple groups of Lie type, only then C depends on the rank, since δ does.

Before [**Hel08**], $\Gamma(G, A)$ was known to be an expander for some particular sets of generators A of $G = \mathrm{SL}_2(\mathbb{F}_q)$. In those cases, then, the diameter bound $\operatorname{diam}\Gamma(G, A) \ll \log |G|$ was also known. The main element of the proof came from modular forms (Selberg's spectral gap [**Sel65**]).

Impatient readers may now jump to the body of the text and leave the rest of the introduction for later. They should certainly read §6.1, on applications of Theorem 1.1 to expander graphs.

1.2.3. *The simple group* $\mathrm{Alt}(n)$. For $G = \mathrm{Alt}(n)$, we have a statement that is somewhat weaker than Babai's conjecture.

THEOREM 1.3. (*Helfgott-Seress,* [**HS14**]) *Let $G = \mathrm{Sym}(n)$ or $G = \mathrm{Alt}(n)$. Let $A \subset G$ be a set of generators of G. Then*

$$\operatorname{diam}(G, A) = e^{O\left((\log n)^4 (\log\log n)\right)} = e^{O_\epsilon\left((\log\log |G|)^{4+\epsilon}\right)} \tag{1.6}$$

for $\epsilon > 0$ arbitrary.

In fact, the bound $\operatorname{diam}(G, A) = \exp(O((\log n)^4 (\log\log n)))$ holds for all transitive groups $G < \mathrm{Sym}(n)$, and can be deduced from Thm. 1.3. We could state this result as follows: let us be given a permutation puzzle with n pieces that has a solution and satisfies transitivity (that is, any piece can be sent to any other one by some succession of moves). Then there is always a short solution, starting from any reachable position. Incidentally, non-transitive puzzles, such as Rubik's cube, can be reduced to transitive ones at some cost, by means of Schreier generators.

We cannot have a product theorem just like Thm. 1.1 in $\mathrm{Alt}(n)$ or $\mathrm{Sym}(n)$.

COUNTEREXAMPLE 1 (Pyber, Spiga). *Let H be the subgroup of $\mathrm{Sym}(n)$ consisting of all permutations of $\{1,\ldots,m\}$. Let σ be the cycle taking i to $i+1$ $(i \leq n-1)$ and n to 1. Let $A = H \cup \{\sigma, \sigma^{-1}\}$. Then*

$$|A^3| = \left|\{\sigma,\sigma^{-1},e\} \cdot H \cdot \{\sigma,\sigma^{-1},e\} \cup H\sigma^{\pm 1}H\right| \leq 9m! + 2(m+1)! \leq (2m+11)|A|.$$

The factor $(2m+11)$ compared to $|A|$ for A large; if we set, say, $m \sim n/2$, then $(2m+11) \ll |A|^{3/n}$.

It might be that one of $|A^{O(n^C)}| \geq |A|^{1+\delta}$ or $A^{O(n^C)} = G$ always holds. Even having one of $\left|A^{O(n^C)}\right| \geq |A|^{1+\delta/\log n}$ or $A^{O(n^C)} = G$ would be a definite improvement over Thm. 1.3. The exponents 4 in (1.6) would become 3, and, at any rate, as we shall later see, product theorems have consequences other than diameter bounds.

It would be natural to hope that some ideas in 1.3, or its later version [**Hel19**], or future strengthenings thereof, will be useful in addressing Babai's conjecture over groups of Lie type of unbounded rank. It is not just that the known counterexamples to strong product theorems over $\mathrm{Sym}(n)$ and SL_n are related. There are ways to define the "field with one element" $\mathbb{F}_{\mathrm{un}}$, and objects over it; then one generally obtains that $\mathrm{Sym}(n) \sim \mathrm{SL}_n(\mathbb{F}_{\mathrm{un}})$. See, e.g., [**Lor18**].

1.2.4. *Solvable and nilpotent groups.* A group G is *solvable* if it has a subnormal series

$$\{e\} = H_0 \triangleleft H_1 \triangleleft \cdots \triangleleft H_k = G \tag{1.7}$$

all of whose quotients H_{i+1}/H_i are abelian. As we said before, questions on growth behave well under quotients, but such a reduction does not help us as much as we would like, since the best results available for the abelian case are considerably less strong than $|A \cdot A \cdot A| \geq |A|^{1+\delta}$.

A solvable group is *nilpotent* if it has a subnormal series (1.7) with G_{i+1}/G_i contained in the center of G/G_i for every $0 \leq i < k$. Nilpotent groups can often be seen as "almost abelian", and our context is no exception. One should not hope to get stronger results on growth in nilpotent groups than for abelian groups – and, on the positive side, one can study nilpotent groups with Freiman's and Ruzsa's tools, supplemented by a Lie-algebra framework ([**Toi14**]; see also [**FKP10**] and Tao [**Tao10**]).

What one can aim for is to show that, given a set A in a solvable group, either A grows rapidly, or we are really in a nilpotent case. We can make such a statement precise as follows.

CONJECTURE 2. *Let $A \subset \mathrm{GL}_n(K)$, K a field. Assume that the group $\langle A \rangle$ generated by A is solvable. Then, for any $C \geq 1$, either*

$$|A^3| \geq C|A| \tag{1.8}$$

or there are subgroups $N \triangleleft G_0 \triangleleft \langle A \rangle$ such that G_0/N is nilpotent and

$$N \subset (A \cup A^{-1} \cup \{e\})^k, \qquad \left|(A \cup A^{-1} \cup \{e\})^k \cap G_0\right| \geq C^{-O_n(1)}|A|, \tag{1.9}$$

where k depends only on n.

We can, of course, set $C = |A|^\delta$, so that (1.8) has the familiar form $|A^3| \geq |A|^{1+\delta}$.

Gill and Helfgott proved Conjecture 2 for $K = \mathbb{F}_p$ [**GH14**]. The case $K = \mathbb{F}_q$ remains open. The case $K = \mathbb{C}$ is relatively straightforward [**BG11**]; in that case, the group N can be taken to be trivial.

Putting the result for $K = \mathbb{F}_p$ together with [**PS16**], it is simple to show that the same result holds for $A \subset \mathrm{GL}_n(\mathbb{F}_p)$ general, without the assumption that the group $\langle A \rangle$ generated by A be solvable. (What [**PS16**] does is reduce the general case to the solvable case.) Again, the same conclusion is believed to hold over $\mathbb{F}_q$. Breuillard, Green and Tao have proved [**BGT12**] that, if one is willing to replace $C^{-O_n(1)}$ in (1.9) by a factor dependent in an unspecified way on C (but still independent of $|A|$), one does not even need to assume that A is contained in $\mathrm{GL}_n(K)$; they start from a completely general, abstract group. They kindly gave the name *Helfgott-Lindenstrauss conjecture* to the statement they proved, though I would personally give that name to Conj. 2.

We shall study what is arguably the simplest interesting solvable case, namely, the *affine group*

$$\left\{ \begin{pmatrix} r & x \\ 0 & 1 \end{pmatrix} : r \in K^*, x \in K \right\}. \tag{1.10}$$

over a field K. As we shall see, the question of growth in it is essentially equivalent to the *sum-product theorem* over a field. Indeed, our treatment (§3.2) will show how to take one of the ideas of proofs of the sum-product theorem over finite fields (as in [**BKT04**] or [**BGK06**]) and reinterpret it in the context of groups ("pivoting"). A version of the same idea (really just a form of induction) will appear again in our treatment of $\mathrm{SL}_2(K)$.

1.2.5. *Groups over* $\mathbb{R}$ *or* $\mathbb{C}$. The proof we shall give of Theorem 1.1 also works for K infinite. Even the first proof worked for $K = \mathbb{R}$, indeed more easily than over $\mathbb{Z}/p\mathbb{Z}$. Actually the statement of Theorem 1.1 turns out to have already been known over $\mathbb{R}$: the proof of [**EK01**, Thm. 2] suffices to establish it.

Some results in combinatorics – such as the sum-product theorem, which underlay the first proof [**Hel08**] of Thm. 1.1, or Beck's theorem [**Bec83**], on which [**EK01**] relies – are both stronger and easier to prove over the reals than over finite fields. In fact, some results are known only over $\mathbb{R}$, or were known only over $\mathbb{R}$ for many years. The reason is that, over $\mathbb{R}$, the topology of the real plane can be used in the solution of geometrical problems. A line divides the real plane into two halves; such a statement does not hold or even make sense over $\mathbb{Z}/p\mathbb{Z}$.

As it turns out, for many applications, we need to know not just a statement such as Theorem 1.1 for a linear group over the reals, but a stronger version thereof. To be precise: one needs to show that the maximal number $n_\delta(A)$ of points in A separated by δ in the real or complex metric grows: $n_\delta(A^3) \geq n_\delta(A)^{1+\delta}$.

Fortunately, as Bourgain and Gamburd first made clear [**BG08a**], existing proofs of Theorem 1.1 and its generalizations can be modified to yield such stronger variants. They worked with the proof in [**Hel08**], but the same should hold of later proofs. The applications they found consisted in or involved expander graphs. We will discuss results on expander graphs in §6.1.

1.3. Notation. By $f(n) \ll g(n)$, $g(n) \gg f(n)$ and $f(n) = O(g(n))$ we mean the same thing, namely, that there are $N > 0$, $C > 0$ such that $|f(n)| \leq C \cdot g(n)$ for all $n \geq N$. We write $\ll_a$, $\gg_a$, O_a if N and C depend on a (say).

As usual, $f(n) = o(g(n))$ means that $|f(n)|/g(n)$ tends to 0 as $n \to \infty$. We write $O^*(x)$ to mean any quantity at most x in absolute value. Thus, if $f(n) = O^*(g(n))$, then $f(n) = O(g(n))$ (with $N = 1$ and $C = 1$).

Given a subset $A \subset X$, we let $1_A : X \to \mathbb{C}$ be the characteristic function of A:

$$1_A(x) = \begin{cases} 1 & \text{if } x \in A, \\ 0 & \text{otherwise.} \end{cases}$$

2. Elementary tools

2.1. Additive combinatorics. Some of additive combinatorics can be described as the study of *sets that grow slowly*. In abelian groups, results are often stated so as to classify sets A such that $|A^2|$ is not much larger than $|A|$; in non-abelian groups, works starting with [**Hel08**] classify sets A such that $|A^3|$ is not much larger than $|A|$. Why?

In an abelian group, if $|A^2| < K|A|$, then $|A^k| < K^{O(k)}|A|$ – i.e., if a set does not grow after one multiplication with itself, it will not grow under several. This is a result of Plünnecke [**Plü70**] and Ruzsa [**Ruz89**]. (Petridis [**Pet12**] recently gave a purely additive-combinatorial proof.)

In a non-abelian group G, there can be sets A breaking this rule.

EXERCISE 2.1. *Let G be a group. Let $H < G$, $g \in G \setminus H$ and $A = H \cup \{g\}$. Then $|A^2| < 3|A|$, but $A^3 \supset HgH$, and HgH may be much larger than A. Give an example with $G = \mathrm{SL}_2(\mathbb{F}_p)$.* Hint: let H is the subgroup of G consisting of the elements $g \in G$ leaving the basis vector $e_1 = (1, 0)$ fixed.

However, Ruzsa's ideas do carry over to the non-abelian case, as was pointed out in [**Hel08**] and [**Tao08**]. We must assume that $|A^3|$ is small, not just $|A^2|$, and then it does follow that $|A^k|$ is small. The formal statement is Exercise 2.3, below. To prove it, we need the following lemma.

LEMMA 2.2 (Ruzsa triangle inequality). *Let A, B and C be finite subsets of a group G. Then*

$$|AC^{-1}||B| \leq |AB^{-1}||BC^{-1}|. \tag{2.1}$$

Commutativity is not needed. In fact, what is being used is in some sense more basic than a group structure; as shown in [**GHR15**], the same argument works naturally in any abstract projective plane endowed with the little Desargues axiom.

PROOF. We will construct an injection $\iota : AC^{-1} \times B \hookrightarrow AB^{-1} \times BC^{-1}$. For every $d \in AC^{-1}$, choose $(f_1(d), f_2(d)) = (a, c) \in A \times C$ such that $d = ac^{-1}$. Define $\iota(d, b) = (f_1(d)b^{-1}, b(f_2(d))^{-1})$. We can recover $d = f_1(d)(f_2(d))^{-1}$ from $\iota(d, b)$; hence we can recover $(f_1, f_2)(d) = (a, c)$, and thus b as well. Therefore, ι is an injection. □

EXERCISE 2.3. *Let G be a group. Prove that*

$$\frac{|(A \cup A^{-1} \cup \{e\})^3|}{|A|} \leq \left(3\frac{|A^3|}{|A|}\right)^3 \tag{2.2}$$

for every finite subset A of G. Show as well that, if $A = A^{-1}$ (i.e., if $g^{-1} \in A$ for every $g \in A$), then

$$\frac{|A^k|}{|A|} \leq \left(\frac{|A^3|}{|A|}\right)^{k-2}. \tag{2.3}$$

for every $k \geq 3$. Conclude that

$$\frac{|(A \cup A^{-1} \cup \{e\})^k|}{|A|} \leq 3^{k-2} \left(\frac{|A^3|}{|A|}\right)^{3(k-2)} \tag{2.4}$$

for every $A \subset G$ and every $k \geq 3$.

Inequalities (2.2)–(2.4) go back to Ruzsa (or Ruzsa-Turjányi [**RT85**]), at least for G abelian.

This means that, from now on, we can generally focus on studying when $|A^3|$ is or isn't much larger than $|A|$. Thanks to (2.2), we can also assume in many contexts that $e \in A$ and $A = A^{-1}$ without loss of generality.

2.2. The orbit-stabilizer theorem for sets. A theme recurs in work on growth in groups: results on subgroups can often be generalized to subsets. This is especially the case if the proofs are quantitative, constructive, or, as we shall later see, probabilistic.

The *orbit-stabilizer theorem for sets* is a good example, both because of its simplicity (it should really be called a lemma) and because it underlies a surprising number of other results on growth. It also helps to put forward a case for seeing group actions, rather than groups themselves, as the main object of study.

We recall that an *action* $G \curvearrowright X$ is a homomorphism from a group G to the group of automorphisms of a set X. (The automorphisms of a set X are just the bijections from X to X; we will see actions on objects with richer structures later.) For $A \subset G$ and $x \in X$, the *orbit* Ax is the set $Ax = \{g \cdot x : g \in A\}$. The *stabilizer* $\mathrm{Stab}(x) \subset G$ is given by $\mathrm{Stab}(x) = \{g \in G : g \cdot x = x\}$.

The statement we are about to give is as in [**HS14**, §3.1].

LEMMA 2.4 (Orbit-stabilizer theorem for sets). *Let G be a group acting on a set X. Let $x \in X$, and let $A \subseteq G$ be non-empty. Then*

$$|(A^{-1}A) \cap \mathrm{Stab}(x)| \geq \frac{|A|}{|Ax|}. \tag{2.5}$$

Moreover, for every $B \subseteq G$,

$$|BA| \geq |A \cap \mathrm{Stab}(x)||Bx|. \tag{2.6}$$

The usual orbit-stabilizer theorem – usually taught as part of a first course in group theory – states that, for H a subgroup of G,

$$|H \cap \mathrm{Stab}(x)| = \frac{|H|}{|Hx|}.$$

This the special case $A = B = H$ of the Lemma we (or rather you) are about to prove.

EXERCISE 2.5. *Prove Lemma 2.4. Suggestion: for (2.5), use the pigeonhole principle.*

If we try to apply Lemma 2.4 to the (left) action of the group G on itself by left multiplication

$$g \mapsto (h \mapsto g \cdot h)$$

or to the (left) action by right multiplication

$$g \mapsto (h \mapsto h \cdot g^{-1}),$$

we do not get anything interesting: the stabilizer of any element is trivial. The same is of course true of the right actions $g \mapsto (h \mapsto g^{-1}h)$ and $g \mapsto (h \mapsto hg)$. However, we also have the action by conjugation

$$g \mapsto (h \mapsto ghg^{-1}).$$

The stabilizer of a point $h \in G$ is its *centralizer*

$$C(h) = C_G(h) = \{g \in G : gh = hg\};$$

the orbit of a point $h \in G$ under the action of the group G is the *conjugacy class*

$$\mathrm{Cl}(h) = \{ghg^{-1} : g \in G\}.$$

Thus, we obtain the following result, which will show itself to be crucial later. Its importance resides in making upper bounds on intersections of A (or rather A^{l+2}) with $\mathrm{Cl}(g)$ imply lower bounds on intersections of A^2 with $C(g)$. In other words, the plan is to show that there are not too many elements of A^{l+2} of a special form, and then Lemma 2.6 will imply that there are many elements of A^2 of another special form. Having many elements of a special form will be very useful.

LEMMA 2.6. *Let $A \subset G$ be a non-empty set with $A = A^{-1}$. Then, for every $g \in A^l$, $l \geq 1$,*

$$|A^2 \cap C(g)| \geq \frac{|A|}{|A^{l+2} \cap \mathrm{Cl}(g)|}.$$

PROOF. Let $G \curvearrowright G$ be the action of G on itself by conjugation. Apply (2.5) with $x = g$; the orbit of g under conjugation by A is contained in $A^{l+2} \cap \mathrm{Cl}(g)$. □

It is instructive to see some other consequences of Lemma 2.4.

EXERCISE 2.7. *Let G be a group and H a subgroup thereof. Let $A \subset G$ be a set with $A = A^{-1}$. Then*

$$|A^2 \cap H| \geq \frac{|A|}{r}, \tag{2.7}$$

where r is the number of cosets of H intersecting A.

Hint: Consider the action $G \curvearrowright X = G/H$ by left multiplication, that is, $g \mapsto (aH \mapsto gaH)$. Then apply (2.5).

The following exercise tells us that, if we show that the intersection of A with a subgroup H grows rapidly, then we know that A itself grows rapidly.

EXERCISE 2.8. *Let G be a group and H a subgroup thereof. Let $A \subset G$ be a non-empty set with $A = A^{-1}$. Prove that, for any $k > 0$,*

$$|A^{k+1}| \geq \frac{|A^k \cap H|}{|A^2 \cap H|}|A|. \tag{2.8}$$

Hint: Consider the action $G \curvearrowright G/H$ again, and apply both (2.6) and (2.5).

EXERCISE 2.9. *Let G be a group and H a subgroup thereof. Write $\pi_{G/H} : G \to G/H$ for the quotient map. Let $A \subseteq G$ be a non-empty set with $A = A^{-1}$. Then, for any $k > 0$,*

$$|A^{k+2}| \geq \frac{|\pi_{G/H}(A^k)|}{|\pi_{G/H}(A)|}|A|.$$

3. Growth in a solvable group

3.1. Remarks on abelian groups. Let G be an abelian group and A be a finite subset of G. This is the classical setup for what nowadays is called *additive combinatorics* – a field that may be said to have started to split off from additive number theory with Roth [**Rot53**] and Freiman [**Fre73**].

In general, for G abelian, $A \subset G$ may be such that $|A+A|$ is barely larger than $|A|$, and that is the case even if we assume that A generates G. For instance, take A to be a segment of an arithmetic progression: $A = \{2, 5, 8, \ldots, 3m-1\}$. Then $|A| = m$ and $|A+A| = 2m-1 < 2|A|$.

Freiman's theorem [**Fre73**] (generalized first to abelian groups of bounded torsion [**Ruz99**] and then to arbitrary abelian groups [**GR07**]) tells us that, in a very general sense, this is the only kind of set that grows slowly. We have to start by giving a generalization of what we just called a segment of an arithmetic progression.

DEFINITION 3.1. *Let G be a group. A* centered convex progression *of dimension d is a set $P \subset G$ such that there exist*

(a) *a convex subset $Q \subset \mathbb{R}^d$ that is also symmetric ($Q = -Q$),*
(b) *a homomorphism $\phi : \mathbb{Z}^d \to G$,*

for which $\phi(\mathbb{Z}^d \cap Q) = P$. We say P is proper *if $\phi|_{\mathbb{Z}^d \cap Q}$ is injective.*

PROPOSITION 3.1 (Freiman; Ruzsa-Green). *Let G be an abelian group. Let $A \subset G$ be finite. Assume that $|A+A| \leq K|A|$ for some K. Then A is contained in at most $c_{K,1}$ copies of $P+H$ for some proper, centered convex progression P of dimension $\leq c_{K,2}$ and some finite subgroup $H < G$ such that $|P+H| \ll e^{c_{K,2}}|A|$. Here $c_{K,1}, c_{K,2} > 0$ depend only on K.*

The best known bounds are essentially those of Sanders [**San12**], as improved by Konyagin (see [**San13**]): $c_{K,1}, c_{K_2} \ll (\log K)^{3+o(1)}$.

This is a broad field into which we will not venture further. Notice just that, in spite of more than forty years of progress, we do not yet have what is conjectured to be the optimal result, namely, the above with $f(K), g(K) \ll \log K$ (the "polynomial Freiman-Ruzsa conjecture"). Thus the state of our knowledge here is in some sense less satisfactory than in the case of simple groups, as will later become clear.

The situation for nilpotent groups is much like the situation for abelian groups: there is a generalization of the Freiman-Ruzsa theorem to the nilpotent case, due to Tointon [**Toi14**] (see also Tessera-Tointon [**TT16**]), based on groundwork laid by Fisher-Katz-Peng [**FKP10**] and Tao [**Tao10**].

Brief excursus. There is of course also the matter of the role of nilpotent groups in the study of growth in a different if related sense, within geometric group theory: for A a subset of an infinite group G, how does $|A^k|$ behave as $k \to \infty$? It is easy to see that, if G is nilpotent, then $|A^k|$ grows polynomially on k. Gromov's theorem [**Gro81**], a deep and celebrated result, states the converse: if $|A^k|$ is bounded by a polynomial on k, then $\langle A\rangle$ has a nilpotent subgroup of finite index. There are several clearly distinct proofs of Gromov's theorem by now; of them, the one closest to the study of "growth" in the sense of the present paper is clearly [**Hru12**]. See [**BGT12**] for further work in that direction.

3.2. The affine group.

3.2.1. *Growth in the affine group.* We defined the *affine group* G over a field K in (1.10). (If we were to insist on using language in exactly the same way as later, we would say that the affine group is an algebraic group G (a variety with morphisms defining the group operations) and that (1.10) describes the group $G(K)$ consisting of its rational points. For the sake of simplicity, we avoid this sort of distinction here. We will go over most of these terms once the time to use them has come.)

Consider the following subgroups of G:

$$U=\left\{\begin{pmatrix}1 & a\\ 0 & 1\end{pmatrix} : a\in K\right\},\qquad T=\left\{\begin{pmatrix}r & 0\\ 0 & 1\end{pmatrix} : r\in K^*\right\}. \tag{3.1}$$

These are simple examples of a *solvable* group G, of a maximal *unipotent* subgroup U and of a maximal torus T. In general, in SL_n, a maximal torus is just the group of matrices that are diagonal with respect to some fixed basis of $\overline{K}^n$, or, what is the same, the centralizer of any element that has n distinct eigenvalues. Here, in our group G, the centralizer $C(g)$ of any element g of G not in U is a maximal torus.

When we are looking at what elements of the group G do to each other by the group operation, we are actually looking at two actions: that of U on itself (by the group operation) and that of T on U (by conjugation; U is a normal subgroup of G). They turn out to correspond to addition and multiplication in K, respectively:

$$\begin{pmatrix}1 & a_1\\ 0 & 1\end{pmatrix}\cdot\begin{pmatrix}1 & a_2\\ 0 & 1\end{pmatrix}=\begin{pmatrix}1 & a_1+a_2\\ 0 & 1\end{pmatrix}$$
$$\begin{pmatrix}r & 0\\ 0 & 1\end{pmatrix}\cdot\begin{pmatrix}1 & a\\ 0 & 1\end{pmatrix}\cdot\begin{pmatrix}r^{-1} & 0\\ 0 & 1\end{pmatrix}=\begin{pmatrix}1 & ra\\ 0 & 1\end{pmatrix}.$$

Thus, we see that growth in U under the actions of U and T is tightly linked to growth in K under addition and multiplication. This can be seen as motivation for studying growth in the affine group G. Perhaps we need no such motivation: we are studying growth in general, through a series of examples, and the affine group is arguably the simplest interesting example of a solvable group.

At the same time, the study of growth in a field under addition and multiplication was historically important in the passage from the study of problems in commutative groups (additive combinatorics) to the study of problems in noncommutative groups by related tools. (Growth in noncommutative groups had of course been studied before, but from very different perspectives, e.g., that of geometric group theory.) Some of the ideas we are about to see in the context of groups come ultimately from [**BKT04**] and [**GK07**], which are about finite fields, not about groups.

Of course, the way we choose to develop matters emphasizes what the approach to the affine group has in common with the approach to other, not necessarily solvable groups. The idea of *pivoting* will appear again when we study SL_2.

LEMMA 3.2. *Let G be the affine group over $\mathbb{F}_p$. Let U be the maximal unipotent subgroup of G, and $\pi : G\to G/U$ the quotient map.*

Let $A \subset G$, $A = A^{-1}$. Assume $A \not\subset U$; let x be an element of A not in U. Then

$$|A^2 \cap U| \geq \frac{|A|}{|\pi(A)|}, \qquad |A^2 \cap T| \geq \frac{|A|}{|A^5|}|\pi(A)| \tag{3.2}$$

for $T = C(x)$.

Recall U is given by (3.1). Since $x \notin U$, its centralizer $T = C(x)$ is a maximal torus.

PROOF. By (2.7), $A_u := A^2 \cap U$ has at least $|A|/|\pi(A)|$ elements. Consider the action of G on itself by conjugation. Then, by Lemma 2.4, $|A^2 \cap T| \geq |A|/|A(x)|$. (Here $A(x)$ is the orbit of x under the action of A by conjugation, and $\mathrm{Stab}(x) = C(g) = T$ is the stabilizer of g under conjugation.) We set $A_t := A^2 \cap T$. Clearly, $|A(x)| = |A(x)x^{-1}|$. Since the derived group of G is U (meaning, in particular, that $axa^{-1}x^{-1} \in U$ for any a and x), we see that $A(x)x^{-1} \subset A^4 \cap U$, and so $|A(x)| \leq |A^4 \cap U|$. At the same time, by (2.6) applied to the action $G \curvearrowright G/U$ by left multiplication, $|A^5| = |A^4 A| \geq |A^4 \cap U| \cdot |\pi(A)|$. Hence

$$|A_t| \geq \frac{|A|}{|A^4 \cap U|} \geq \frac{|A|}{|A^5|}|\pi(A)|.$$

□

The proof of the following proposition will proceed essentially by induction. This may be a little unexpected, since we are in a group G, not in, say, $\mathbb{Z}$, which has a natural ordering. However, as the proof will make clear, one can do induction on a group with a finite set of generators, even in the absence of an ordering.

PROPOSITION 3.3. *Let G be the affine group over $\mathbb{F}_p$, U the maximal unipotent subgroup of G, and T a maximal torus. Let $A_u \subset U$, $A_t \subset T$. Assume $A_u = A_u^{-1}$, $e \in A_t, A_u$ and $A_u \neq \{e\}$. Then*

$$|(A_t^2(A_u))^6| \geq \min(|A_u||A_t|, p). \tag{3.3}$$

To be clear: here

$$A_t^2(A_u) = \{t_1(u_1) : t_1 \in A_t^2, u_1 \in A_u\},$$

where $t(u) = tut^{-1}$, since T acts on U by conjugation.

PROOF. Call $a \in U$ a *pivot* if the function $\phi_a : A_u \times A_t \to U$ given by

$$(u,t) \mapsto ut(a) = utat^{-1}$$

is injective.

Case (a): There is a pivot a in A_u. Then $|\phi_a(A_u, A_t)| = |A_u||A_t|$, and so

$$|A_u A_t(a)| \geq |\phi_a(A_u, A_t)| = |A_u||A_t|.$$

This is the motivation for the name "pivot": the element a is the pivot on which we build an injection ϕ_a, giving us the growth we want.

Case (b): There are no pivots in U. As we are about to see, this case can arise only if either A_u or A_t is large with respect to p. Say that $(u_1, t_1), (u_2, t_2) \in A_u \times A_t$ *collide* for $a \in U$ if $\phi_a(u_1, t_1) = \phi_a(u_2, t_2)$. Saying that there are no pivots in U is the same as saying that, for every $a \in U$, there are at least two distinct $(u_1, t_1), (u_2, t_2) \in A_u \times A_t$ that collide for a. Now, two distinct (u_1, t_1), (u_2, t_2) can collide for at most one $a \in U \setminus \{e\}$. (As one can easily see, such an a corresponds to a solution to a non-trivial linear equation, which can have at most one solution.) Hence, if there are no pivots, $|A_u|^2|A_t|^2 \geq |U \setminus \{e\}| = p-1$, i.e., $|A_u| \cdot |A_t|$ is large ($\geq \sqrt{p-1}$). This fact already hints that this case will not be hard.

Let κ_a denote the number of collisions for a given $a \in U$:

$$\kappa_a = |\{u_1, u_2 \in A_u, t_1, t_2 \in A_t : \phi_a(u_1, t_1) = \phi_a(u_2, t_2)\}|.$$

As we were saying, two distinct (u_1, t_1), (u_2, t_2) collide for at most one $a \in U \setminus \{e\}$. Hence the total number of collisions $\sum_{a \in U \setminus \{e\}} \kappa_a$ is $\leq |A_u|^2|A_t|^2$, and so there is an $a \in U \setminus \{e\}$ such that

$$\kappa_a \leq \frac{|A_u|^2|A_t|^2}{p-1}.$$

Now,

$$\begin{aligned}(|A_u||A_t|)^2 &= \left(\sum_{x \in \phi_a(A_u, A_t)} |\{(u,t) \in A_u \times A_t : \phi_a(u,t) = x\}|\right)^2 \\ &\leq |\phi_a(A_u, A_t)| \sum_{x \in \phi_a(A_u, A_t)} |\{(u,t) \in A_u \times A_t : \phi_a(u,t) = x\}|^2 \\ &= |\phi_a(A_u, A_t)| \cdot \kappa_a,\end{aligned}$$

where the inequality is just Cauchy-Schwarz. Thus, $|\phi_a(A_u, A_t)| \geq |A_u|^2|A_t|^2/\kappa_a$, and so

$$|\phi_a(A_u, A_t)| \geq \frac{|A_u|^2|A_t|^2}{\frac{|A_u|^2|A_t|^2}{p-1}} - p - 1.$$

We are not quite done, since a may not be in A. Since a is *not* a pivot (as there are none), there exist distinct (u_1, t_1), (u_2, t_2) such that $\phi_a(u_1, t_1) = \phi_a(u_2, t_2)$. Then $t_1 \neq t_2$ (why?), and so the map $\psi_{t_1,t_2} : U \to U$ given by $u \mapsto t_1(u)(t_2(u))^{-1}$ is injective. The idea is that the very non-injectivity of ϕ_a gives an implicit definition of it, much like a line that passes through two distinct points is defined by them.

What follows may be thought of as the "unfolding" step, in that we wish to remove an element a from an expression, and we do so by applying to the expression a map that will send a to something known. We will be using the commutativity of T here.

For any $u \in U$, $t \in T$, since T is abelian,

$$\begin{aligned}\psi_{t_1,t_2}(\phi_a(u,t)) &= t_1(ut(a))(t_2(ut(a)))^{-1} = t_1(u)t(t_1(a)(t_2(a))^{-1})(t_2(u))^{-1} \\ &= t_1(u)t(\psi_{t_1,t_2}(a))(t_2(u))^{-1} = t_1(u)t(u_1^{-1}u_2)(t_2(u))^{-1},\end{aligned} \tag{3.4}$$

where $\psi_{t_1,t_2}(a) = u_1^{-1}u_2$ holds because $\phi_a(u_1, t_1) = \phi_a(u_2, t_2)$. Note that a has disappeared from the last expression in (3.4). We obtain

$$\psi_{t_1,t_2}(\phi_a(A_u, A_t)) \subset A_t(A_u)A_t(A_u^2)A_t(A_u) \subset (A_t(A_u))^4.$$

Since ψ_{t_1,t_2} is injective, we conclude that

$$|(A_t(A_u))^4| \geq |\psi_{t_1,t_2}(\phi_a(A_u, A_t))| = |\phi_a(A_u, A_t)| \geq p-1,$$

that is to say, at most a single element of U is missing from $(A_t(A_u))^4$. Since A_u contains at least one element besides e, we obtain immediately that

$$(A_t(A_u))^6 \supset (A_t(A_u))^4 A_u = U.$$

There is an idea here that we are about to see again: any element a that is not a pivot can, by this very fact, be given in terms of some $u_1, u_2 \in A_u$, $t_1, t_2 \in A_t$, and so an expression involving a can often be transformed into one involving only elements of A_u and A_t.

Case (c): There are pivots and non-pivots in U. Here comes what we can think of as the inductive step. Since $A_u \neq \{e\}$, A_u generates U. Thus, there is a non-pivot $a \in U$ and a $g \in A_u$ such that ga is a pivot. Then $\phi_{ag} : A_u \times A_t \to U$ is injective. Much as in (3.4), we unfold:

$$\begin{aligned}\psi_{t_1,t_2}(\phi_{ga}(u,t)) &= t_1(ut(g)t(a))(t_2(ut(g)t(a)))^{-1} \\ &= t_1(ut(g))t(u_1^{-1}u_2)(t_2(ut(g)))^{-1},\end{aligned} \tag{3.5}$$

where (u_1, t_1), (u_2, t_2) are distinct pairs such that $\phi_a(u_1, t_1) = \phi_a(u_2, t_2)$. Just as before, ψ_{t_1,t_2} is injective. Hence

$$|A_t(A_u)A_t^2(A_u)A_t(A_u^2)A_t^2(A_u)A_t(A_u)| \geq |\psi_{t_1,t_2}(\phi_{ga}(u,t))| = |A_u||A_t|$$

and we are done.

The idea to recall here is that, if S is a subset of an orbit $\mathcal{O} = \langle A \rangle x$ such that $S \neq \emptyset$ and $S \neq \mathcal{O}$, then there is an $s \in S$ and a $g \in A$ such that $gs \notin S$. It is in this fashion that we can use induction even in the absence of a natural ordering of $\langle A \rangle$. □

We are using the fact that G is the affine group over $\mathbb{F}_p$ (and not over some other field) only at the beginning of case (c), when we say that, for $A_u \subset U$, $A_u \neq \{e\}$ implies $\langle A_u \rangle = U$.

PROPOSITION 3.4. *Let G be the affine group over $\mathbb{F}_p$. Let U be the maximal unipotent subgroup of G, and $\pi : G \to G/U$ the quotient map.*

Let $A \subset G$, $A = A^{-1}$, $e \in A$. Assume A is not contained in any maximal torus. Then either

$$|A^{73}| \geq \sqrt{|\pi(A)|} \cdot |A| \tag{3.6}$$

or

$$U \subset A^{72}. \tag{3.7}$$

The exponents 72, 73 in (3.6) are not optimal. For instance, one can obtain 52, 53 by looking closer at the proof of Prop. 3.3.

PROOF. We can assume $A \not\subset U$, as otherwise what we are trying to prove is trivial. Let g be an element of A not in U; its centralizer $C(g)$ is a maximal torus T. By assumption, there is an element h of A not in T. Then $hgh^{-1}g^{-1} \neq e$. At the same time, $hgh^{-1}g^{-1}$ does lie in $A^4 \cap U$, and so $A^4 \cap U$ is not $\{e\}$.

Let $A_u = A^4 \cap U$, $A_t = A^2 \cap T$; their size is bounded from below by (3.2). Applying Prop. 3.3, we obtain

$$|A^{72} \cap U| \geq \min(|A_u||A_t|, p) \geq \min\left(\frac{|A|^2}{|A^5|}, p\right).$$

By (2.6), $|A^{73}| \geq |A^{72} \cap U| \cdot |\pi(A)|$. Clearly, if $|A|/|A^5| < 1/\sqrt{|\pi(A)|}$, then $|A^{57}| \geq |A^5| > \sqrt{|\pi(A)|} \cdot |A|$. If $|A|/|A^5| \geq 1/\sqrt{|\pi(A)|}$, then either $|A^{72} \cap U| \geq |A|/\sqrt{|\pi(A)|}$ and so $|A^{73}| \geq \sqrt{|\pi(A)|} \cdot |A|$, or $|A^{72} \cap U| = p$ and so $U \subset A^{72}$. □

For $A \subset U$, getting a better-than-trivial lower bound on $|A^k|$, k a constant, amounts to Freiman's theorem in $\mathbb{F}_p$, and getting a growth factor of the form $|\pi(A)|^\delta$, $\delta > 0$, would involve proving a version of Freiman's theorem of polynomial strength. As we discussed before, that is a difficult open problem.

3.2.2. *Brief remarks on a generalization and an application.* We can see Prop. 3.4 as a very simple result of the "classification of approximate subgroups" kind. If a set A (with $A = A^{-1}$, $e \in A$) in the affine group over $\mathbb{F}_p$ grows slowly ($|A^k| \leq |A|^{1+\delta}$, $k = 73$, δ small) then either (i) A is contained in a maximal torus, (ii) A is contained in a few cosets of the maximal unipotent subgroup U (that is, $|\pi(A)| \leq |A|^{2\delta}$), or (iii) A^k contains a subgroup (namely, U) such that $\langle A \rangle / H$ is nilpotent (here, in fact, abelian).

EXERCISE 3.5. *Give examples of subsets A of the affine group over $\mathbb{F}_p$ that fail to grow for each of the reasons above: a set contained in a maximal torus, a set almost contained in U, and a set containing U.*

The following more general statement has been proved for $K = \mathbb{F}_p$ [**GH11**]. (It remains open for general finite K.) Let $A \subset G = \mathrm{GL}_n(K)$ ($A = A^{-1}$, $e \in A$) be such that $\langle A \rangle$ is solvable. Then, for any $\delta > 0$, if $|A^3| < |A|^{1+\delta}$, there are a subgroup $S \triangleleft \langle A \rangle$ and a unipotent subgroup $U \triangleleft S$ such that (a) S/U is nilpotent, (b) $U \subset A^k$, where $k = O_n(1)$, (c) A is contained in $|A|^{O_n(\delta)}$ cosets of U.

EXERCISE 3.6. *Verify that each of the cases (i)-(iii) enumerated above in the case of the affine group satisfies this description, i.e., there are S and U such that (a)–(c) are fulfilled.*

What is also interesting is that the results we have proved on growth in the affine linear group can be interpreted as a *sum-product theorem.*

EXERCISE 3.7. *Let $X \subset \mathbb{F}_p$, $Y \subset \mathbb{F}_p^*$ be given with $X = -X$, $0 \in X$, $1 \in Y$. Using Prop. 3.3, show that*

$$|6Y^2X| \geq \min(|X||Y|, p-1). \tag{3.8}$$

This is almost exactly [**GK07**], Corollary 3.5], say.

Using (3.8), or any estimate like it, one can prove the following.

THEOREM 3.8 (Sum-product theorem [**BKT04**], [**BGK06**]; see also [**EM03**]). *For any $A \subset \mathbb{F}_p^*$ with $|A| \leq p^{1-\epsilon}$, $\epsilon > 0$, we have*

$$\max(|A \cdot A|, |A + A|) \geq |A|^{1+\delta},$$

where $\delta > 0$ depends only on ϵ.

In fact, the proof we have given of Prop. 3.3 takes its ideas from proofs of the sum-product theorem. In particular, the idea of *pivoting* is already present in them. We will later see how to apply it in a broader context.

3.2.3. *Diameter bounds in a remaining case.* We have proved that growth occurs in SL_2 under some weak conditions. This leaves open the question of what happens with A^k, k unbounded, for A not obeying those conditions. In particular: what happens when A, while not contained in the maximal unipotent group U, is contained in the union of few cosets of U?

One thing that is certainly relevant here is that, in general, there is no vertex expansion in the affine group, and thus no expansion. Indeed, the purpose of this subsection is to give a glimpse of the issue of diameter bounds in situations in which neither expansion nor rapid growth hold.

Let us state the lack of vertex expansion in elementary terms.

PROPOSITION 3.9. *For any $\lambda_1, \ldots, \lambda_k \in \mathbb{Z}$, and any $\epsilon > 0$, there is a constant C depending on ϵ such that, for every prime $p > C$, there is a set $S \subset \mathbb{F}_p$, $0 < |S| \leq p/2$, such that*

$$|S \cup (S+1) \cup \lambda_1 S \cup \ldots \cup \lambda_k S| \leq (1+\epsilon)|S|. \tag{3.9}$$

EXERCISE 3.10. *Prove Proposition* 3.9. Hints: *prove this for $k = 1$ first; you can assume $\lambda = \lambda_1$ is ≥ 2. Here is a plan. We want to show that $|S \cup (S+1) \cup \lambda S| \leq (1+\epsilon)|S|$. For $|S \cup (S+1)|$ to be $\leq (1+\epsilon/2)|S|$, it is enough that S be a union of intervals of length $> 2/\epsilon$. (By an* interval *we mean the image of an interval $[a,b] \cap \mathbb{Z}$ under the map $\mathbb{Z} \to \mathbb{Z}/p\mathbb{Z} \sim \mathbb{F}_p$.) We also want $|S \cup \lambda S| \leq (1+\epsilon)|S|$; this will be the case if S is the union of disjoint sets of the form V, $\lambda^{-1}V$, $\ldots$, $\lambda^{-r}V$, $r \geq \epsilon/2$. Now, in $\mathbb{F}_p$, if I is an interval of length ℓ, then $\lambda^{-1}I$ is the union of λ intervals (why? of what length?). Choose V so that $V, \lambda^{-1}V, \ldots, \lambda^{-r}V$ are disjoint. Let S be the union of these sets; verify that it fulfills (3.9).*

The following exercise shows that Prop. 3.9 is closely connected to the fact that a certain group is *amenable.*

EXERCISE 3.11. *Let $\lambda \geq 2$ be an integer. Define the* Baumslag-Solitar group $\mathrm{BS}(1,\lambda)$ *by*

$$\mathrm{BS}(1,\lambda) = \langle a_1, a_2 | a_1 a_2 a_1^{-1} = a_2^{\lambda} \rangle.$$

(a) *A group G with generators $a_1, \ldots, a_\ell$ is called* amenable *if, for every $\epsilon > 0$, there is a finite $S \subset G$ such that*

$$|F \cup a_1 F \cup \ldots \cup a_\ell F| \leq (1+\epsilon)|F|.$$

Show that $\mathrm{BS}(1,\lambda)$ *is amenable.* Hint: *to construct F, take your inspiration from Exercise* 3.10.

(b) *Express the subgroup of the affine group over $\mathbb{F}_p$ generated by the set*

$$A_\lambda = \left\{ \begin{pmatrix} \lambda & 0 \\ 0 & 1 \end{pmatrix}, \begin{pmatrix} 1 & 1 \\ 0 & 1 \end{pmatrix} \right\} \tag{3.10}$$

as a quotient of $\mathrm{BS}(1,\lambda)$, *i.e., as the image of a homomorphism π_p defined on* $\mathrm{BS}(1,\lambda)$.

(c) *Displace or otherwise modify your sets F so that, for each of them, $\pi_p|_F$ is injective for p larger than a constant. Conclude that $S = \pi_p(F)$ satisfies (3.9), thus giving a (slightly) different proof of exercise* 3.10.

Amenability is not good news when we are trying to prove that a diameter is small, in that it closes a standard path towards showing that it is logarithmic in the size of the group. However, it does not imply that the diameter is not small.

Let us first be clear about what we can prove or rather about what we cannot hope to prove. We should not aim at a bound on the diameter of the affine group G with respect to an arbitrary set of generators A: it is easy to choose A so that the diameter of $\Gamma(G,A)$ is very large.

EXERCISE 3.12. *Let A_λ be as in (3.10) for λ a generator of $\mathbb{F}_p^*$. Let $A = A_\lambda \cup A_\lambda^{-1}$. Then A generates the affine group G over $\mathbb{F}_p$. Show that* $\operatorname{diam} \Gamma(G,A) = (p-1)/2$.

Rather, we should aim for a bound on the diameter of the *Schreier graph* of the action of the affine group G by conjugation on its maximal unipotent subgroup U. In general, the Schreier graph of an action $G \curvearrowright X$ of a group G on a set X with respect to a set of generators A of G is the graph having X as its set of vertices and $\{(x, ax) : x \in X, a \in A\}$ as its set of edges. In our case ($X = U$, $A = A_\lambda \cup A_\lambda^{-1}$, $\lambda \in \mathbb{F}_p^*$), the Schreier graph is isomorphic to the graph $\Gamma_{p,\lambda}$ with vertex set $\mathbb{F}_p$ and edge set

$$\{(x,x+1) : x \in \mathbb{F}_p\} \cup \{(x,x-1) : x \in \mathbb{F}_p\} \cup \{(x,\lambda x) : x \in \mathbb{F}_p\} \cup \{(x,\lambda^{-1}x) : x \in \mathbb{F}_p\}.$$

We are not avoiding the problem posited by the fact that the Baumslag-Solitar group $\mathrm{BS}(1,\lambda)$ is amenable, since what amenability impedes is precisely a natural approach to prove logarithmic diameter bounds on $\Gamma_{p,\lambda}$. If Proposition 3.9 were not true, then the diameter of $\Gamma_{p,\lambda}$ would be $O(\log p)$. (Why?)

If λ is the projection of a fixed integer λ_0, then it is possible, and easy, to give a logarithmic diameter bound nevertheless.

EXERCISE 3.13. *Let $\lambda_0 \geq 2$ be an integer. Let $\lambda = \lambda_0 \bmod p$, which lies in $\mathbb{F}_p^*$ for $p > \lambda_0$. Show that the diameter of the graph $\Gamma_{p,\lambda}$ is $O(\lambda_0 \log p)$.* Hint: *lift elements of $\mathbb{F}_p$ to $\mathbb{Z} \cap [0, p-1]$, and write them out in base λ_0.*

It turns out to be possible to give a *polylogarithmic* bound for general $\lambda \in \mathbb{F}_p^*$:

$$\operatorname{diam} \Gamma_{p,\lambda} \ll (\log p)^{O(1)}, \tag{3.11}$$

where the implied constants are independent of p and λ. Here we need not assume that λ generates $\mathbb{F}_p^*$, but we do assume that the order of λ is $\gg \log p$. (Indeed, if the order of λ is very small, viz., $o((\log p)/\log\log p)$, then (3.11) cannot hold; why?)

The proof of (3.11) was the outcome of a series of discussions among B. Bukh, A. Harper, E. Lindenstrauss and the author. It is essentially an exercise in Fourier analysis using bounds on exponential sums due to Konyagin [**Kon92**].

EXERCISE 3.14. *Let p be a prime, $\lambda \in \mathbb{F}_p^*$. Assume λ has order $\geq \log p$. Write $e(t) = e^{2\pi i t}$ and $e_p(t) = e^{2\pi i t/p}$. Konyagin* [**Kon92**, Lemma 6] *showed that, for any $\epsilon > 0$, there is a $c_\epsilon > 0$ such that, for any $p \geq c_\epsilon$ prime and $\alpha, \lambda \in (\mathbb{Z}/p\mathbb{Z})^*$ with λ of order $\geq c_\epsilon (\log p)/(\log\log p)^{1-\epsilon}$ in the group $(\mathbb{Z}/p\mathbb{Z})^*$,*

$$\sum_{j=0}^{J} |\{\alpha\lambda^j/p\}|^2 \geq \frac{1}{(\log p)^{3^{\epsilon/4}}}, \tag{3.12}$$

where $J = \lfloor c_\epsilon \log p (\log\log p)^4 \rfloor$ and $\{x\}$ is the element of $(-1/2, 1/2]$ such that $x - \{x\}$ is an integer.

(a) *Show that* (3.12) *implies that $S(\alpha) = \sum_{j=0}^{J} e_p(\alpha\lambda^j)$ satisfies $|S(\alpha)| \leq J + 1 - 1/(\log p)^{3^{\epsilon/4}/2}$ for every $\alpha \in (\mathbb{Z}/p\mathbb{Z})^*$.*

(b) *Deduce that every element of $\mathbb{Z}/p\mathbb{Z}$ can be written as a sum $\sum_{i=1}^{K} \lambda^{j_i}$, where $0 \leq j_i \leq J$ and K is bounded by*

$$K \ll J(\log p)^{3^{\epsilon/4}/2}(\log p) \ll_\epsilon (\log p)^{2+3^{\epsilon/4}/2}(\log\log p)^4 \ll_\epsilon (\log p)^{5/2+\epsilon}.$$

To do so, show first that for any sequence $r_0, \ldots, r_J \in \mathbb{Z}/p\mathbb{Z}$, the number of ways of expressing $x \in \mathbb{Z}/p\mathbb{Z}$ as a sum of K elements (not necessarily *distinct) of a subset $A \subset \mathbb{Z}/p\mathbb{Z}$ equals*

$$\frac{1}{p}\sum_{\alpha\in\mathbb{Z}/p\mathbb{Z}} S_A(\alpha)^K e_p(-\alpha x),$$

where $S_A(\alpha) = \sum_{a\in A} e(\alpha a)$. This approach is the circle method *over $\mathbb{Z}/p\mathbb{Z}$.*

(c) *Conclude that the graph $\Gamma_{p,\lambda}$ with vertex set $\mathbb{F}_p$ and edge set*

$$\{(x, x+1) : x \in \mathbb{F}_p\} \cup \{(x, \lambda x) : x \in \mathbb{F}_p\}$$

has diameter $\ll_\epsilon (\log p)^{5/2+\epsilon}$.

4. Intersections with varieties

Let G a linear algebraic group defined over a field K. Let A be a finite set of generators of the set of points of G over K.

We will first show that, unless all the points of G over K lie in V, there are (plenty of) elements of A^k, k bounded, that do not lie on V (*escape from subvarieties*). Here the constant k depends only on some invariants of V (its number of components, their degree and their dimension), not on K or on other properties of V.

Our main aim will then be to show that, if A grows slowly, then A is truly a beautiful object, very regular from many points of view. Of course, this is a strategy for showing in the following section that A does not exist (or is almost all of G).

"Very regular" here means "behaving well with respect to the algebraic geometry of the ambient group G". To be precise: the intersection of a slowly growing set A with any variety V will be bounded by not much more than $|A|^{\dim(V)/\dim(G)}$ (Theorem 4.4; the *dimensional estimate*).

Here is an intuitive image. Thinking for a moment in three dimensions (that is, $\dim(G) = 3$), one might say that this estimate means that A is very regular in the sense of being a roughly spherical blob, as its intersection with any line, or any curve of bounded degree, is bounded by $O(|A|^{1/3})$, and its intersection with any plane, or any surface of bounded degree, is bounded by $O(|A|^{2/3})$.

Finally, we will see that for some kinds of varieties V – namely, centralizers – we can give a *lower* bound on the intersection of A with V, roughly of the same order as the upper bound above. This fact will be a crucial tool in §5.

4.1. Preliminaries from algebraic geometry and algebraic groups.

We will have the choice of working sometimes over linear algebraic groups and sometimes over Lie algebras (as in [**Hel15**], following [**Hel11**]) or solely over linear algebraic groups (as in [**Tao15**], which follows [**BGT11**]). We will follow the first path. Naturally, we will need some preliminaries on varieties, their behavior under mappings, the derivatives of such mappings, and so forth. It will all be a quick review for some readers. When it comes to basic algebraic geometry, we will cite mainly [**Mum99**] and [**Har77**], as they are standard sources for English speakers. In the case of either source, we will limit ourselves to the first chapter, that is, to classical foundations. Our definitions for terms related to algebraic groups come mostly from [**Spr98**] and [**Bor91**]; basic facts on finite groups of Lie type come from [**MT11**, ch. 21 and 24].

4.1.1. *Basic definitions.* We will need some basic terms from algebraic geometry. Let K be a field; denote by $\overline{K}$ an algebraic closure of K. For us, a variety V will simply be an affine or a projective variety – that is, the algebraic set consisting of the solutions in $\mathbb{A}^n$ to a system of polynomial equations, or the solutions in $\mathbb{P}^n$ to a system of homogeneous polynomial equations. We say V is defined over K if V can be described by polynomial equations with coefficients in K. Given a field L containing K, we write $V(L)$ for the set of solutions with coordinates in L. When we simply say "points on V", we mean elements of $V(\overline{K})$.

Abstract algebraic varieties (as in, say, [**Mum99**, Def. I.6.2]) will not really be needed, although they do give a very natural way to handle a variety that parametrizes a family of varieties, among many other things. For instance, we will tacitly refer to the variety of all d-dimensional planes in projective space, and, while that variety (a *Grassmanian*) can indeed be defined as an algebraic set in projective space, that is a non-obvious though standard fact.

The *Zariski topology* on $\mathbb{A}^n$ or $\mathbb{P}^n$ is the topology whose open sets are the complements of varieties (affine ones if we work in $\mathbb{A}^n$, projective ones if we work in $\mathbb{P}^n$). It induces a topology, also called Zariski topology, on any variety V; its open sets are the complements $V \setminus W$ of subvarieties W of V. (A *subvariety* of V is a variety contained in V.) The *Zariski closure* $\overline{S}$ of a subset S of V is its closure in the Zariski topology.

A variety V is *irreducible* if it is not the union of two varieties $V_1, V_2 \neq \emptyset, V$. (Note that many authors call an algebraic set a variety only if it is irreducible.) Every variety V can be written as a finite union of irreducible varieties V_i, with $V_i \not\subset V_j$ for $i \neq j$; they are called the *irreducible components* (or simply the components) of V.

When we say "property P holds for a generic point in the variety V", we simply means that there is a dense open subset $U \subset V$ such that property P holds for every point on U. It is easy to see that a non-empty open subset of an irreducible variety is always dense.

The *dimension* $\dim V$ of an irreducible variety V is the largest d such that there exists a chain of irreducible varieties

$$V_0 \subset V_1 \subset \cdots \subset V_d = V.$$

The union of several irreducible varieties of dimension d is called a *pure-dimensional* variety of dimension d. If W is a pure-dimensional proper subvariety of an irreducible variety V, then $\dim W < \dim V$ [**Mum99**, Cor. I.7.1]. (A subvariety $W \subset V$ is *proper* if $W \neq V$.)

The direct product $V \times W$ of irreducible varieties V, W is an irreducible variety of dimension is $\dim V + \dim W$ ([**Har77**, Exer. I.3.15 and I.2.14] or [**Mum99**, Prop. I.6.1, Thm. I.6.3 and Prop. I.7.5]).

4.1.2. *Degrees. Bézout's theorem.* The degree of a pure-dimensional variety V in $\mathbb{A}^n$ or $\mathbb{P}^n$ of dimension d is its number of points of intersection with a generic plane of dimension $n - d$. (See? We just referred tacitly to...)

Bézout's theorem, in its classical formulation, states that, for any two distinct irreducible curves C_1, C_2 in $\mathbb{A}^2$, the number of points of intersection $(C_1 \cap C_2)(\overline{K})$ is at most $d_1 d_2$. (In fact, for C_1 and C_2 generic, the number of points of intersection is exactly $d_1 d_2$; the same is true for *all* distinct C_1, C_2 if we count points of intersection with multiplicity.)

In general, if V_1 and V_2 are irreducible varieties, and we write $V_1 \cap V_2$ as a union of irreducible varieties $W_1, W_2, \ldots, W_k$ with $W_i \not\subset W_j$ for $i \neq j$, a generalization of Bézout's theorem tells us that

$$\sum_{i=1}^{k} \deg(W_k) \leq \deg(V_1) \deg(V_2). \tag{4.1}$$

See, for instance, [**DS98**, p.251], where Fulton and MacPherson are mentioned in connection to this and even more general statements.

Inequality (4.1) implies immediately that, if a variety V is defined by at most m equations of degree at most d, then the number and degrees of the irreducible components of V are bounded in terms of m and d alone.

4.1.3. *Morphisms.* A *morphism* from a variety $V_1 \subset \mathbb{A}^m$ to a variety $V_2 \subset \mathbb{A}^n$ is simply a map $f : V_1 \to V_2$ of the form

$$(x_1, \ldots, x_m) \mapsto (P_1(x_1, \ldots, x_m), \ldots, P_n(x_1, \ldots, x_m)),$$

where $P_1, \ldots, P_n$ are polynomials. It is clear that the preimage $f^{-1}(W)$ of a subvariety $W \subset V_2$ is a subvariety of V_1.

What is not at all evident a priori is that, for $W \subset V_1$ a subvariety, the image $\phi(W)$ is a *constructible set*, meaning a finite union of terms of the form $W \setminus W'$, where W and $W' \subset W$ are varieties. (For instance, if $V \subset \mathbb{A}^2$ is the variety given by $x_1 x_2 = 1$ (a hyperbola), then its image under the morphism $\phi(x_1, x_2) = x_1$ is the constructible set $\mathbb{A}^1 \setminus \{0\}$.) This result is due to Chevalley [**Mum99**, Cor. I.8.2].[2]

Let V be irreducible and let $f : V \to \mathbb{A}^n$ be a morphism. It is easy to see that the Zariski closure $\overline{f(V)}$ must be irreducible, and that $\dim \overline{f(V)} \leq \dim V$. Let $d = \dim V - \overline{f(V)}$. Then there is a Zariski open subset $U \subset \overline{f(V)}$ such that, for every $x \in U$, the preimage $f^{-1}(\{x\})$ is a pure-dimensional variety of dimension d [**Mum99**, Thm. I.8.3].

It is easy to see (by Bézout (4.1)) that the degree of $f^{-1}(\{x\})$ is bounded in terms of $\deg(V)$, n and the degrees of the polynomials $P_1, \ldots, P_n$ defining f. If $\dim V = \overline{f(V)}$, $f^{-1}(\{x\})$ is 0-dimensional, and so its number of points is bounded by its degree, by the definition of degree.

4.1.4. *Tangent spaces and derivatives.* Let $V \subset \mathbb{A}^n$ be a variety of dimension d defined by equations $P_i(x_1, \ldots, x_n) = 0$, $1 \leq i \leq k$. The *tangent space* $T_x V$ of V at x is the kernel of the linear map from $\mathbb{A}^n$ to $\mathbb{A}^k$ given by the matrix $\mathscr{P}|_x = (\partial P_i / \partial x_j)_{1 \leq i \leq k, 1 \leq j \leq n}$. (These are formal partial derivatives.) A point x on V is *non-singular* if $\dim T_x V = \dim V$, and *singular* otherwise. The set of singular points is a proper subvariety of V [**Har77**, Thm. I.5.3].

Let $V \subset \mathbb{A}^n$, $W \subset \mathbb{A}^m$ be varieties and let $f : V \to W$ be a morphism. At any point $\vec{x}$ on V, the linear map given by the matrix $J|_x = \left(\frac{\partial f_i}{\partial x_j}\right)_{1 \leq i \leq m, 1 \leq j \leq n}$ restricts to a linear map $Df|_x : T_x V \to T_x W$ (as follows from the chain rule). For any $r \geq 0$, the set of non-singular points on V such that the rank of $Df|_x$ is at least r is Zariski-open in V. This fact is easy to see for $V = \mathbb{A}^n$: the rank is then $< r$ if and only if every r-by-r minor of $J|_x$ is 0, a condition that defines a subvariety. For V general, define a new matrix by putting the matrix $\mathscr{P}|_x$ on top of the matrix

[2] As R. Vakil says of the closely related statement that the image of a projective variety under a morphism is a projective variety: "a great deal of classical algebra and geometry is contained in this theorem as special cases." In model-theoretical terms, we are talking of quantifier elimination.

$J|_x$, and note that the new matrix will have rank at least $n - \dim(V) + r$ if and only if $Df|_x$ has rank at least r; thus we can proceed as for $V = \mathbb{A}^n$.

EXERCISE 4.1. *Let V, W be varieties, V irreducible, $f : V \to W$ a morphism, and x a non-singular point on V. Prove that, if the rank of $Df|_x$ is at least r, then the dimension of $\overline{f(V)}$ is at least r.*

4.1.5. *Linear algebraic groups.* A *linear algebraic group* over a field K is a subvariety G of GL_n, defined over K, that is closed under multiplication and inversion.[3] We thus have morphisms $\cdot : G \times G \mapsto G$ and ${}^{-1} : G \to G$. An *algebraic* or *closed* subgroup of G is a subvariety H of G that is also closed under multiplication and inversion.

We will assume that the field of definition K is *perfect*, meaning that every finite extension of k is separable; this assumption will save us from possible trouble. Finite fields, fields of characteristic 0 and algebraically closed fields are always perfect fields.

A linear algebraic group G is *semisimple* if it has no connected, non-trivial and solvable normal algebraic subgroups, even defined over $\overline{K}$. ("Connected" means "connected in the Zariski topology"; an algebraic group is connected if and only if it is irreducible [**Spr98**, Prop. 2.2.1]. For algebraic groups, being *solvable* is defined analogously as for groups [**Bor91**, §2.4].) We say G is *simple* (over K) if it is semisimple, connected and has no connected, proper and non-trivial normal algebraic subgroups defined over K.[4]

Let G be an arbitrary linear algebraic group over a field K. An element $g \in G(\overline{K})$. is *semisimple* if it is diagonalizable over $\overline{K}$. Note that, by [**Bor91**, §4.3, Prop.] and the first definition in [**Bor91**, §4.5], the semisimplicity of g is invariant under isomorphisms of G, i.e., it does not actually depend on the embedding of G into GL_n.

A *torus* $T < \mathrm{GL}_n$ is an algebraic group isomorphic to GL_1^r over $\overline{K}$ for some $r \geq 1$. A torus defined over K is always diagonalizable over $\overline{K}$ [**Bor91**, §8.5, Prop.]; that is, there exists $g \in \mathrm{GL}_n(\overline{K})$ such that gTg^{-1} is a subgroup of the group of diagonal matrices in GL_n. A *maximal torus* of a connected linear algebraic group G is a torus $T < G$ with r maximal. We call r the *rank* of G. If G is connected, then every semisimple $g \in G(\overline{K})$ lies in a maximal torus [**Spr98**, Thm. 6.4.5(ii)].

The centralizer $C(g)$ of a semisimple point g in G has dimension at least $r = \mathrm{rank}(G)$; if $\dim C(g) = \mathrm{rank}(G)$, we say g is *regular*. When G is semisimple, a semisimple element $g \in G(\overline{K})$ is regular if and only if the connected component $C(g)^\circ$ of $C(g)$ containing the identity is a maximal torus ([**Bor91**, §12.2, Prop., and §13.17, Cor. 2(c)]). A regular semisimple element $g \in G(\overline{K})$ lies in exactly one maximal torus [**Bor91**, §12.2, Prop.]. For G semisimple, regular semisimple elements form a non-empty open subset of G [**Ste65**, §2.14].

4.1.6. *Lie algebras.* A *Lie algebra* is a vector space $\mathfrak{g}$ over a field K together with a bilinear map $[\cdot,\cdot] : \mathfrak{g} \times \mathfrak{g} \to \mathfrak{g}$ satisfying the identities

$$[x, y] = -[y, x], \qquad [x, [y, z]] + [y, [z, x]] + [z, [x, y]] = 0. \tag{4.2}$$

[3]Alternatively, we could define a linear algebraic group G to be an affine variety with two morphisms $\cdot : G \times G \mapsto G$ and ${}^{-1} : G \to G$ satisfying the usual rules, and then prove that G is isomorphic to a subvariety of GL_n with the multiplication and inversion morphisms it inherits from GL_n [**Bor91**, Prop. 1.10].

[4]Some sources (e.g., [**Bor91**, §22.8]) give the name *almost-simple* to what we call *simple*.

An *ideal* of a Lie algebra is a subspace $\mathfrak{v} \subsetneq \mathfrak{g}$ such that $[g, \mathfrak{v}] \subset \mathfrak{v}$. We say a Lie algebra is *simple* if it has no ideals other than (0).

A linear algebraic group G acts on its tangent space $\mathfrak{g} = T_e G$ at the origin by conjugation: for $g \in G$, we define the linear map $\mathrm{Ad}_g : \mathfrak{g} \to \mathfrak{g}$ to be the derivative of $y \mapsto gyg^{-1}$. The derivative of Ad_g with respect to g can be written as a bilinear map $\mathfrak{g} \times \mathfrak{g} \to \mathfrak{g}$, which we call $[\cdot,\cdot]$; it is fairly straightforward to check that it satisfies the identities in (4.2), and thus makes $\mathfrak{g}$ into a Lie algebra.

It is easy to see that, if a subspace $\mathfrak{v}$ of the Lie algebra $\mathfrak{g}$ of a linear algebraic group G is invariant under Ad_g for every $g \in G$, then $\mathfrak{v}$ is an ideal. Thus, if G is not simple, then $\mathfrak{g}$ is not simple.

It would be convenient if G simple implied $\mathfrak{g}$ simple, but that is not quite true[5]. However, there are only a few exceptions, all in small characteristic. To summarize: for $G = \mathrm{SL}_n$, the Lie algebra $\mathfrak{g} = \mathfrak{sl}_n$ is simple provided that the characteristic p of the field K does not divide n. (If $p|n$, then $\mathfrak{sl}_n$ has non-trivial center, namely, the multiples of the diagonal matrix I.) For almost simple Lie groups G such that $\mathfrak{g}$ is not isomorphic to $\mathfrak{sl}_n$, we have that $\mathfrak{g}$ is simple provided that $\mathrm{char}(K) > 3$ [**Hog82**, Table 1]. (The assumption in [**Hog82**] that the ground field is algebraically closed is harmless, as, if $\mathfrak{g}$ is simple over $\overline{K}$, it follows trivially that $\mathfrak{g}$ is simple over K: a decomposition over K would also be valid over $\overline{K}$.) In fact, $\mathrm{char}(K) > 2$ is enough for all Lie algebras of type other than A_n (corresponding to SL_n), E_6 and G_2, by the same table.

In spite of this small-characteristic phenomenon, we will nevertheless descend from the algebraic groups to Lie algebra at an important step (proof of Lemma 4.6), as then matters arguably become particularly clear and straightforward.

4.1.7. *Finite groups of Lie type.* The general definition of a finite group of Lie type is that it is the group G^F of points on a semisimple algebraic group G defined over a finite field $\mathbb{F}_q$ that are left fixed by a *Steinberg endomorphism* $F : G \to G$. A Steinberg endomorphism is an endomorphism $F : G \to G$ such that, for some $m \geq 1$, F^m is the *Frobenius map* with respect to $\mathbb{F}_q$. The Frobenius map with respect to $\mathbb{F}_q$ is the map sending every element $g \in G(\overline{\mathbb{F}_q})$ with entries $g_{i,j}$ to the element with entries $g_{i,j}^q$. It fixes precisely the elements of $G(\mathbb{F}_q)$.

The most familiar finite groups of Lie type (classical groups and Chevalley groups) are of the form $G(\mathbb{F}_q)$, G a semisimple algebraic group; they correspond to the case $m = 1$. The groups that require $m > 1$ are called *twisted groups*.

We will work out growth in $G(K)$, $G = \mathrm{SL}_2$, K finite (or, more generally, perfect) in a way that generalizes easily to other groups of Lie type with G simple. It is possible to include twisted groups, as was shown in [**PS16**]; however, our notation will be of the form $G(K)$, as is appropriate for $m = 1$.

Requiring G to be simple is not quite the same as requiring the group of Lie type $G^F = G(K)$ to be simple. The simple groups coming from groups of Lie type are of the form $G^F/Z(G^F)$, G simple.[6] The center $Z(G^F)$ is described in

[5]To the contrary of what was carelessly stated in the proof of Prop. 5.3 in the survey [**Hel15**].

[6]Two comments for the sake of precision are in order. (a) There is one group in the classification of finite simple groups that is almost but not quite of the type $G^F/Z(G^F)$: the *Tits group* [**MT11**, p. 213]. As we said before, we need not care about individual groups in the classification, since we aim at asymptotic statements. (b) By a result of Tits [**MT11**, Thm. 24.17], given G simple and *simply connected* [**MT11**, Def. 9.14], the group $G^F/Z(G^F)$ will be simple, provided we are not in a finite list of exceptions. Notably, SO_n is not simply connected; one uses a simply-connected finite cover of SO_n in its stead.

[**MT11**, Table 24.2]. It is very easy to pass from statements on growth in G^F to statements on growth in $G^F/Z(G^F)$, as we will see in the case for $G = \mathrm{SL}_2$, where $Z(G^F) = \{I, -I\}$.

4.2. Escape from subvarieties. We are working with a finite subset A of a group G. At some points in the argument, we will need to make sure that we can find an element $g \in A^k$ (k small) that is *not* special: for example, we want to be able to use a g that is not unipotent, that does not have a given $\vec{v}$ as an eigenvector, that *is* regular semisimple, etc.

It is possible to give a completely general argument of this form. Let us first set the framework. Let G be a group acting by linear transformations on n-dimensional space $\mathbb{A}^n$ over a field K. In other words, we are given a homomorphism $\phi : G \to \mathrm{GL}_n(K)$ from G to the group of invertible matrices $\mathrm{GL}_n(K)$. Let W be a proper subvariety of $\mathbb{A}^n$. We may think of points on W as being *special*, and points outside W as being generic. We start with a point x of $\mathbb{A}^n$, and a subset A of G. The following proposition ensures us that, if, starting from x and acting on it repeatedly by A, we can eventually escape from W, then we can escape from it in a bounded number of steps, and in many ways.

The proof[7] proceeds by induction on the dimension, with the degree kept under control. What is crucial for us is that the dimension is an integer, and thus can be used as a counter for induction. (Alternatively, we could say that the kind of induction we are about to undertake works because the ring $K[x_1, \ldots, x_n]$ is Noetherian.)

PROPOSITION 4.2. *Let us be given*

- G *a group acting linearly on affine space* $\mathbb{A}^n$ *over a field* K,
- $W \subsetneq \mathbb{A}^n$, *a subvariety,*
- A *a set of generators of* G *with* $A = A^{-1}$, $e \in A$,
- $x \in \mathbb{A}^n$ *such that the orbit* $G \cdot x$ *of* x *is not contained in* W.

Then there are constants k, c *depending only the number, dimension and degree of the irreducible components of* W *such that there are at least* $\max(1, c|A|)$ *elements* $g \in A^k$ *for which* $gx \notin W(K)$.

PROOF FOR A SPECIAL CASE. Let us first do the special case of W an irreducible linear subvariety. We will proceed by induction on the dimension of W. If $\dim(W) = 0$, then W consists of a single point, and the statement is clear: since $G \cdot x \not\subset \{x\}$ and A generates G, there exists a $g \in A$ such that $gx \neq x$; if there are fewer than $|A|/2$ such elements of A, we let g_0 be one of them, and note that any product $g^{-1} g_0$ with $gx = x$ satisfies $g^{-1} g_0 x \neq x$; there are $> |A|/2$ such products.

Assume, then, that $\dim(W) > 0$, and that the statement has been proven for all W' with $\dim(W') < \dim(W)$. If $gW = W$ for all $g \in A$, then either (a) gx does not lie on W for any $g \in A$, proving the statement, or (b) gx lies on W for every $g \in G = \langle A \rangle$, contradicting the assumption. Assume that $gW \neq W$ for some $g \in A$; then $W' = gW \cap W$ is an irreducible linear variety with $\dim(W') < \dim(W)$. Thus, by the inductive hypothesis, there are at least $\max(1, c'|A|)$ elements $g' \in A^{k'}$ (c', k' depending only on $\dim(W')$) such that $g'x$ does not lie on $W' = gW \cap W$. Hence, for each such g', either $g^{-1} g' x$ or $g'x$ does not lie on W. We have thus proven the statement with $c = c'/2$, $k = k' + 1$. □

[7] The statement of the proposition is as in [**Hel11**], based closely on [**EMO05**], but the idea is probably older.

EXERCISE 4.3. *Generalize the proof so that it works without the assumptions that W be linear or irreducible. Sketch: work first towards removing the assumption of irreducibility. Let W be the union of r components, not necessarily all of the same dimension. The intersection $W' = gW \cap W$ may also have several components, but no more than r^2; this is what we meant by "keeping the degree under control". Now pay attention to d, the maximum of the dimensions of the components of a variety, and m, the number of components of maximal dimension. Show that either (1) d is lower for $W' = gW \cap W$ than for W, or (2) d is the same in both cases, but m is lower for W' than for W, or (3) x does not lie in any component of W of dimension d, and thus we may work instead with W with those components removed. Use this fact to carry out the inductive process.*

Now note that you never really used the fact that W is linear. Instead of keeping track of the number of components r, keep track of the sum of their degrees. Control that using the generalized form (4.1) of Bézout's theorem.

4.3. Dimensional estimates. By a *dimensional estimate* we mean a lower or upper bound on an intersection of the form $A^k \cap V$, where $A \subset G(K)$, V is a subvariety of G and G/K is an algebraic group. As you will notice, the bounds that we obtain will be meaningful when A grows relatively slowly. However, no assumption on A is made, other than that it generate $G(K)$.

Of course, Proposition 4.2 may already be seen as a dimensional estimate of sorts, in that it tells us that $\gg |A|$ elements of A^k, k bounded, lie outside W. We are now aiming at much stronger bounds; Proposition 4.2 will be a useful tool along the way.

We aim for the estimates whose most general form is as follows.

THEOREM 4.4. *Let $G < \mathrm{GL}_n$ be a simple linear algebraic group over a finite field K. Let $A \subset G(K)$ be a finite set of generators of $G(K)$. Assume $A = A^{-1}$, $e \in A$. Let V be a pure-dimensional subvariety of G. Then*

$$|A \cap V(K)| \ll |A^k|^{\frac{\dim V}{\dim G}}, \tag{4.3}$$

where k and the implied constant depend only on n and on $\deg(V)$.

Estimates of this form can be traced in part to [**LP11**] (A a subgroup, V general) and in part to [**Hel08**] y [**Hel11**] (A an arbitrary set, but V special). We now have Theorem 4.4, thanks to [**BGT11**] and [**PS16**]. In fact, [**PS16**] gives a more general statement, in that twisted groups of Lie type are covered. Actually, one can state Theorem 4.4 in an even more general form, in that the assumption that K is finite can be dropped, and the condition that A generate $G(K)$ can be replaced by a condition that $\langle A \rangle$ be "Zariski-dense enough", meaning not contained in a union of $\leq C$ varieties of degree $\leq C$, where C depends only on n and $\deg(V)$.

We will show how to prove the estimate (4.3) in the case we actually need, but in a way that can be generalized to arbitrary V and arbitrary simple G. We will give a detailed outline of how to obtain the generalization.

Actually, as a first step towards the general strategy, let us study a particular V that we will not use in the end; it was crucial in earlier versions of the proof, and, more importantly, it makes several of the key ideas clear quickly. The proof is basically the same as in [**Hel08**, §4]. In particular, it will not look as if we used any algebraic geometry; however, the concrete procedure we follow here will then lead us naturally to a general procedure that will ask for the language and the basic tools of algebraic geometry.

LEMMA 4.5. *Let $G = \mathrm{SL}_2$, K a field. Let $A \subset G(K)$ be a finite set of generators of $G(K)$. Assume $A = A^{-1}$, $e \in A$. Let T be a maximal torus of G. Then*

$$|A \cap T(K)| \ll |A^k|^{1/3}, \tag{4.4}$$

where k and the implied constant are absolute.

PROOF. We can assume without loss of generality that $|K|$ and $|A|$ are greater than a constant, as otherwise the statement is trivial. We can also write the elements of T as diagonal matrices, by conjugation by an element of $\mathrm{SL}_2(\overline{K})$.

Let

$$g = \begin{pmatrix} a & b \\ c & d \end{pmatrix} \tag{4.5}$$

be any element of $\mathrm{SL}_2(\overline{K})$ with $abcd \neq 0$. Consider the map $\phi : T(K) \times T(K) \times T(K) \to G(K)$ given by

$$\phi(x, y, z) = x \cdot gyg^{-1} \cdot z.$$

We would like to show that this map is in some sense almost injective. (What for? If the map were injective, and we had $g \in A^\ell$, ℓ bounded by a constant, we would have

$$|A \cap T(K)|^3 = |\phi(A \cap T(K), A \cap T(K), A \cap T(K))| \leq |AA^\ell AA^{-\ell}A| = |A^{2\ell+3}|,$$

which would imply immediately the result we are trying to prove. Here we are simply using the fact that the image $\phi(D)$ of an injection ϕ has the same number of elements as the domain D.)

Multiplying matrices, we see that, for

$$x = \begin{pmatrix} r & 0 \\ 0 & r^{-1} \end{pmatrix}, \quad y = \begin{pmatrix} s & 0 \\ 0 & s^{-1} \end{pmatrix}, \quad z = \begin{pmatrix} t & 0 \\ 0 & t^{-1} \end{pmatrix},$$

$\phi((x, y, z))$ equals

$$\begin{pmatrix} rt(sad - s^{-1}bc) & rt^{-1}(s^{-1} - s)ab \\ r^{-1}t(s - s^{-1})cd & r^{-1}t^{-1}(s^{-1}ad - sbc) \end{pmatrix}. \tag{4.6}$$

Let $s \in \overline{K}$ be such that $s^{-1} - s \neq 0$ and $sad - s^{-1}bc \neq 0$. A brief calculation shows then that $\phi^{-1}(\{\phi((x, y, z))\})$ has at most 16 elements: we have

$$rt^{-1}(s^{-1} - s)ab \cdot r^{-1}t(s - s^{-1})cd = -(s - s^{-1})^2 abcd,$$

and, since $abcd \neq 0$, at most 4 values of s can give the same value $-(s - s^{-1})^2 abcd$ (the product of the top right and bottom left entries of ((4.6)); for each such value of s, the product and the quotient of the upper left and upper right entries of (4.6) determine r^2 and t^2, respectively, and obviously there are at most 2 values of r and 2 values of t for r^2, t^2 given.

Now, there are at most 4 values of s such that $s^{-1} - s = 0$ or $sad - s^{-1}bc = 0$. Hence,

$$|\phi(A \cap T(K), A \cap T(K), A \cap T(K))| \geq \frac{1}{16} |A \cap T(K)| (|A \cap T(K)| - 4) |A \cap T(K)|,$$

and, at the same time, $\phi(A \cap T(K), A \cap T(K), A \cap T(K)) \subset AA^\ell AA^{-\ell}A = A^{3+2\ell}$, as we said before. If $|A \cap T(K)|$ is less than 8 (or any other constant), conclusion (4.4) is trivial. Therefore,

$$|A \cap T(K)|^3 \leq 2|A \cap T(K)|(|A \cap T(K)| - 4)|A \cap T(K)| \leq 32|A^{2\ell+3}|,$$

i.e., (4.4) holds.

It only remains to verify that there exists an element (4.5) of A^ℓ with $abcd \neq 0$. Now, $abcd = 0$ defines a subvariety W of $\mathbb{A}^4$, where $\mathbb{A}^4$ is identified with the space of 2-by-2 matrices. Moreover, for $|K| > 2$, there are elements of $G(K)$ outside that variety. Hence, the conditions of Prop. 4.2 hold (with $x = e$). Thus, we obtain that there is a $g \in A^\ell$ (ℓ a constant) such that $g \notin W(K)$, and that was what we needed. □

Let us abstract the essence of what we have just done, so that we can then generalize the result to an arbitrary variety V instead of working just with T. For the sake of convenience, we will do the case $\dim V = 1$, which is, at any rate, the case we will need. The strategy of the proof of Lemma 4.5 is to construct a morphism $\phi : V \times V \times \cdots \times V \to G$ (r copies of V, where $r = \dim(G)$) of the form

$$\phi(v_1, \ldots, v_r) = v_1 g_1 v_2 g_2 \cdots v_{r-1} g_{r-1} v_r, \tag{4.7}$$

where $g_1, g_2, \ldots, g_{r-1} \in A^\ell$, in such a way that, for $v = (v_1, \ldots, v_r)$ a generic point in $V \times V \times \cdots \times V$, the preimage $\phi^{-1}(\{\phi(v)\})$ has dimension 0. Actually, as we have just seen, it is enough to prove that this is true for $(g_1, g_2, \ldots, g_{r-1})$ a generic element of G^{r-1}; the escape argument (Prop. 4.2) takes care of the rest.

The following lemma is the same as [**Tao15**, Prop. 5.5.3], which, in turn, is the same as [**LP11**, Lemma 4.5]. We will give a proof valid for $\mathfrak{g}$ simple.

LEMMA 4.6. *Let $G < \mathrm{SL}_n$ be a simple algebraic group defined over a field K. Let $V, V' \subsetneq G$ be irreducible subvarieties with $\dim(V) < \dim(G)$ and $\dim(V') > 0$. Then, for every $g \in G(\overline{K})$ outside a subvariety $W \subsetneq G$ depending on V and V', the variety $\overline{VgV'}$ has dimension $> \dim(V)$.*

Moreover, the number and degrees of the irreducible components of W are bounded by a constant that depends only on n and $\deg(V)$ and $\deg(V')$.

In fact, the proof we will now see bounds the number and degrees of the components of W in terms of n alone.

PROOF FOR $\mathfrak{g}$ SIMPLE. We can assume without loss of generality – replacing V and V' by varieties Vh and $h'V$, $h, h' \in G(\overline{K})$, if necessary – that V and V' go through the origin, and that the origin is a non-singular point for V and V'. We may also assume without loss of generality that K is algebraically closed.

Let $\mathfrak{v}$ and $\mathfrak{v}'$ be the tangent spaces to V and V' at the origin. The tangent space to $\overline{VgV'g^{-1}}$ at the identity is $\mathfrak{v} + \mathrm{Ad}_g \mathfrak{v}'$. Thus, for $\overline{V'gV}$ to have dimension $> \dim(V)$, it is enough that $\mathfrak{v} + \mathrm{Ad}_g \mathfrak{v}'$ have dimension $> \dim(\mathfrak{v}) = \dim(V)$.

Suppose that this is not the case for any g on G. Then the space $\mathfrak{w}$ spanned by all spaces $\mathrm{Ad}_g \mathfrak{v}'$, for all g, is contained in $\mathfrak{v}$. Since $\dim(V) < \dim(G)$, $\mathfrak{v} \subsetneq \mathfrak{g}$. Clearly, $\mathfrak{w}$ is non-empty and invariant under Ad_g for every g. Hence it is an ideal. However, we are assuming $\mathfrak{g}$ to be simple. Contradiction.

Thus, $\mathfrak{v} + \mathrm{Ad}_g \mathfrak{v}'$ has dimension greater than $\dim(\mathfrak{v})$ for some g. It is easy to see that the points g where that is not the case are precisely those such that all $(\dim(\mathfrak{v}) + 1) \times (\dim(\mathfrak{v}) + 1)$ minors of a matrix – whose entries are polynomial on the entries of g – vanish. We let W be the subvariety of V where those minors all vanish. The claim on the number and degrees of components of W follows by Bézout (4.1). □

We can now generalize our proof of Lemma 4.5, and thus prove (4.3) for all varieties of dimension 1. Before we start, we need a basic counting lemma, left as an exercise.

EXERCISE 4.7. *Let $W \subset \mathbb{A}^n$ be a variety defined over K such that every component of W has dimension $\leq d$. Let S be a finite subset of K. Then the number of points $(x_1, \ldots, x_n) \in S \times S \times \cdots \times S$ (n times) lying on W is $\ll |S|^d$, where the implied constant depends only on n and on the number and degrees of the components of W.*

PROPOSITION 4.8. *Let $G \subset \mathrm{SL}_n$ be an simple algebraic group over a finite field K. Assume that $|G(K)| \geq c|K|^{\dim(G)}$, $c > 0$. Let $Z \subset G$ be a variety of dimension 1. Let $A \subset G(K)$ be a set of generators of $G(K)$ such that $A = A^{-1}$, $e \in A$.*

Then

$$|A \cap Z(K)| \ll |A^k|^{1/\dim(G)}, \tag{4.8}$$

where k and the implied constant depend only on n, c, $\deg(G)$ and the number and degrees of the irreducible components of Z.

Obviously, $G = \mathrm{SL}_n$ is a valid choice, since it is simple and $|\mathrm{SL}_n(K)| \gg |K|^{n^2-1} = |K|^{\dim(G)}$.

PROOF. We will use Lemma 4.6 repeatedly. When we apply it, we get a subvariety $W \subsetneq G$ such that, for every g outside W, some component of $\overline{VgV'}$ has dimension $> \dim(V)$ (where V and V' are varieties satisfying the conditions of Lemma 4.6). Since G is irreducible, every component of W has dimension less than $\dim(G)$. By Exercise 4.7 (with $S = K$) and the assumption $|G(K)| \geq c|K|^{\dim(G)}$, there is at least one point of $G(K)$ not on W, provided that $|K|$ is larger than a constant, as we can indeed assume. Hence, we can use escape from subvarieties (Prop. 4.2) to show that there is a $g \in (A \cup A^{-1} \cup \{e\})^\ell$, where ℓ depends only on the number and degrees of components of W, that is to say – by Lemma 4.6 – only on n and $\deg(G)$.

So: first, we apply Lemma 4.6 with $V = V' = Z$; we obtain a variety $V_2 = \overline{Vg_1V'} = \overline{Zg_1Z}$ with $g \in (A \cup A^{-1} \cup \{e\})^\ell$ such that V_2 has at least one component of dimension 2. (We might as well assume V is irreducible from now on; then V_2 is irreducible.) We apply Lemma 4.6 again with $V = V_2$, $V' = Z$, and obtain a variety $V_3 = \overline{V_2g_2Z} = \overline{Zg_1Zg_2Z}$ of dimension 3. We go on and on, and get that there are $g_1, \ldots, g_{m-1} \in (A \cup A^{-1} \cup \{e\})^{\ell'}$, $r = \dim(G)$, such that $\overline{Zg_1Zg_2 \ldots Zg_{r-1}Z}$ has dimension r.

Hence, the variety W of singular points of the map f from $Z^r = Z \times Z \times \cdots \times Z$ (r times) to G given by

$$f(z_1, \ldots, z_m) = z_1 g_1 z_2 g_2 \ldots z_{r-1} g_{r-1} z_r$$

cannot be all of $Z \times \ldots \times Z$. Thus, since $Z \times \ldots \times Z$ is irreducible, every component of W is of dimension less than $\dim V$. Again by Exercise 4.7 (with $S = A \cap Z(K)$), at most $O(|A \cap Z(K)|^{r-1})$ points of $(A \cap Z(K)) \times \cdots \times (A \cap Z(K))$ (r times) on W. The number of points of $(A \cap Z(K)) \times \cdots \times (A \cap Z(K))$ not on W is at most the degree of f times the number of points on $f(A\cap Z(K),\ldots,A\cap Z(K))$, which is contained in A^k for $k=r+(r-1)\ell'$. Therefore,

$$|A\cap Z(K)|^r \leq \deg(f)|A^k| + O\left(|A\cap Z(K)|^{r-1}\right),$$

and so we are done. □

In general, one can prove (4.3) for $\dim(V)$ arbitrary using very similar arguments, together with an induction on the dimension of the variety V in (4.3). We will demonstrate the basic procedure doing things in detail for $G=\mathrm{SL}_2$ and for the kind of variety V for which we really need to prove estimates.

We mean the variety V_t defined by

$$\det(g)=1, \mathrm{tr}(g)=t \tag{4.9}$$

for $t\neq \pm 2$. Such varieties are of interest to us because, for any regular semisimple $g\in \mathrm{SL}_2(K)$ (meaning: any matrix in $\mathrm{SL}_2(K)$ having two distinct eigenvalues), the conjugacy class $\mathrm{Cl}(g)$ is contained in $V_{\mathrm{tr}(g)}$.

PROPOSITION 4.9. *Let K be a finite field. Let $A\subset \mathrm{SL}_2(K)$ be a set of generators of $\mathrm{SL}_2(K)$ with $A=A^{-1}$, $e\in A$. Let V_t be given by (4.9).*

Then, for every $t\in K$ other than ± 2,

$$|A\cap V_t(K)| \ll |A^k|^{\frac{2}{3}}, \tag{4.10}$$

where k and the implied constant are absolute.

Needless to say, $\dim(\mathrm{SL}_2)=3$ and $\dim(V_t)=2$, so this is a special case of (4.3).

PROOF. Consider the map $\phi: V_t(K)\times V_t(K)\to \mathrm{SL}_2(K)$ given by

$$\phi(y_1,y_2)=y_1y_2^{-1}.$$

It is clear that

$$\phi(A\cap V_t(K), A\cap V_t(K))\subset A^2.$$

Thus, if ϕ were injective, we would obtain immediately that $|A\cap V_t(K)|^2\leq |A^2|$. Now, ϕ is not injective, not even nearly so. The preimage of $\{h\}$, $h\in \mathrm{SL}_2(K)$, is

$$\phi^{-1}(\{h\})=\{(w,h^{-1}w): \mathrm{tr}(w)=t, \mathrm{tr}(h^{-1}w)=t\}.$$

We should thus ask ourselves how many elements of A lie on the subvariety $Z_{t,h}$ of G defined by

$$Z_{t,h}=\{(w,hw): \mathrm{tr}(w)=t, \mathrm{tr}(h^{-1}w)=t\}.$$

For $h\neq \pm e$, $\dim(Z_{t,h})=1$, and the number and degrees of irreducible components of $Z_{t,h}$ are bounded by an absolute constant. Thus, applying Proposition 4.8, we get that, for $h\neq \pm e$,

$$|A\cap Z_{t,h}(K)|\ll |A^{k'}|^{1/3},$$

where k' and the implied constant are absolute.

Now, for every $y_1\in V_t(K)$, there are at least $|V_t(K)|-2$ elements $y_2\in V_t(K)$ such that $y_1y_2^{-1}\neq \pm e$. We conclude that

$$|A\cap V(K)|(|A\cap V(K)|-2)\leq |A^2|\cdot \max_{g\neq \pm e}|A\cap Z_{t,h}(K)|\ll |A^2||A^{k'}|^{1/3}.$$

We can assume that $|A \cap V(K)| \geq 3$, as otherwise the desired conclusion is trivial. We obtain, then, that

$$|A \cap V(K)| \ll |A^k|^{2/3}$$

for $k = \max(2, k')$, as we wanted. □

Now we can finally prove the result we needed.

Corollary 4.10. *Let $G = \mathrm{SL}_2$, K a finite field. Let A be a set of generators of $G(K)$ with $A = A^{-1}$, $e \in A$. Let $g \in A^\ell$ ($\ell \geq 1$) be regular semisimple. Then*

$$|A^2 \cap C(g)| \gg \frac{A}{|A^{k\ell}|^{2/3}}, \tag{4.11}$$

where k and the implied constant are absolute.

In particular, if $|A^3| \leq |A|^{1+\delta}$, then

$$|A^2 \cap C(g)| \gg_\ell |A|^{1/3 - O(\ell\delta)}. \tag{4.12}$$

Proof. Proposition 4.9 and Lemma 2.6 imply (4.11) immediately, and (4.12) follows readily from (4.11) via (2.4). □

Let us now see two problems whose statements we will not use; they are, however, essential if one wishes to work in SL_n for n arbitrary, or in an arbitrary simple algebraic group. The first problem is challenging, but we have already seen and applied the main ideas involved in its solution. In essence, it is a matter of setting up a recursion properly.

Exercise 4.11. *Generalize Proposition 4.8 to pure-dimensional varieties Z of arbitrary dimension; that is, prove Theorem 4.4.*

The following exercise is easy. In part (b), follow the proof of Corollary 4.10, using Exercise 4.11.

Exercise 4.12. *Let G be a simple algebraic group over a finite field K. Let $A \subset G(K)$, $A = A^{-1}$, $e \in A$, $\langle A \rangle = G(K)$. Let $g \in A^\ell$, $\ell \geq 1$.*

(a) *Using the material in §4.1.3, show that* $\dim G - \dim \overline{\mathrm{Cl}(g)} = \dim C(g)$.

(b) *Show that, if $|A|^3 \leq |A|^{1+\delta}$,*

$$|A^2 \cap C(g)| \gg |A|^{\frac{\dim(C(g))}{\dim(G)} - O(\ell\delta)}, \tag{4.13}$$

where the implied constants depend only on n.

If g is regular semisimple, then, as we know, $C(g)$ is a maximal torus.

5. Growth and diameter in $\mathrm{SL}_2(K)$

5.1. Growth in $\mathrm{SL}_2(K)$, K arbitrary. We come to the proof of our main result. Here we will be closer to newer treatments (in particular, [**PS16**]) than to what was the first proof, given in [**Hel08**]; these newer versions generalize more easily. We will give the proof only for SL_2, and point out the couple of places in the proof where one would has to be especially careful when generalizing matters to SL_n, $n > 2$, or other linear algebraic groups.

The proof in [**Hel08**] used the sum-product theorem (Thm. 3.8). We will not use it, but the idea of "pivoting" will reappear. It is also good to note that, just as before, there is an inductive process here, carried out on a group G, even though G does not have a natural order $(1, 2, 3, \ldots)$. All we need for the induction to work is a set of generators A of G.

THEOREM 5.1 (Helfgott [**Hel08**]). *Let K be a finite field. Let $A \subset \mathrm{SL}_2(K)$ be a set of generators of $\mathrm{SL}_2(K)$ with $A = A^{-1}$, $e \in A$. There either*

$$|A^3| \geq |A|^{1+\delta}, \tag{5.1}$$

where $\delta > 0$ is an absolute constant, or

$$A^3 = \mathrm{SL}_2(K). \tag{5.2}$$

Actually, [**Hel08**] proved this result (with A^k, k a constant, instead of A^3 in (5.2)) for $K = \mathbb{F}_p$; the first generalization to a general finite field K was given by [**Din11**]. The proof we are about to see works for K general without any extra effort. It works, incidentally, for K infinite as well, dropping the condition $|A| < |\mathrm{SL}_2(K)|^{1-\epsilon}$, which becomes trivially true. The case of characteristic 0 is actually easier than the case $K = \mathbb{F}_p$; the proof in [**Hel08**] was already valid for $K = \mathbb{R}$ or $K = \mathbb{C}$, say. However, for applications, the "right" result for $K = \mathbb{R}$ or $K = \mathbb{C}$ is not really Thm. 5.1, but a statement counting how many elements there can be in A and $A \cdot A \cdot A$ that are separated by a given small distance from each other; that was proven in [**BG08a**], adapting the techniques in [**Hel08**].

PROOF. We may assume that $|A|$ is larger than an absolute constant, since otherwise the conclusion would be trivial. Let $G = \mathrm{SL}_2$.

Suppose that $|A^3| < |A|^{1+\delta}$, where $\delta > 0$ is a small constant to be determined later. By escape (Prop. 4.2), there is an element $g_0 \in A^c$ that is regular semisimple (that is, $\mathrm{tr}(g_0) \neq \pm 2$), where c is an absolute constant. (Easy exercise: show we can take $c = 2$.) Its centralizer in $G(K)$ is $\mathbf{T} := C(g) = T(\overline{K}) \cap G(K)$ for some maximal torus T.

Call $\xi \in G(K)$ a *pivot* if the map $\phi_\xi : A \times \mathbf{T} \to G(K)$ defined by

$$(a, t) \mapsto a\xi t \xi^{-1} \tag{5.3}$$

is injective as a function from $(\pm e \cdot A)/\{\pm e\} \times \mathbf{T}/\{\pm e\}$ to $G(K)/\{\pm e\}$.

Case (a): There is a pivot ξ in A. By Corollary 4.10, there are $\gg |A|^{1/3 - O(c\delta)}$ elements of $\mathbf{T}$ in $A^{-1}A$. Hence, by the injectivity of ϕ_ξ,

$$|\phi_\xi(A, A^2 \cap \mathbf{T})| \geq \frac{1}{4}|A||A^2 \cap \mathbf{T}| \gg |A|^{\frac{4}{3} - O(c\delta)}.$$

At the same time, $\phi_\xi(A, A^2 \cap \mathbf{T}) \subset A^5$, and thus

$$|A^5| \gg |A|^{4/3 - O(c\delta)}.$$

For $|A|$ larger than a constant and $\delta > 0$ less than a constant, this inequality gives us a contradiction with $|A^3| < |A|^{1+\delta}$ (by Ruzsa (2.3)).

Case (b): There are no pivots ξ in $G(K)$. Then, for every $\xi \in G(K)$, there are $a_1, a_2 \in A$, $t_1, t_2 \in \mathbf{T}$, $(a_1, t_1) \neq (\pm a_2, \pm t_2)$ such that $a_1 \xi t_1 \xi^{-1} = \pm e \cdot a_2 \xi t_2 \xi^{-1}$, and that gives us that

$$a_2^{-1} a_1 = \pm e \cdot \xi t_2 t_1^{-1} \xi^{-1}.$$

In other words, for each $\xi \in G(K)$, A^2 has a non-trivial intersection with the torus $\xi T \xi^{-1}$:

$$A^2 \cap \xi \mathbf{T} \xi^{-1} \not\subset \{\pm e\}. \tag{5.4}$$

(Note this means that case (b) never arises for K infinite. Why?)

Choose any $g \in A^2 \cap \xi \mathbf{T} \xi^{-1}$ with $g \neq \pm e$. Then g is regular semisimple. (This fact is peculiar to SL_2, or rather to groups of rank 1. This is one place in the proof that requires some work when you generalize it to other groups.)

The centralizer $C(g)$ of g equals $\xi \mathbf{T} \xi^{-1}$ (why?). Hence, by Corollary 4.10, we obtain that there are $\geq c'|A|^{1/3-O(\delta)}$ elements of $\xi \mathbf{T} \xi^{-1}$ in A^2, where c' and the implied constant are absolute.

At least $(1/2)|G(K)|/|\mathbf{T}|$ maximal tori of G are of the form $\xi T \xi^{-1}$, $\xi \in G(K)$ (check this yourself!). Every semisimple element of G that is not $\pm e$ is regular (again, something peculiar to SL_2); thus, every element of G that is not $\pm e$ can lie on at most one maximal torus. Hence

$$|A^2| \geq \frac{1}{2} \frac{|G(K)|}{|\mathbf{T}|} (c|A|^{1/3-O(\delta)} - 2) \gg |G(K)|^{2/3} |A|^{1/3-O(\delta)}.$$

Therefore, either $|A^2| > |A|^{1+2\delta}$ (say) or $|A| \geq |G|^{1-O(\delta)}$. In the first case, we have obtained a contradiction. In the second case, Proposition 5.6 implies that $A^3 = G$.

Case (c): There are pivots and non-pivots in $G(K)$. Since $\langle A \rangle = G(K)$, this implies that there exists a non-pivot $\xi \in G$ and an $a \in A$ such that $a\xi \in G$ is a pivot. Since ξ is not a pivot, (5.4) holds, and thus there are $|A|^{1/3-O(\delta)}$ elements of $\xi \mathbf{T} \xi^{-1}$ in A^k.

At the same time, $a\xi$ is a pivot, i.e., the map $\phi_{a\xi}$ defined in (5.3) is injective (considered as an application from $A/\{\pm e\} \times \mathbf{T}/\{\pm e\}$ to $G(K)/\{\pm e\}$). Therefore,

$$\left|\phi_{a\xi}(A, \xi^{-1}(A^2 \cap \xi T \xi^{-1})\xi)\right| \geq \frac{1}{4}|A||A^2 \cap \xi T \xi^{-1}| \geq \frac{1}{4}|A|^{\frac{4}{3}-O(\delta)}.$$

Since $\phi_{a\xi}(A, \xi^{-1}(A^2 \cap \xi T \xi^{-1})\xi) \subset A^5$, we obtain that

$$|A^5| \geq \frac{1}{4}|A|^{4/3-O(\delta)}. \tag{5.5}$$

Thanks again to Ruzsa (2.3), this inequality contradicts $|A^3| \leq |A|^{1+\delta}$ for $\delta > 0$ smaller than a constant. □

The following is a trivial exercise.

EXERCISE 5.2. *Using Theorem 5.1, show that the statement of Thm. 5.1 is also true with PSL_2 in place of SL_2. This step finishes the proof of Thm. 1.1.*

For SL_n, $n > 2$, or for general algebraic groups, there is, as we have seen, one difficulty in generalizing the above proof: a semisimple element other than $\pm e$ is not necessarily regular. The key to circumventing this difficulty is to use Theorem 4.4 to bound the number of elements on non-maximal subtori of a maximal torus T, and, in that way, bound the number of non-semisimple elements of A^k on T.

EXERCISE 5.3. *Using this observation, modify the proof of Thm. 5.1 so as to work for any simple linear algebraic group G.*

There remains the question of what the optimal value of δ in Thm. 5.1 could be. Kowalski [**Kow13**] proves Thm. 5.1 with $\delta = 1/3024$ (under the assumption $A = A^{-1}$). Button and Roney-Dougal prove (under the same assumption) that one cannot do better than $\delta = (\log_2 7 - 1)/6 \approx 0.3012$ [**BRD15**].

To obtain a good value of δ, it seems best to aim for a statement with a conclusion of the form

$$|A^3| \geq c|A|^{1+\delta}$$

instead of (5.1). It may be even better to aim for a result of the form, say,

$$|AA_0^kAA_0^kAA_0^kA| \geq c|A|^{1+\delta},$$

where A_0 is an arbitrary set of generators of $\mathrm{SL}_2(K)$. Then, when using our result to prove a diameter bound (as in exercise 1.2), we can set A_0 to be our initial set of generators S, whereas we set A equal to increasing powers of S. The resulting constant C in the exponent of the bound $\operatorname{diam}\Gamma(G,S) \ll (\log|G|)^C$ should then improve substantially over the value $C = 3323$ given in [**Kow13**].

Of course, we still need to prove Prop. 5.6. Let us do so.

5.2. The case of large subsets. Let us first see how A grows when $A \subset \mathrm{SL}_2(\mathbb{F}_q)$ is large with respect to $G = \mathrm{SL}_2(\mathbb{F}_q)$. In fact, it is not terribly hard to show that, if $|A| \geq |G|^{1-\delta}$, $\delta > 0$ a small constant, then $(A \cup A^{-1} \cup \{e\})^k = G$, where k is an absolute constant. To proceed as in [**Hel08**]: we can use (2.7) to pass to the solvable group of upper- or lower-triangular matrices, then go on as in §3.2 to show that the subgroups $U^{\pm}$ of upper- or lower-triangular matrices are contained in $(A \cup A^{-1} \cup \{e\})^{k'}$, k' a constant; we are then done by $G = U^-U^+U^-U^+$.

We will prove a stronger and nicer result: $A^3 = G$. The proof is due to Nikolov and Pyber [**NP11**]; it is based on a classical idea, brought to bear to this particular context by Gowers [**Gow08**]. It will give us the opportunity to revisit the adjacency operator $\mathscr{A}$ and its spectrum.

Recall that a *complex representation* of a group G is just a homomorphism $\phi : G \to \mathrm{GL}_d(\mathbb{C})$; by the *dimension* of the representation we just mean d. A representation ϕ is *trivial* if $\phi(g) = e$ for every $g \in G$.

The following result is due to Frobenius (1896), at least for q prime. It can be proven simply by examining a character table, as in [**Sha99**]. The same procedure gives analogues of the same result for other groups of Lie type. Alternatively, there is a very nice elementary proof for q prime, to be found, for example, in [**Tao15**, Lemma 1.3.3].

PROPOSITION 5.4. *Let $G = \mathrm{SL}_2(\mathbb{F}_q)$, $q = p^\alpha$. Then every non-trivial complex representation of G has dimension $\geq (q-1)/2$.*

We recall that the adjacency operator $\mathscr{A}$ on a Cayley graph $\Gamma(G,A)$ is the linear operator that takes a function $f : V \to \mathbb{C}$ to the function $\mathscr{A}f : V \to \mathbb{C}$ given by

$$\mathscr{A}f(g) = \frac{1}{|A|}\sum_{a\in A} f(ag). \tag{5.6}$$

Assume, as usual, that $A = A^{-1}$. Then $\mathscr{A}$ is symmetric and all its eigenvalues are real:

$$\ldots \leq \nu_2 \leq \nu_1 \leq \nu_0 = 1.$$

The largest eigenvalue ν_0 corresponds to the eigenspace of constant functions.

EXERCISE 5.5. *Show that no eigenvalue ν can be larger than 1. Hint: assume $\nu > 1$, and show, using (5.6), that, for g such that $|f(g)|$ is maximal, the equation $\mathscr{A}f(g) = \nu f(g)$ leads to a contradiction.*

By an *eigenspace* of $\mathscr{A}$ we mean, of course, the vector space consisting of functions f such that $\mathscr{A}f = \nu f$ for some fixed eigenvalue ν. It is clear from the definition that every eigenspace of $\mathscr{A}$ is invariant under the action of G by multiplication on the right. Hence, an eigenspace of $\mathscr{A}$ is a complex representation of G – and it can be trivial only if it is the eigenspace of constant functions, i.e., the eigenspace corresponding to ν_0. Thus, by Prop. 5.4, all other eigenvalues have multiplicity $\geq (q-1)/2$.

The idea now is to obtain a spectral gap, i.e., a non-trivial upper bound on ν_j, $j > 0$. It is standard to use the fact that the trace of a power $\mathscr{A}^r$ of an adjacency operator $\mathscr{A}$ can be expressed in two ways: as a the number of cycles of length r in the graph $\Gamma(G, A)$ (multiplied by $1/|A|^r$), and as the sum of the rth powers of the eigenvalues of $\mathscr{A}$. In our case, for $r = 2$, this gives us

$$\frac{|G||A|}{|A|^2} = \sum_j \nu_j^2 \geq \frac{q-1}{2}\nu_j^2, \tag{5.7}$$

for any $j \geq 1$, and hence

$$|\nu_j| \leq \sqrt{\frac{|G|/|A|}{(q-1)/2}}. \tag{5.8}$$

This is a very low upper bound when $|A|$ is large. This means that a few applications of the operator $\mathscr{A}$ are enough to render any function almost uniform, since any component orthogonal to the space of constant functions is multiplied by some ν_j, $j \geq 1$, at every step. The following proof puts in practice this observation efficiently.

PROPOSITION 5.6 ([**NP11**]). *Let* $G = \mathrm{SL}_2(\mathbb{F}_q)$, $q = p^\alpha$. *Let* $A \subset G$, $A = A^{-1}$. *Assume* $|A| \geq 2|G|^{8/9}$. *Then*

$$A^3 = G.$$

Actually, [**NP11**] proves this result without the assumption $A = A^{-1}$. We need $A = A^{-1}$ for $\mathscr{A}$ to be a symmetric operator, but, thanks to [**Gow08**], essentially the same argument works in the case $A \neq A^{-1}$.

PROOF. Suppose there is a $g \in G$ such that $g \notin A^3$. Then the scalar product

$$\begin{aligned}\langle \mathscr{A}1_A, 1_{gA}\rangle = \langle \mathscr{A}1_A, 1_{gA}\rangle &= \sum_{x\in G} (\mathscr{A}1_A)(x)\cdot 1_{gA}(x)\\ &= \frac{1}{|A|}\sum_{x\in G}\sum_{a\in G} 1_A(ax)\cdot 1_{gA}(x)\end{aligned}$$

equals 0, as otherwise there is an $x \in gA$ and an $a \in A$ such that $ax \in A$, and that would imply $g \in A^{-1}AA^{-1} = A^3$.

Since $\mathscr{A}$ is symmetric, it has full spectrum, that is, there exists a system of $n = |G|$ orthonormal eigenvectors $v_0, v_1, \ldots$ of $\mathscr{A}$. Here v_0 is the constant function satisfying $\langle v_0, v_0\rangle = 1$, that is, the constant function taking the value $1/\sqrt{|G|}$ everywhere. Then

$$\begin{aligned}\langle \mathscr{A}1_A, 1_{gA}\rangle &= \langle \sum_{j\geq 0} \nu_j \langle 1_A, v_j\rangle v_j, 1_{gA}\rangle\\ &= \nu_0\langle 1_A, v_0\rangle\langle v_0, 1_{gA}\rangle + \sum_{j>0}\nu_j\langle 1_A, v_j\rangle\langle v_j, 1_{gA}\rangle.\end{aligned}$$

Now

$$\nu_0\langle 1_A, v_0\rangle\langle v_0, 1_{gA}\rangle = 1\cdot\frac{|A|}{\sqrt{|G|}}\cdot\frac{|gA|}{\sqrt{|G|}} = \frac{|A|^2}{|G|}.$$

At the same time, by (5.8) and Cauchy-Schwarz,

$$\left|\sum_{j>0} \nu_j \langle 1_A, v_j\rangle\langle v_j, 1_{gA}\rangle\right| \leq \sqrt{\frac{2|G|/|A|}{q-1}} \sqrt{\sum_{j\geq 1} |\langle 1_A, v_j\rangle|^2} \sqrt{\sum_{j\geq 1} |\langle v_j, 1_{gA}\rangle|^2}$$
$$\leq \sqrt{\frac{2|G|/|A|}{q-1}} |1_A|_2 |1_{gA}|_2 = \sqrt{\frac{2|G||A|}{q-1}}.$$

Since $|G| = q(q^2-1)$, we see that $|A| \geq 2|G|^{8/9}$ implies

$$\frac{|A|^2}{|G|} > \sqrt{\frac{2|G||A|}{q-1}},$$

and thus $\langle \mathscr{A} 1_A, 1_{gA}\rangle > 0$. Contradiction. □

6. Further perspectives and open problems

6.1. Expansion, random walks and the affine sieve. Let G be a group, $A \subset G$, $A = A^{-1}$. As we saw in §1.1, the adjacency operator $\mathscr{A}$ has full real spectrum, and we can define what it means for the graph $\Gamma(G, A)$ to be a *δ-spectral expander*, or simply an *δ-expander*. An infinite family of graphs $\Gamma(G_i, A_i)$ is called an *expander family* if there is an $\epsilon > 0$ such that every $\Gamma(G_i, A_i)$ is an ϵ-expander. Of particular interest are expander families with $|A_i|$ bounded.

Using Thm. 5.1, Bourgain and Gamburd proved the following result [**BG08b**].

THEOREM 6.1. *Let $A_0 \subset \mathrm{SL}(\mathbb{Z})$. Assume that A_0 is not contained in any proper algebraic subgroup of* SL_2. *Then*

$$\{\Gamma(\mathrm{SL}_2(\mathbb{Z}/p\mathbb{Z}), A_0 \bmod p)\}_{p>C, p\ prime} \tag{6.1}$$

is an expander family for some constant C.

The proof also involves Proposition 5.4 (applied as in [**SX91**]) as well as a non-commutative version [**Tao08**] of the Balog-Gowers-Szemerédi theorem from additive combinatorics. There are by now wide-ranging generalizations of Thm. 6.1; see, e.g., [**GV12**].

A *random walk* on a graph is what it sounds like: we start at a vertex v_0, and at every step we move to one of the d neighbors of the vertex we are at – choosing any one of them with probability $1/d$. For convenience we work with a *lazy random walk*: at every step, we decide to stay where we are with probability $1/2$, and to move to a neighbor with probability $1/2d$. The *mixing time* is the number of steps it takes for ending point of a lazy random walk to become almost equidistributed (where "almost" is understood in any reasonable metric). In an ϵ-expander graph $\Gamma(G, A)$, the mixing time is $O_\epsilon(\log |G|)$, i.e., about as small as it could be: it is easy to see that, for $|A|$ bounded, the mixing time (and even the diameter) has to be $\gg \log |G|$.

EXERCISE 6.2. *Let G be a group, $A \subset G$, $A = A^{-1}$, $\langle A\rangle = G$. Let $\mathscr{A}$ be the adjacency operator on the Cayley graph.*

(a) *Take a lazy random walk with k steps on the Cayley graph, starting at the identity e. Show that the probability of your final position is given by the function $\phi_k = ((\mathscr{A} + I)/2)^k \delta_e$, where $\delta_e : G \to \mathbb{C}$ is the function taking the value 1 at e and 0 elsewhere.*

(b) *Write* δ_e *as a linear combination* $\delta_e = \sum_j c_j v_j$*, where each* v_j *is an eigenvector of* $\mathscr{A}$*. What is the coefficient in front of the constant eigenvector* v_0*? What is* $((\mathscr{A}+I)/2)^k \delta_e$*, as a linear combination of the eigenvectors* v_j*?*

(c) *Assume* $\Gamma(G, A)$ *is a* δ*-expander. Show that, for* $k \geq (2C/\delta) \log |G|$, $C \geq 1$*, the probability distribution* ϕ_k *is nearly uniform in both the* ℓ^2*- and the* ℓ^∞*-norms:*

$$\sum_{g \in G} \left| \phi_k(g) - \frac{1}{|G|} \right|^2 \leq \frac{1}{|G|^C},$$

$$\max_{g \in G} \left| \phi_k(g) - \frac{1}{|G|} \right| \leq \frac{1}{|G|^{C-1}}.$$

That is to say, the mixing time with respect to either the ℓ^2*- or the* ℓ^∞*-norms is* $\ll (1/\delta) \log |G|$.

Thus, Thm. 6.1 gives us small mixing times. This fact has made the *affine sieve* possible [**BGS10**]. The affine sieve is an analogue of classical sieve methods; they are recast as sieves based on the natural action of $\mathbb{Z}$ on $\mathbb{Z}$, whereas a general affine sieve considers the actions of other groups, such as $\mathrm{SL}_2(\mathbb{Z})$.

Expansion had been shown before for some specific A_0. In particular, when A_0 generates $\mathrm{SL}_2(\mathbb{Z})$ (or a subgroup of finite index before) then the fact that (6.1) is an expander graph can be derived from the *Selberg spectral gap* [**Sel65**], i.e., the fact that the Laplacian on the quotient $\mathrm{SL}_2(\mathbb{Z})\backslash\mathbb{H}$ of the upper half plane $\mathbb{H}$ has a spectral gap. Nowadays, one can go in the opposite direction: spectral gaps on more general quotients can be proven using Thm. 6.1 [**BGS11**].

Let us finish this discussion by saying that it is generally held to be plausible that the family of *all* Cayley graphs of $\mathrm{SL}_2(\mathbb{Z}/p\mathbb{Z})$, for all p, is an expander family; in other words, there may be an $\epsilon > 0$ such that, for every prime p and every generator A of $\mathrm{SL}_2(\mathbb{Z}/p\mathbb{Z})$, the graph $\Gamma(\mathrm{SL}_2(\mathbb{Z}/p\mathbb{Z}), A)$ is an ϵ-expander. This statement has seemed plausible at least since [**LR92**], but proving it is an open problem believed to be very hard. It has been shown that there exists a thin family of primes such that the statement is true if those primes are omitted [**BG10**].

6.2. Algorithmic and probabilistic questions. It is one thing to show that the diameter of a group G is small, that is, to show that every element of G can be written as short word on any set of generators A. (By a *word* on A we mean a product of elements of $A \cup A^{-1}$.) It is quite another to be able to find that word – reasonably quickly, it is understood.

Larsen [**Lar03**] gave a probabilistic algorithm that expresses an arbitrary $g \in \mathrm{SL}_2(\mathbb{Z}/p\mathbb{Z})$ as a word of length $O(\log p \log\log p)$ in the generators

$$A = \left\{ \begin{pmatrix} 1 & 1 \\ 0 & 1 \end{pmatrix}, \begin{pmatrix} 1 & 0 \\ 1 & 1 \end{pmatrix} \right\}$$

in time $(\log p)^{O(1)}$. No algorithm is known for arbitrary generators of $\mathrm{SL}_2(\mathbb{Z}/p\mathbb{Z})$. Neither do we have an algorithm for finding short words on arbitrary generators of finite simple groups in any other family.

Another question is what happens when g_1, g_2 are random elements of a group G. For several kinds of groups (linear algebraic, $\mathrm{Alt}(n)$) it is known that, with probability tending to one, g and h generate G. What is the diameter of the Cayley graph of G with respect to $\{g, h\}$ likely to be? For $G = \mathrm{SL}_2(\mathbb{F}_p)$, it is known that it is $O(\log|G|)$ with probability tending to one (by [**GHS**$^+$**09**] taken together with Thm. 5.1). For $\mathrm{Alt}(n)$, it is known to be $O(n^2(\log n)^{O(1)})$ with probability tending to one [**HSZ15**]. Is it actually $O(n(\log n)^{O(1)})$, or even $O(n\log n)$, with probability tending to one?

One can combine algorithmic and probabilistic questions. The proof in [**BBS04**] (supplemented by [**BH05**]) yields a probabilistic algorithm that, for a proportion $\to 1$ (as $n \to \infty$) of all pairs of elements g_1, g_2 of $\mathrm{Alt}(n)$, expresses any given element g of $\mathrm{Alt}(n)$ as a word of polynomial length on g_1 and g_2, and does so in (Las Vegas) polynomial time. (If the algorithm will fail for a given pair (g_1, g_2), it states so at an initial stage taking polynomial time.) The procedure in [**HSZ15**] gives a probabilistic algorithm that finds a word of length $O(n^2(\log n)^{O(1)})$ in time $O(n^2(\log n)^{O(1)})$ for a proportion $\to 1$ of all pairs g_1, g_2 and g arbitrary, as is sketched in [**HSZ15**, App. B].

No analogous algorithm is known over $\mathrm{SL}_2(\mathbb{F}_q)$, or for any other simple group of Lie type; we do not know how to express an arbitrary element of $\mathrm{SL}_2(\mathbb{F}_q)$ as a word of length $(\log q)^{O(1)}$ on a random pair of generators of G in time $(\log q)^{O(1)}$.

6.3. Final remarks. Let us briefly mention some links with other areas.

Group classification. It is by now clear that it is useful to look at a particular kind of result in group classification: the kind that was developed so as to avoid casework, and to do without the Classification of Finite Simple Groups. (The Classification is now generally accepted, but this was not always the case, and it is still sometimes felt to be better to prove something without it than with it; what we are about to see gives itself some validation to this viewpoint.) While results proven without the Classification are sometimes weaker than others, they are also more robust. Classifying subgroups of a finite group G is the same as classifying subsets $A \subset G$ such that $e \in A$ and $|AA| = |A|$. Some Classification-free classification methods can be adapted to help in classifying subsets $A \subset G$ such that $e \in A$ and $|AAA| \leq |A|^{1+\delta}$ – in other words, precisely what we are studying. It is in this way that [**LP11**] was useful in [**BGT11**], and [**Bab82**], [**Pyb93**] were useful in [**HS14**].

Model theory. Model theory is essentially a branch of logic with applications to algebraic structures. Hrushovski and his collaborators [**HP95**], [**HW08**], [**Hru12**] have used model theory to study subgroups of algebraic groups. This was influenced by Larsen-Pink [**LP11**], and also served to explain it. In turn, [**Hru12**] influenced later work, especially [**BGT12**].

Permutation-group algorithms. Much work on permutation groups has been algorithmic in nature. Here a standard reference is [**Ser03**]. A good example is a problem we mentioned before – that of bounding the diameter of $\mathrm{Sym}(n)$ with respect to a random pair of generators; the approach in [**BBS04**] combines probabilistic and algorithmic ideas – as does [**HSZ15**], which builds on [**BBS04**], and as, for that matter, does [**HS14**]. The reference [**LPW09**] treats several of the relevant probabilistic tools.

Geometric group theory. Here much work remains to be done. Geometric group theory, while still a relatively new field, is considerably older than the approach followed in these notes. It is clear that there is a connection, but it has not yet been fully explored. Here it is particularly worth remarking that [**Hru12**] gave a new proof of Gromov's theorem by means of the study of sets A that grow slowly in the sense used in these notes.

Acknowledgments

I was supported by ERC Consolidator grant 648329 (codename GRANT) and by funds from my Humboldt professorship. Many thanks are due to a helpful and spirited anonymous referee. Thanks are due as well to Lifan Guan, for providing a useful reference and catching several typos, and to the audiences both at the Arizona Winter School and at the Hausdorff Institute (HIM), for real-time feedback.

References

[AM85] N. Alon and V. D. Milman, λ_1, *isoperimetric inequalities for graphs, and superconcentrators*, J. Combin. Theory Ser. B **38** (1985), no. 1, 73–88, DOI 10.1016/0095-8956(85)90092-9. MR782626

[Bab82] L. Babai, *On the order of doubly transitive permutation groups*, Invent. Math. **65** (1981/82), no. 3, 473–484, DOI 10.1007/BF01396631. MR643565

[BBS04] L. Babai, R. Beals, and Á. Seress, *On the diameter of the symmetric group: polynomial bounds*, Proceedings of the Fifteenth Annual ACM-SIAM Symposium on Discrete Algorithms, ACM, New York, 2004, pp. 1108–1112. MR2291003

[Bec83] J. Beck, *On the lattice property of the plane and some problems of Dirac, Motzkin and Erdős in combinatorial geometry*, Combinatorica **3** (1983), no. 3-4, 281–297, DOI 10.1007/BF02579184. MR729781

[BG08a] J. Bourgain and A. Gamburd, *On the spectral gap for finitely-generated subgroups of* SU(2), Invent. Math. **171** (2008), no. 1, 83–121, DOI 10.1007/s00222-007-0072-z. MR2358056

[BG08b] J. Bourgain and A. Gamburd, *Uniform expansion bounds for Cayley graphs of* $\mathrm{SL}_2(\mathbb{F}_p)$, Ann. of Math. (2) **167** (2008), no. 2, 625–642, DOI 10.4007/annals.2008.167.625. MR2415383

[BG10] E. Breuillard and A. Gamburd, *Strong uniform expansion in* $\mathrm{SL}(2,p)$, Geom. Funct. Anal. **20** (2010), no. 5, 1201–1209, DOI 10.1007/s00039-010-0094-3. MR2746951

[BG11] E. Breuillard and B. Green, *Approximate groups, II: The solvable linear case*, Q. J. Math. **62** (2011), no. 3, 513–521, DOI 10.1093/qmath/haq011. MR2825469

[BGK06] J. Bourgain, A. A. Glibichuk, and S. V. Konyagin, *Estimates for the number of sums and products and for exponential sums in fields of prime order*, J. London Math. Soc. (2) **73** (2006), no. 2, 380–398, DOI 10.1112/S0024610706022721. MR2225493

[BGS10] J. Bourgain, A. Gamburd, and P. Sarnak, *Affine linear sieve, expanders, and sum-product*, Invent. Math. **179** (2010), no. 3, 559–644, DOI 10.1007/s00222-009-0225-3. MR2587341

[BGS11] J. Bourgain, A. Gamburd, and P. Sarnak, *Generalization of Selberg's* $\frac{3}{16}$ *theorem and affine sieve*, Acta Math. **207** (2011), no. 2, 255–290, DOI 10.1007/s11511-012-0070-x. MR2892611

[BGT11] E. Breuillard, B. Green, and T. Tao, *Approximate subgroups of linear groups*, Geom. Funct. Anal. **21** (2011), no. 4, 774–819, DOI 10.1007/s00039-011-0122-y. MR2827010

[BGT12] E. Breuillard, B. Green, and T. Tao, *The structure of approximate groups*, Publ. Math. Inst. Hautes Études Sci. **116** (2012), 115–221, DOI 10.1007/s10240-012-0043-9. MR3090256

[BH05] L. Babai and T. P. Hayes, *Near-independence of permutations and an almost sure polynomial bound on the diameter of the symmetric group*, Proceedings of the Sixteenth Annual ACM-SIAM Symposium on Discrete Algorithms, ACM, New York, 2005, pp. 1057–1066. MR2298365

[Bil99] Y. Bilu. Structure of sets with small sumset. In *Structure theory of set addition*, pages 77–108. Paris: Société Mathématique de France, 1999.

[BKT04] J. Bourgain, N. Katz, and T. Tao, *A sum-product estimate in finite fields, and applications*, Geom. Funct. Anal. **14** (2004), no. 1, 27–57, DOI 10.1007/s00039-004-0451-1. MR2053599

[Bor91] A. Borel. *Linear algebraic groups.* 2nd enlarged edition, Springer-Verlag, New York. 1991.

[BRD15] J. Button and C. M. Roney-Dougal, *An explicit upper bound for the Helfgott delta in* $\mathrm{SL}(2,p)$, J. Algebra **421** (2015), 493–511, DOI 10.1016/j.jalgebra.2014.09.001. MR3272393

[BS92] L. Babai and Á. Seress, *On the diameter of permutation groups*, European J. Combin. **13** (1992), no. 4, 231–243, DOI 10.1016/S0195-6698(05)80029-0. MR1179520

[Che70] J. Cheeger. A lower bound for the smallest eigenvalue of the Laplacian. In *Problems in analysis (Papers dedicated to Solomon Bochner, 1969)* Princeton Univ. Press, Princeton, N. J., 1970, pp. 195–199, 1970.

[CS10] E. Croot and O. Sisask, *A probabilistic technique for finding almost-periods of convolutions*, Geom. Funct. Anal. **20** (2010), no. 6, 1367–1396, DOI 10.1007/s00039-010-0101-8. MR2738997

[Din11] O. Dinai, *Growth in* SL_2 *over finite fields*, J. Group Theory **14** (2011), no. 2, 273–297, DOI 10.1515/JGT.2010.056. MR2788087

[DS98] V. I. Danilov and V. V. Shokurov, *Algebraic curves, algebraic manifolds and schemes*, Springer-Verlag, Berlin, 1998. Translated from the 1988 Russian original by D. Coray and V. N. Shokurov; Translation edited and with an introduction by I. R. Shafarevich; Reprint of the original English edition from the series Encyclopaedia of Mathematical Sciences [*Algebraic geometry. I*, Encyclopaedia Math. Sci., 23, Springer, Berlin, 1994; MR1287418 (95b:14001)]. MR1658464

[EK01] G. Elekes and Z. Király, *On the combinatorics of projective mappings*, J. Algebraic Combin. **14** (2001), no. 3, 183–197, DOI 10.1023/A:1012799318591. MR1869409

[EM03] G. A. Edgar and C. Miller, *Borel subrings of the reals*, Proc. Amer. Math. Soc. **131** (2003), no. 4, 1121–1129, DOI 10.1090/S0002-9939-02-06653-4. MR1948103

[EMO05] A. Eskin, S. Mozes, and H. Oh, *On uniform exponential growth for linear groups*, Invent. Math. **160** (2005), no. 1, 1–30, DOI 10.1007/s00222-004-0378-z. MR2129706

[FKP10] D. Fisher, N. H. Katz, and I. Peng, *Approximate multiplicative groups in nilpotent Lie groups*, Proc. Amer. Math. Soc. **138** (2010), no. 5, 1575–1580, DOI 10.1090/S0002-9939-10-10078-1. MR2587441

[Fre73] G. A. Freĭman. *Foundations of a structural theory of set addition.* American Mathematical Society, Providence, R. I., 1973. Translated from the Russian, Translations of Mathematical Monographs, Vol 37.

[GH11] N. Gill and H. A. Helfgott, *Growth of small generating sets in* $\mathrm{SL}_n(\mathbb{Z}/p\mathbb{Z})$, Int. Math. Res. Not. IMRN **18** (2011), 4226–4251. MR2836020

[GH14] N. Gill and H. A. Helfgott, *Growth in solvable subgroups of* $\mathrm{GL}_r(\mathbb{Z}/p\mathbb{Z})$, Math. Ann. **360** (2014), no. 1-2, 157–208, DOI 10.1007/s00208-014-1008-8. MR3263161

[GHR15] N. Gill, H. A. Helfgott, and M. Rudnev, *On growth in an abstract plane*, Proc. Amer. Math. Soc. **143** (2015), no. 8, 3593–3602, DOI 10.1090/proc/12309. MR3348800

[GHS+09] A. Gamburd, S. Hoory, M. Shahshahani, A. Shalev, and B. Virág, *On the girth of random Cayley graphs*, Random Structures Algorithms **35** (2009), no. 1, 100–117, DOI 10.1002/rsa.20266. MR2532876

[GK07] A. A. Glibichuk and S. V. Konyagin, *Additive properties of product sets in fields of prime order*, Additive combinatorics, CRM Proc. Lecture Notes, vol. 43, Amer. Math. Soc., Providence, RI, 2007, pp. 279–286. MR2359478

[Gow08] W. T. Gowers, *Quasirandom groups*, Combin. Probab. Comput. **17** (2008), no. 3, 363–387, DOI 10.1017/S0963548307008826. MR2410393

[GR07] B. Green and I. Z. Ruzsa, *Freiman's theorem in an arbitrary abelian group*, J. Lond. Math. Soc. (2) **75** (2007), no. 1, 163–175, DOI 10.1112/jlms/jdl021. MR2302736

[Gro81] M. Gromov, *Groups of polynomial growth and expanding maps*, Inst. Hautes Études Sci. Publ. Math. **53** (1981), 53–73. MR623534

[GV12] A. S. Golsefidy and P. P. Varjú, *Expansion in perfect groups*, Geom. Funct. Anal. **22** (2012), no. 6, 1832–1891, DOI 10.1007/s00039-012-0190-7. MR3000503

[Har77] R. Hartshorne, *Algebraic geometry*, Graduate Texts in Mathematics, vol. 52, Springer-Verlag, New York-Heidelberg, 1977. MR0463157

[Hela] H. A. Helfgott. Crecimiento y expansión en SL_2. To appear in *Actas de la escuela AGRA II: Aritmética, grupos y análisis.*

[Hel08] H. A. Helfgott, *Growth and generation in* $\mathrm{SL}_2(\mathbb{Z}/p\mathbb{Z})$, Ann. of Math. (2) **167** (2008), no. 2, 601–623, DOI 10.4007/annals.2008.167.601. MR2415382

[Hel11] H. A. Helfgott, *Growth in* $SL_3(\mathbb{Z}/p\mathbb{Z})$, J. Eur. Math. Soc. (JEMS) **13** (2011), no. 3, 761–851, DOI 10.4171/JEMS/267. MR2781932

[Hel15] H. A. Helfgott, *Growth in groups: ideas and perspectives*, Bull. Amer. Math. Soc. (N.S.) **52** (2015), no. 3, 357–413, DOI 10.1090/S0273-0979-2015-01475-8. MR3348442

[Hel19] H. A. Helfgott. Growth in linear algebraic groups and permutation groups: towards a unified perspective. In *Groups St Andrews 2017 in Birmingham*, volume 455 of *London Math. Soc. Lecture Note Ser.*, pages 300–345. Cambridge Univ. Press, Cambridge, 2019.

[Hog82] G. M. D. Hogeweij, *Almost-classical Lie algebras. I, II*, Nederl. Akad. Wetensch. Indag. Math. **44** (1982), no. 4, 441–452, 453–460. MR683531

[HP95] E. Hrushovski and A. Pillay, *Definable subgroups of algebraic groups over finite fields*, J. Reine Angew. Math. **462** (1995), 69–91. MR1329903

[Hru12] E. Hrushovski, *Stable group theory and approximate subgroups*, J. Amer. Math. Soc. **25** (2012), no. 1, 189–243, DOI 10.1090/S0894-0347-2011-00708-X. MR2833482

[HS14] H. A. Helfgott and Á. Seress, *On the diameter of permutation groups*, Ann. of Math. (2) **179** (2014), no. 2, 611–658, DOI 10.4007/annals.2014.179.2.4. MR3152942

[HSZ15] H. A. Helfgott, Á. Seress, and A. Zuk, *Random generators of the symmetric group: diameter, mixing time and spectral gap*, J. Algebra **421** (2015), 349–368, DOI 10.1016/j.jalgebra.2014.08.033. MR3272386

[HW08] E. Hrushovski and F. Wagner, *Counting and dimensions*, Model theory with applications to algebra and analysis. Vol. 2, London Math. Soc. Lecture Note Ser., vol. 350, Cambridge Univ. Press, Cambridge, 2008, pp. 161–176, DOI 10.1017/CBO9780511735219.005. MR2436141

[Kon92] S. V. Konyagin, *Estimates for Gaussian sums and Waring's problem modulo a prime* (Russian), Trudy Mat. Inst. Steklov. **198** (1992), 111–124; English transl., Proc. Steklov Inst. Math. **1(198)** (1994), 105–117. MR1289921

[Kow13] E. Kowalski, *Explicit growth and expansion for* SL_2, Int. Math. Res. Not. IMRN **24** (2013), 5645–5708, DOI 10.1093/imrn/rns214. MR3144176

[Lar03] M. Larsen, *Navigating the Cayley graph of* $SL_2(\mathbb{F}_p)$, Int. Math. Res. Not. **27** (2003), 1465–1471, DOI 10.1155/S1073792803130383. MR1976231

[Lor18] O. Lorscheid, $\mathbb{F}_1$ *for everyone*, Jahresber. Dtsch. Math.-Ver. **120** (2018), no. 2, 83–116, DOI 10.1365/s13291-018-0177-x. MR3798149

[LP11] M. J. Larsen and R. Pink, *Finite subgroups of algebraic groups*, J. Amer. Math. Soc. **24** (2011), no. 4, 1105–1158, DOI 10.1090/S0894-0347-2011-00695-4. MR2813339

[LPW09] D. A. Levin, Y. Peres, and E. L. Wilmer, *Markov chains and mixing times*, American Mathematical Society, Providence, RI, 2009. With a chapter by James G. Propp and David B. Wilson. MR2466937

[LR92] J. D. Lafferty and D. Rockmore, *Fast Fourier analysis for* SL_2 *over a finite field and related numerical experiments*, Experiment. Math. **1** (1992), no. 2, 115–139. MR1203870

[MT11] G. Malle and D. Testerman, *Linear algebraic groups and finite groups of Lie type*, Cambridge Studies in Advanced Mathematics, vol. 133, Cambridge University Press, Cambridge, 2011. MR2850737

[Mum99] D. Mumford, *The red book of varieties and schemes*, Second, expanded edition, Lecture Notes in Mathematics, vol. 1358, Springer-Verlag, Berlin, 1999. Includes the Michigan lectures (1974) on curves and their Jacobians. With contributions by Enrico Arbarello. MR1748380

[NP11] N. Nikolov and L. Pyber, *Product decompositions of quasirandom groups and a Jordan type theorem*, J. Eur. Math. Soc. (JEMS) **13** (2011), no. 4, 1063–1077, DOI 10.4171/JEMS/275. MR2800484

[Pet12] G. Petridis, *New proofs of Plünnecke-type estimates for product sets in groups*, Combinatorica **32** (2012), no. 6, 721–733, DOI 10.1007/s00493-012-2818-5. MR3063158

[Plü70] H. Plünnecke, *Eine zahlentheoretische Anwendung der Graphentheorie* (German), J. Reine Angew. Math. **243** (1970), 171–183, DOI 10.1515/crll.1970.243.171. MR0266892

[PS16] L. Pyber and E. Szabó, *Growth in finite simple groups of Lie type*, J. Amer. Math. Soc. **29** (2016), no. 1, 95–146, DOI 10.1090/S0894-0347-2014-00821-3. MR3402696

[Pyb93] L. Pyber, *On the orders of doubly transitive permutation groups, elementary estimates*, J. Combin. Theory Ser. A **62** (1993), no. 2, 361–366, DOI 10.1016/0097-3165(93)90053-B. MR1207742

[Rot53] K. F. Roth, *On certain sets of integers*, J. London Math. Soc. **28** (1953), 104–109, DOI 10.1112/jlms/s1-28.1.104. MR0051853

[RT85] I. Z. Ruzsa and S. Turjányi, *A note on additive bases of integers*, Publ. Math. Debrecen **32** (1985), no. 1-2, 101–104. MR810596

[Ruz89] I. Z. Ruzsa, *An application of graph theory to additive number theory*, Sci. Ser. A Math. Sci. (N.S.) **3** (1989), 97–109. MR2314377

[Ruz99] I. Z. Ruzsa. An analog of Freiman's theorem in groups. In *Structure theory of set addition*, pages 323–326. Paris: Société Mathématique de France, 1999.

[San12] T. Sanders, *On the Bogolyubov-Ruzsa lemma*, Anal. PDE **5** (2012), no. 3, 627–655, DOI 10.2140/apde.2012.5.627. MR2994508

[San13] T. Sanders, *The structure theory of set addition revisited*, Bull. Amer. Math. Soc. (N.S.) **50** (2013), no. 1, 93–127, DOI 10.1090/S0273-0979-2012-01392-7. MR2994996

[Sel65] A. Selberg, *On the estimation of Fourier coefficients of modular forms*, Proc. Sympos. Pure Math., Vol. VIII, Amer. Math. Soc., Providence, R.I., 1965, pp. 1–15. MR0182610

[Ser03] Á. Seress, *Permutation group algorithms*, Cambridge Tracts in Mathematics, vol. 152, Cambridge University Press, Cambridge, 2003. MR1970241

[Sha99] Y. Shalom, *Expander graphs and amenable quotients*, Emerging applications of number theory (Minneapolis, MN, 1996), IMA Vol. Math. Appl., vol. 109, Springer, New York, 1999, pp. 571–581, DOI 10.1007/978-1-4612-1544-8_23. MR1691549

[Spr98] T. A. Springer, *Linear algebraic groups*, 2nd ed., Progress in Mathematics, vol. 9, Birkhäuser Boston, Inc., Boston, MA, 1998. MR1642713

[Ste65] R. Steinberg, *Regular elements of semisimple algebraic groups*, Inst. Hautes Études Sci. Publ. Math. **25** (1965), 49–80. MR0180554

[SX91] P. Sarnak and X. X. Xue, *Bounds for multiplicities of automorphic representations*, Duke Math. J. **64** (1991), no. 1, 207–227, DOI 10.1215/S0012-7094-91-06410-0. MR1131400

[Tao08] T. Tao, *Product set estimates for non-commutative groups*, Combinatorica **28** (2008), no. 5, 547–594, DOI 10.1007/s00493-008-2271-7. MR2501249

[Tao10] T. Tao, *Freiman's theorem for solvable groups*, Contrib. Discrete Math. **5** (2010), no. 2, 137–184. MR2791295

[Tao15] T. Tao, *Expansion in finite simple groups of Lie type*, Graduate Studies in Mathematics, vol. 164, American Mathematical Society, Providence, RI, 2015. MR3309986

[Toi14] M. C. H. Tointon, *Freiman's theorem in an arbitrary nilpotent group*, Proc. Lond. Math. Soc. (3) **109** (2014), no. 2, 318–352, DOI 10.1112/plms/pdu005. MR3254927

[TT16] R. Tessera and M. C. H. Tointon, *Properness of nilprogressions and the persistence of polynomial growth of given degree*, Discrete Anal. (2018), Paper No. 17, 38. MR3877012

Mathematisches Institut, Georg-August Universität Göttingen, Bunsenstrasse 3-5, D-37073 Göttingen, Germany –and– IMJ-PRG, UMR 7586, 58 avenue de France, Bâtiment S. Germain, case 7012, 75013 Paris CEDEX 13, France

Email address: harald.helfgott@gmail.com

Contemporary Mathematics
Volume **740**, 2019
https://doi.org/10.1090/conm/740/14903

Lectures on applied ℓ-adic cohomology

Étienne Fouvry, Emmanuel Kowalski, Philippe Michel, and Will Sawin

Abstract. We describe how a systematic use of the deep methods from ℓ-adic cohomology pioneered by Grothendieck and Deligne and further developed by Katz and Laumon help make progress on various classical questions from analytic number theory. This text is an extended version of a series of lectures given during the 2016 Arizona Winter School.

Contents

1. Introduction

One of the most basic question in number theory is to understand how various sets of integers behave when restricted to (i.e. intersected with) *congruence classes*, a notion that goes back at least to Euclid and was exposed systematically by Gauss

2010 *Mathematics Subject Classification.* Primary 11F03, 11L05 14F20.

in his 1801 *Disquisitiones Arithmeticae* (following works of Fermat, Euler, Wilson, Lagrange, Legendre and their predecessors from the middle ages and antiquity), and which is fundamental to number theory.

Let us recall that given an integer $q \in \mathbf{Z} - \{0\}$, a *congruence class* (a.k.a. an *arithmetic progression*) modulo q is a subset of $\mathbf{Z}$ of the shape

$$a \,(\mathrm{mod}\, q) = a + q\mathbf{Z} \subset \mathbf{Z}$$

for some integer a. The set of congruence classes modulo q is denoted $\mathbf{Z}/q\mathbf{Z}$; it is a finite ring of cardinality q (with addition and multiplication induced by that of $\mathbf{Z}$).

In number theory, especially analytic number theory, one is interested in studying the behaviour of some given arithmetic function along congruence classes, for instance to determine whether a set of integers has finite or infinite intersection with some congruence class. The analysis of such problem, which may involve quite sophisticated manipulations, often involves certain specific classes of functions on $\mathbf{Z}/q\mathbf{Z}$.

When studying such functions, it is natural to invoke the *Chinese Remainder Theorem*

$$\mathbf{Z}/q\mathbf{Z} \simeq \prod_{p^\alpha \| q} \mathbf{Z}/p^\alpha\mathbf{Z}$$

which largely reduces the study to the case of prime power moduli; then, in many instances, the deepest case is when q is a prime; the ring $\mathbf{Z}/q\mathbf{Z}$ is then a finite field, denoted $\mathbf{F}_q$, and often the functions that occur are what we will call *trace functions.*

The objective of these lectures is utilitarian: our aim is to describe these trace functions, many examples, their theory and most importantly how they are handled when they occur in analytic number theory. Indeed the mention of "étale" or "ℓ-adic cohomology", "sheaves", "purity", "functors", "local systems" or "vanishing cycles" sounds forbidding to the working analytic number theorist and often prevents him/her to embrace the subject and make full use of the powerful methods that Deligne, Katz, Laumon have developed for us. It is our hope that after these introductory lectures, any of the remaining readers will feel ready for and at ease with more serious activities such as the reading of the wonderful series of orange books by Katz, and eventually will be able to tackle by him/herself any trace function that nature has laid in front of him/her.

2. Examples of trace functions

Unless stated otherwise, we now assume that q is a prime number.

2.1. Characters. *Trace functions modulo* q are special classes of $\mathbf{C}$-valued functions on $\mathbf{F}_q$ of geometric origin. Perhaps the first significant example, beyond the constant function 1, is the *Legendre symbol* (for $q \geqslant 3$)

$$\left(\frac{\cdot}{q}\right) : x \in \mathbf{F}_q \to \begin{cases} 0 & \text{if } x = 0 \\ +1 & \text{if } x \in (\mathbf{F}_q^\times)^2 \\ -1 & \text{if } x \in \mathbf{F}_q^\times - (\mathbf{F}_q^\times)^2 \end{cases}$$

which detects the squares modulo q, and whose arithmetic properties (especially the *quadratic reciprocity law*) were studied by Gauss in the *Disquisitiones.*

The class of trace functions was further enriched by P. G. Dirichlet: on his way to proving his famous theorem on primes in arithmetic progressions, he introduced what are now called *Dirichlet characters*, i.e. the homomorphisms of the multiplicative group

$$\chi : (\mathbf{Z}/q\mathbf{Z})^\times \to \mathbf{C}^\times$$

(with $\chi(0)$ defined to be 0 for χ non-trivial).

Another significant class of trace functions are the additive characters

$$\psi : (\mathbf{Z}/q\mathbf{Z}, +) \to \mathbf{C}^\times.$$

These are all of the shape

$$x \in \mathbf{Z}/q\mathbf{Z} \mapsto \mathrm{e}_q(ax) := \exp\left(2\pi i \frac{\tilde{a}\tilde{x}}{q}\right)$$

(say) for some $a \in \mathbf{Z}/q\mathbf{Z}$, where $\tilde{a}$ and $\tilde{x}$ denote elements (lifts) of the congruence classes $a \pmod q$ and $x \pmod q$. Both additive and multiplicative characters satisfy the important *orthogonality relations*

$$\frac{1}{q}\sum_{x\in\mathbf{F}_q} \psi(x)\overline{\psi'(x)} = \delta_{\psi=\psi'}, \quad \frac{1}{q-1}\sum_{x\in\mathbf{F}_q^\times} \chi(x)\overline{\chi'(x)} = \delta_{\chi=\chi'};$$

and we will see later a generalization of these relations to arbitrary trace functions.

Additive and multiplicative characters can be combined together (by means of a Fourier transform) to form the (normalized) *Gauss sums*

$$\varepsilon_\chi(a) = \frac{1}{q^{1/2}}\sum_{x\in\mathbf{F}_q^\times} \chi(x)\mathrm{e}_q(ax),$$

but these are not really new functions of a: by a simple change of variable, one has

$$\varepsilon_\chi(a) = \overline{\chi}(a)\varepsilon_\chi(1)$$

for $a \in \mathbf{F}_q^\times$. For χ non-trivial, Gauss proved that

$$|\varepsilon_\chi(1)| = 1.$$

2.2. Algebraic exponential sums. Another important source of trace functions comes from the study of the diophantine equations

$$Q(\boldsymbol{x}) = 0,\ \boldsymbol{x} = (x_1, \ldots, x_n) \in \mathbf{Z}^n,\ Q(X_1, \ldots, X_n) \in \mathbf{Z}[X_1, \ldots, X_n]. \tag{2.1}$$

For instance, the analysis of the *major arcs* in the *circle method* of Hardy–Littlewood (cf. [**Vau97**, Chap. 4]) leads to the following algebraic exponential sums on $(\mathbf{Z}/q\mathbf{Z})^n$ obtained by Fourier transform

$$(a, \boldsymbol{x}) \in (\mathbf{Z}/q\mathbf{Z})^{n+1} \mapsto \frac{1}{q^{n/2}}\sum_{\boldsymbol{y}\in(\mathbf{Z}/q\mathbf{Z})^n} \mathrm{e}_q(aQ(\boldsymbol{y}) + \boldsymbol{x}.\boldsymbol{y}).$$

In the 1926's, while studying the case of a positive definite homogeneous polynomial Q of degree 2 in four variables (a positive definite integral quaternary quadratic form), and introducing a new variant of the circle method, Kloosterman [**Klo27**], defined the so-called (normalized) *Kloosterman sums*

$$\mathrm{Kl}_2(a; q) = \frac{1}{q^{1/2}}\sum_{\substack{x,y\in\mathbf{F}_q^\times\\ xy=a}} \mathrm{e}_q(x + y).$$

This is another example of a trace function, and indeed one that is defined via Fourier transform.

By computing their fourth moment (see [**Iwa97**, (4.26)]), Kloosterman was able to obtain the first non-trivial bound for Kloosterman sums, namely

$$|\operatorname{Kl}_2(a;q)| \leqslant 2q^{1/4}.$$

This estimate proved crucial for the study of equation (2.1) in the case of quaternary positive definite quadratic forms. In the 1940's, this bound was improved by A. Weil, who as a consequence of his proof of the Riemann hypothesis for curves over finite fields proved the best individual upper bound (see [**IK04**, §11.7]):

$$|\operatorname{Kl}_2(a;q)| \leqslant 2.$$

In 1939, Kloosterman sums appeared again in the work of Petersson who related them to Fourier coefficients of modular forms.[1] Since then, via the works of Selberg, Kuznetsov, Deshouillers-Iwaniec and many others, Kloosterman sums play a fundamental role in the analytic theory of automorphic forms[2].

A further important example of trace functions are the (normalized) *hyper-Kloosterman sums.* These are higher dimensional generalisations of Kloosterman sums, and are given, for any integer $k \geqslant 1$ by

$$\operatorname{Kl}_k(a;q) = \frac{1}{q^{(k-1)/2}} \sum_{\substack{x_1,\ldots,x_k \in \mathbf{F}_q^\times \\ x_1.x_2.\ldots.x_k = a}} e_q(x_1 + x_2 + \ldots + x_k).$$

Hyper-Kloosterman sums were introduced by P. Deligne, who also established the following generalization of the Weil bound:

$$|\operatorname{Kl}_k(a;q)| \leqslant k.$$

Hyper-Kloosterman sums can be interpreted as inverse (discrete) Mellin transforms of powers of Gauss sums, and therefore can be used to study the distribution of Gauss sums. As was denoted by Katz in [**Kat80**], this fact and Deligne's bound imply the following[3]

THEOREM 2.1. *As $q \to \infty$, the set of (normalized) Gauss sums*

$$\{\varepsilon_\chi(1),\ \chi \,(\mathrm{mod}\, q) \text{ non trivial}\}$$

become equidistributed on the unit circle $\mathbf{S}^1 \subset \mathbf{C}^\times$ with respect to the uniform (Haar) probability measure.

Hyper-Kloosterman sums also occur in the theory of automorphic forms; for instance, Luo, Rudnick and Sarnak used the fact that powers of Gauss sums occur in the root number of the functional equation of certain automorphic L-functions, the inverse Mellin transform property and Deligne's bound, to obtain non-trivial estimates for the Langlands parameters of automorphic representations on GL_n (giving in particular the first improvement of Selberg's famous 3/16 bound for the Laplace eigenvalues of Maass cusp forms).

In addition, just as for the classical Kloosterman sums, hyper-Kloosterman sums also occur in the spectral theory of GL_k automorphic forms.

There are many more examples of trace functions, and we will describe some below along with ways to construct new trace functions from older ones.

---

[1]In fact, Poincaré had already written them down in one of his last papers, published posthumously.

[2]The double occurence of Kloosterman sums in the context of quadratic forms and of modular forms is explained by the theta correspondence

[3]See [**Kat12**] for a considerable generalisation of this theorem.

3. Trace functions and Galois representations

Let $\mathbf{P}^1_{\mathbf{F}_q}$ be the projective line and $\mathbf{A}^1_{\mathbf{F}_q} \subset \mathbf{P}^1_{\mathbf{F}_q}$ be the affine line and $K = \mathbf{F}_q(X)$ be the field of functions of $\mathbf{P}^1_{\mathbf{F}_q}$.

In the sequel we fix some prime $\ell \neq q$, $\overline{\mathbf{Q}_\ell}$ an algebraic closure of the field of ℓ-adic numbers $\mathbf{Q}_\ell$ and an embedding $\iota\colon \overline{\mathbf{Q}_\ell} \hookrightarrow \mathbf{C}$ into the complex numbers. Trace functions modulo q are $\overline{\mathbf{Q}_\ell}$-valued functions[4] defined on the set of $\mathbf{F}_q$-points of the affine line $\mathbf{A}^1(\mathbf{F}_q) \simeq \mathbf{F}_q$. They are obtained from *constructible* ℓ-adic sheaves (often denoted $\mathcal{F}$) for the étale topology on $\mathbf{P}^1_{\mathbf{F}_q}$. All these notions are quite forbidding at first; fortunately the category of *constructible* ℓ-adic sheaves on $\mathbf{P}^1_{\mathbf{F}_q}$ can be rather conveniently described in terms of the category of representations of the Galois group of K. Following [**Kat80, Kat88**], we will start from this viewpoint.

Let $K^{\mathrm{sep}} \supset K$ be a separable closure of K, and $\overline{\eta}$ the associated geometric generic point (i.e. $\mathrm{Spec}(K^{\mathrm{sep}}) = \overline{\eta}$). Let $\overline{\mathbf{F}_q} \subset K^{\mathrm{sep}}$ denote the separable (or algebraic) closure of $\mathbf{F}_q$ in K^{sep}. We denote

$$G^{\mathrm{geom}} := \mathrm{Gal}(K^{\mathrm{sep}}/\overline{\mathbf{F}_q}.K) \subset G^{\mathrm{arith}} = \mathrm{Gal}(K^{\mathrm{sep}}/K),$$

the *geometric*, resp. *arithmetic*, Galois group. By restricting the action of an element of G^{arith} to $\overline{\mathbf{F}_q}$ we have the exact sequence

$$1 \to G^{\mathrm{geom}} \to G^{\mathrm{arith}} \to \mathrm{Gal}(\overline{\mathbf{F}_q}/\mathbf{F}_q) \to 1. \tag{3.1}$$

DEFINITION 3.1. Let $U \subset \mathbf{A}^1_{\mathbf{F}_q}$ be a non-empty open subset of $\mathbf{A}^1_{\mathbf{F}_q}$ that is defined over $\mathbf{F}_q$. An ℓ-adic sheaf lisse on U, say $\mathcal{F}$, is a continuous finite-dimensional Galois representation

$$\varrho_{\mathcal{F}} : G^{\mathrm{arith}} \to \mathrm{GL}(V_{\mathcal{F}})$$

where $V_{\mathcal{F}}$ is a finite dimensional $\overline{\mathbf{Q}_\ell}$-vector space, which is unramified at every closed point x of U. The dimension $\dim V_{\mathcal{F}}$ is called the rank of $\mathcal{F}$ and is denoted $\mathrm{rk}(\mathcal{F})$. The vector space $V_{\mathcal{F}}$ is also denoted $\mathcal{F}_{\overline{\eta}}$.

3.1. Closed points on the affine line. In this section we spell-out the meaning of the sentence "unramified at every closed point x of U".

Let us recall that the datum of closed point of $\mathbf{P}^1_{\mathbf{F}_q}$ is equivalent to the datum of an embedding $\mathcal{O}_x \hookrightarrow K$ of a local ring[5] $\mathcal{O}_x$ (the ring of rational functions defined in a neighborhood of x) whose field of fractions is K. Given such an embedding, we denote by $\mathfrak{p}_x$ its unique prime ideal, π_x a generator of π_x (an uniformizer) and by $v_x : K \to \mathbf{Z} \cup \{\infty\}$ the associated discrete valuation (normalized so that $v_x(\pi_x) = 1$): we have

$$\mathcal{O}_x = \{f \in K,\ v_x(f) \geqslant 0\} \supset \mathfrak{p}_x = \{f \in K,\ v_x(f) > 0\}.$$

We denote by $k_x = \mathcal{O}_x/\mathfrak{p}_x$ its residue field and by $q_x = |k_x| =: q^{\deg x}$ the size of k_x and $\deg x$ its degree

The set of closed points of the projective line $\mathbf{P}^1_{\mathbf{F}_q}$ is the union of the set of closed points of the affine line $\mathbf{A}^1_{\mathbf{F}_q}$ which is indexed by the set of monic, irreducible (non-constant) polynomials of $\mathbf{F}_q[X]$ and the point ∞.

- For π irreducible, monic and not constant, the local ring $\mathcal{O}_\pi$ is the localization of $\mathbf{F}_q[X]$ at the prime ideal $(\pi) \subseteq \mathbf{F}_q[X]$:

$$\mathcal{O}_\pi = \{P/Q,\ P, Q \in \mathbf{F}_p[X],\ \pi \nmid Q\} \supset \mathfrak{p}_\pi = \{P/Q,\ P, Q \in \mathbf{F}_p[X],\ \pi | P,\ \pi \nmid Q\},$$

[4]Hence $\mathbf{C}$-valued via the fixed embedding ι

[5]A PID with a unique prime ideal [**Ser79**, Chap. 1]

the valuation v_π is the usual valuation: for any polynomial $P \in \mathbf{F}_q[X]$, $v_x(P) = v_\pi(P)$ is the exponent of the highest power of π dividing P which is extended to K by setting $v_x(P/Q) = v_\pi(P) - v_\pi(Q)$, and the degree is $\deg \pi$.

– For ∞,

$$\begin{aligned}\mathcal{O}_\infty &= \{P/Q,\ P, Q \in \mathbf{F}_p[X],\ \deg P \leqslant \deg Q\} \supset \mathfrak{p}_\infty \\ &= \{P/Q,\ P, Q \in \mathbf{F}_p[X],\ \deg P < \deg Q\},\end{aligned}$$

the valuation is minus the degree of the rational fraction $v_\infty(P/Q) = \deg(Q) - \deg(P)$, and the degree of ∞ is 1.

REMARK 3.2. We denote by $\mathbf{P}^1(\mathbf{F}_q)$ the set of closed points of degree 1 and by $\mathbf{A}^1(\mathbf{F}_q) = \mathbf{P}^1(\mathbf{F}_q) - \{\infty\}$. Note that $\mathbf{A}^1(\mathbf{F}_q)$ is identified with $\mathbf{F}_q$ by identifying $x \in \mathbf{F}_q$ with the degree 1 (irreducible) polynomial $X - x$.

Similarly a non-empty open set $U \subset \mathbf{A}^1_{\mathbf{F}_q}$ is the open complement of the closed set $Z_Q \subset \mathbf{A}^1_{\mathbf{F}_q}$ of zeros of some (non-zero) polynomial $Q \in \mathbf{F}_q[X]$, i.e. defined by the equation $Q(x) = 0$.

The "closed points of U" are the closed point associated with the irreducible monic polynomials $\pi \in \mathbf{F}_q[X]$ coprime to Q and the set of closed points of degree 1, is identified with the complement of the set of roots of Q contained in $\mathbf{F}_q$:

$$U(\mathbf{F}_q) \simeq \{x \in \mathbf{F}_q,\ Q(x) \neq 0\} \subset \mathbf{F}_q.$$

3.1.1. *Decomposition group, inertia and Frobenius.* The valuation v_x can be extended (in multiple ways) to a ($\mathbf{Q}$-valued) valuation on K^{sep} and the choice of one such extension (denoted $v_{\{x\}}$) determines a decomposition and an inertia subgroup in the arithmetic Galois group

$$I_{\{x\}} \subset D_{\{x\}} \subset G^{\mathrm{arith}}$$

fitting in the exact sequence

$$1 \to I_{\{x\}} \to D_{\{x\}} \to \mathrm{Gal}(\overline{\mathbf{F}_q}/k_x) \to 1. \tag{3.2}$$

Let also us recall that $\mathrm{Gal}(\overline{\mathbf{F}_q}/k_x)$ is topologically generated by the *arithmetic Frobenius*

$$\mathrm{Frob}_{k_x}^{\mathrm{arith}} : \begin{array}{ccc}\overline{\mathbf{F}_q} & \to & \overline{\mathbf{F}_q} \\ u & \to & u^{q_x}\end{array}.$$

In the sequel we will denote by $\mathrm{Frob}_{k_x}^{\mathrm{geom}}$ its inverse, also called the *geometric Frobenius.* The lifts of the (geometric) Frobenius therefore define a (left) $I_{\{x\}}$-class in the decomposition subgroup which we denote by

$$\mathrm{Frob}_{\{x\}} \subset D_{\{x\}}$$

and which we call the Frobenius class at $\{x\}$.

REMARK 3.3. The choice of a different extension $v_{\{x\}'}$ of v_x yields a priori another decomposition, inertia subgroups and Frobenius class, $D_{\{x\}'}, I_{\{x\}'},\ Fr_{\{x\}'}$, but these are conjugate to $D_{\{x\}}, I_{\{x\}},\ Fr_{\{x\}}$ because G^{arith} acts transitively on the set of extensions. As we will see the various quantities that we will discuss in relation to these sets will be conjugacy-invariant and therefore depend only on x but not of a choice of $\{x\}$ and will use the indice x instead of $\{x\}$. Sometimes, to simplify notations, we will implicitly assume the choice of an $\{x\}$ without mentioning it and will simply write $D_x, I_x, \mathrm{Frob}_x$

We can now explain the term unramified.

DEFINITION 3.4. Given x a closed point of $\mathbf{P}^1_{\mathbf{F}_q}$, a G^{arith}-module V is unramified (or lisse) at x at if for one (or equivalently any) extension $\{x\}$, the corresponding inertia subgroup $I_{\{x\}}$ acts trivially on V. Otherwise V is ramified at x.

If V is unramified at x, all the elements in the Frobenius class $\mathrm{Frob}_{\{x\}}$ act by the same automorphism of V and we will denote this automorphism by $(\mathrm{Frob}_{\{x\}}\,|V)$.

Moreover if we change the extension $\{x\}$ we obtain an automorphism which is G^{arith}-conjugate to $(\mathrm{Frob}_{\{x\}}\,|V)$. We denote by $(\mathrm{Frob}_x\,|V_{\mathcal{F}})$ this conjugacy class.

It follows from this discusion that for any sheaf $\mathcal{F}$ there is a non-empty open subset on which $\mathcal{F}$ is unramified and maximal for this property. We will note this open set $U_{\mathcal{F}}$.

3.2. The trace function attached to a lisse sheaf. Let $\mathcal{F}$ be an ℓ-adic sheaf lisse on $U \subset \mathbf{A}^1_{\mathbf{F}_q}$ and

$$\varrho_{\mathcal{F}} \colon G^{\mathrm{arith}} \to \mathrm{GL}(V_{\mathcal{F}})$$

the corresponding representation.

For $x \in U(\mathbf{F}_q)$ a closed point of degree 1 at which the representation $\varrho_{\mathcal{F}}$ is unramified, we have, in the previous section, associated a Frobenius conjugacy class $(\mathrm{Frob}_x\,|V_{\mathcal{F}})$ namely the union of all the $(\mathrm{Frob}_{\{x\}}\,|V_{\mathcal{F}})$. By conjugacy, the trace of all these automorphisms $(\mathrm{Frob}_{\{x\}}\,|V_{\mathcal{F}})$ is constant within that class: we denote this common value by

$$\mathrm{tr}(\mathrm{Frob}_x\,|V_{\mathcal{F}})$$

and call it the Frobenius trace of $\mathcal{F}$ at x.

DEFINITION 3.5. Given an ℓ-adic sheaf $\mathcal{F}$ lisse on $U \subset \mathbf{A}^1_{\mathbf{F}_q}$; the trace function $K_{\mathcal{F}}$ associated to this situation is the function on $U(\mathbf{F}_q)$ given by

$$x \in U(\mathbf{F}_q) \mapsto K_{\mathcal{F}}(x) = \mathrm{tr}(\mathrm{Frob}_x\,|V_{\mathcal{F}}).$$

This is a priori a $\overline{\mathbf{Q}_\ell}$-valued function but it can be considered complex-valued via the fixed embedding $\iota \colon \overline{\mathbf{Q}_\ell} \hookrightarrow \mathbf{C}$.

REMARK 3.6. As we have seen in Remark 3.2 $U(\mathbf{F}_q)$ is identified with

$$\{x \in \mathbf{F}_q,\ Q(x) \neq 0\} \subset \mathbf{F}_q$$

and therefore $K_{\mathcal{F}}$ can be considered as a function defined on a subset of $\mathbf{F}_q$.

REMARK 3.7. There are several ways by which one could extend $K_{\mathcal{F}}$ to the whole of $\mathbf{A}^1(\mathbf{F}_q)$. The simplest way is the extension by zero outside $U(\mathbf{F}_q)$; another possible extension (called the *middle extension*) would be to set for any $x \in \mathbf{A}^1(\mathbf{F}_q)$,

$$K_{\mathcal{F}}(x) := \mathrm{tr}(\mathrm{Frob}_{\{x\}}\,|V_{\mathcal{F}}^{I_{\{x\}}})$$

where $V_{\mathcal{F}}^{I_{\{x\}}} \subset V_{\mathcal{F}}$ is the subspace of $I_{\{x\}}$-invariant vectors: the action of the Frobenius class $\mathrm{Frob}_{\{x\}}$ on $V_{\mathcal{F}}^{I_{\{x\}}}$ is well-defined and its trace does not depend on $\{x\}$. For our purpose, any of the two extensions would work (cf. Remark 3.12).

3.3. Trace functions over $U(\mathbf{F}_{q^n})$. In fact, an ℓ-adic sheaf, lisse on $U_{\mathbf{F}_q}$ give rise to a whole family of trace functions.

For any $n \geqslant 1$, let us consider the finite extension $\mathbf{F}_{q^n}$ let us and base change the whole situation to that field: this amounts to replace $\mathbf{P}^1_{\mathbf{F}_q}$ by $\mathbf{P}^1_{\mathbf{F}_{q^n}}$, $K = \mathbf{F}_q(X)$ by $K_n = \mathbf{F}_{q^n}(X)$, and the arithmetic Galois group G^{arith} by $G_n^{\mathrm{arith}} = \mathrm{Gal}(K^{\mathrm{sep}}/K_n)$ (notice that the geometric Galois group does not change).

The group G_n^{arith} is a normal subgroup of G^{arith} (whose quotient is $\mathrm{Gal}(\mathbf{F}_{q^n}/\mathbf{F}_q)$, so we may restrict our initial Galois representation to it: in that way we obtain another ℓ-adic sheaf denoted $\mathcal{F}_n$

$$\varrho_{\mathcal{F}_n} : G_n^{\mathrm{arith}} \to \mathrm{GL}(V_{\mathcal{F}})$$

and another trace function

$$K_{\mathcal{F},n} : \begin{array}{ccc} U(\mathbf{F}_{q^n}) & \to & \mathbf{C} \\ x & \to & \mathrm{tr}(\mathrm{Frob}_x \,|V_{\mathcal{F}}) \end{array}$$

where $U(\mathbf{F}_{q^n})$ denotes now the set of closed points of $\mathbf{P}^1_{\mathbf{F}_{q^n}}$ of degree 1 which are contained in U: this set is identified with the set of irreducible monic polynomials of degree 1 coprime with Q and is therefore identified with

$$\{x \in \mathbf{F}_{q^n},\ Q(x) \neq 0\}.$$

As we will see below, the existence of this sequence of auxiliary functions is very important: for instance (the Chebotareff density theorem) the full sequence $(K_{\mathcal{F},n})_{n \geqslant 1}$ characterizes the representation $\varrho_{\mathcal{F}}$ up to semi-simplification.

REMARK 3.8. As we have remarked already one has the identifications

$$U(\mathbf{F}_q) \simeq \{x \in \mathbf{F}_q,\ Q(x) \neq 0\},\ U(\mathbf{F}_{q^n}) \simeq \{x \in \mathbf{F}_{q^n},\ Q(x) \neq 0\}.$$

However the inclusion

$$\{x \in \mathbf{F}_q,\ Q(x) \neq 0\} \subset \{x \in \mathbf{F}_{q^n},\ Q(x) \neq 0\}$$

does NOT imply that the function $K_{\mathcal{F}}$ is "the restriction" of $K_{\mathcal{F},n}$ to $U(\mathbf{F}_q)$. More precisely, if we denote by x the closed point in $U(\mathbf{F}_q)$ associated with the polynomial $X - x \in \mathbf{F}_q[X]$ and by x_n the closed point in $U(\mathbf{F}_{q^n})$ associated with the same polynomial $X - x \in \mathbf{F}_{q^n}[X]$ one has the formula

$$K_{\mathcal{F},n}(x_n) = \mathrm{tr}(\mathrm{Frob}_{x_n} |V_{\mathcal{F}}) = \mathrm{tr}(\mathrm{Frob}_x^n |V_{\mathcal{F}}).$$

More generally, for d dividing n let $\pi \in \mathbf{F}_q[X]$ be a monic irreducible polynomial of degree d and coprime to Q. Then π defines a closed point x_π of U of degree d. Since $d|n$, the polynomial π splits in $\mathbf{F}_{q^n}$

$$\pi(X) = \prod_{i=1}^{d}(X - x_i)$$

and any of its roots x_i defines a closed point in $U(\mathbf{F}_{q^n})$ (corresponding to the polynomial $X - x_i \in \mathbf{F}_{q^n}[X]$); we then have for $i = 1, \ldots, d$

$$K_{\mathcal{F},n}(x_i) = \mathrm{tr}(\mathrm{Frob}_{x_i} |V_{\mathcal{F}}) = \mathrm{tr}(\mathrm{Frob}_\pi^{n/d} |V_{\mathcal{F}}). \tag{3.3}$$

REMARK 3.9. There is, a priori, no reason to limit ourselves to the affine line: if $\mathcal{C}_{\mathbf{F}_q}$ is any smooth geometrically connected curve over $\mathbf{F}_q$ with function field $K_{\mathcal{C}}$ (which is a finite extension of $\mathbf{F}_q(X)$) and any dense open subset $U \subset \mathcal{C}$ defined over $\mathbf{F}_q$, an ℓ-adic sheaf $\mathcal{F}$ on $\mathcal{C}$ lisse on some non-empty open set U is a continuous representation

$$\varrho_{\mathcal{F}}\colon \operatorname{Gal}(K_{\mathcal{C}}^{sep}/K_{\mathcal{C}}) \to \operatorname{GL}(V_{\mathcal{F}})$$

which is unramified at every closed point of U.

3.4. The language of representations. The definition of sheaves and trace functions in terms of Galois representations enable to use consistently the vocabulary from representation theory. For instance

- An ℓ-adic sheaf is *irreducible* (resp. *isotypic*) if the representation $\varrho_{\mathcal{F}}$ is.
- An ℓ-adic sheaf is *geometrically irreducible* (resp. *geometrically isotypic*) if the *restriction* of $\varrho_{\mathcal{F}}$ to the *geometric Galois group* G^{geom} is.
- An ℓ-adic sheaf is *trivial* if the representation $\varrho_{\mathcal{F}}$ is. The trace function is constant, equal to 1.
- An ℓ-adic sheaf is *geometrically trivial* if the *restriction* of $\varrho_{\mathcal{F}}$ to the *geometric Galois group* G^{geom} is. In view of 3.1 its trace function is a constant, say $K_{\mathcal{F}}(x) = \alpha$ and for any $n \geqslant 1$,

$$K_{\mathcal{F},n}(x) = \alpha^n.$$

One can also create new sheaves and trace function from other sheaves.

- The *dual sheaf* $D(\mathcal{F})$ is the contragredient representation $D(\varrho_{\mathcal{F}})$ acting on the dual space $\operatorname{Hom}(V_{\mathcal{F}}, \overline{\mathbf{Q}_\ell})$. This sheaf is also lisse on U and its trace function is given for $x \in U(\mathbf{F}_q)$ by

$$K_{D(\mathcal{F})}(x) = \operatorname{tr}(\mathrm{Frob}_x^{-1} | V_{\mathcal{F}}).$$

- Given another sheaf $\mathcal{G}$ lisse on some U', one can form the *direct sum sheaf* $\mathcal{F} \oplus \mathcal{G}$ whose representation is $\varrho_{\mathcal{F}\oplus\mathcal{G}} = \varrho_{\mathcal{F}} \oplus \varrho_{\mathcal{G}}$; the sheaf is lisse (at least) on $U \cap U'$, of rank the sum of the ranks, and its trace function is given, for $x \in U(\mathbf{F}_q) \cap U'(\mathbf{F}_q)$ by the sum

$$K_{\mathcal{F}\oplus\mathcal{G}}(x) = K_{\mathcal{F}}(x) + K_{\mathcal{G}}(x).$$

- Given another sheaf $\mathcal{G}$ lisse on some U', one can form the *tensor product sheaf* $\mathcal{F} \otimes \mathcal{G}$ whose representation is $\varrho_{\mathcal{F}\otimes\mathcal{G}} = \varrho_{\mathcal{F}} \otimes \varrho_{\mathcal{G}}$; the sheaf is lisse (at least) on $U \cap U'$, of rank the product of the ranks, and its trace function is given, for $x \in U(\mathbf{F}_q) \cap U'(\mathbf{F}_q)$ by the product

$$K_{\mathcal{F}\otimes\mathcal{G}}(x) = K_{\mathcal{F}}(x) K_{\mathcal{G}}(x).$$

- As a special case, one construct the *sheaf of homomorphisms* between $\mathcal{F}$ and $\mathcal{G}$ and the *sheaf of endomorphisms* of $\mathcal{F}$,

$$\operatorname{Hom}(\mathcal{F}, \mathcal{G}) = D(\mathcal{F}) \otimes \mathcal{G},\ \operatorname{End}(\mathcal{F}) = D(\mathcal{F}) \otimes \mathcal{F}.$$

- Let $H \subset \operatorname{GL}(V_{\mathcal{F}})$ be an algebraic group containing $\varrho_{\mathcal{F}}(G^{\mathrm{arith}})$ and let $r\colon H \to \operatorname{GL}(V')$ be a finite-dimensional continuous ℓ-adic representation; the composite representation $r \circ \varrho_{\mathcal{F}}$ defines an ℓ-adic sheaf, denoted $r \circ \mathcal{F}$, which is lisse on U and has rank $\dim V'$. Its trace function is given, for $x \in U(\mathbf{F}_q)$ by

$$K_{r\circ\mathcal{F}}(x) = \operatorname{tr}(r(\mathrm{Frob}_x | V_{\mathcal{F}}) | V').$$

– Let $f \in \mathbf{F}_q(X)$ be non-constant; we can view f as a non-constant morphism $\mathbf{P}^1_{\mathbf{F}_q} \to \mathbf{P}^1_{\mathbf{F}_q}$. The Galois subgroup corresponding to this covering

$$\mathrm{Gal}(K^{\mathrm{sep}}/\mathbf{F}_q(f(X))) \subset G^{\mathrm{arith}}$$

is isomorphic to G^{arith} and therefore the restriction of $\varrho_{\mathcal{F}}$ to $\mathrm{Gal}(K^{\mathrm{sep}}/\mathbf{F}_q(f(X)))$ defines an ℓ-adic sheaf on $\mathbf{P}^1_{\mathbf{F}_q}$ lisse on $f^{-1}(U)$ which is denoted $f^*\mathcal{F}$ and is called the *pull-back* of $\mathcal{F}$ by f. Its rank equals the rank of $\mathcal{F}$ and its trace function is given, for $x \in f^{-1}(U)(\mathbf{F}_q) - \{\infty\}$ by

$$K_{f*\mathcal{F}}(x) = K_{\mathcal{F}}(f(x)).$$

– If the sequel, we will use this pull-back sheaf construction for the following morphisms: This are special cases of *fractional linear transformations*: given $\gamma = \begin{pmatrix} a & b \\ c & d \end{pmatrix} \in \mathrm{PGL}_2(\mathbf{F}_q)$ (the group of automorphisms of $\mathbf{P}^1_{\mathbf{F}_q}$) one defines the automorphism

$$[\gamma]\colon x \to \frac{ax+b}{cx+d}.$$

We denote the pull-back sheaf by $\gamma^*\mathcal{F}$. In particular, for $\gamma = n(b) = \begin{pmatrix} 1 & b \\ 0 & 1 \end{pmatrix}$ we obtain the the additive translation map $[+b]\colon x \to x+b$, and for $\gamma = t(a) = \begin{pmatrix} a & 0 \\ 0 & 1 \end{pmatrix}$, $a \neq 0$ we obtain the multiplicative translation map $[\times a]\colon x \to ax$.

3.5. Purity. We will be interested in the size of trace functions. For this we need the notion of *purity*.

DEFINITION 3.10. Let $w \in \mathbf{Z}$. an ℓ-adic sheaf $\mathcal{F}$, lisse on U is punctually pure of weight w if, for any $x \in U_{\mathbf{F}_q}$, the eigenvalues of $(\mathrm{Frob}_x\,|V_{\mathcal{F}})$ are complex numbers[6] of modulus $q_x^{w/2}$. It is mixed of weights $\leqslant w$ if (as a representation) it is a successive extension of sheaves punctually pure of weights $\leqslant w$.

In particular, if $\mathcal{F}$ is mixed of weights $\leqslant w$, one has for any $x \in U(\mathbf{F}_q)$

$$|K_{\mathcal{F}}(x)| \leqslant \mathrm{rk}(\mathcal{F})q^{w/2}. \tag{3.4}$$

REMARK 3.11. It is always possible to reduce to the case of ℓ-adic sheaves mixed of weight $w = 0$. For any $w \in \mathbf{Z}$ there exist an ℓ-adic sheaf denoted $\overline{\mathbf{Q}_\ell}(w/2)$ of rank 1, lisse on $\mathbf{P}^1_{\mathbf{F}_q}$, whose restriction to G^{geom} is trivial and such that Frob_x acts by multiplication by $q_x^{-w/2}$ (in particular $\overline{\mathbf{Q}_\ell}(w/2)$ is pure of weight $-w$). Given $\mathcal{F}$ of some weight w', the tensor product

$$\mathcal{F}(w/2) := \mathcal{F} \otimes \overline{\mathbf{Q}_\ell}(w/2)$$

has weight $w' - w$ and has trace function given by

$$x \mapsto q^{-w/2}K_{\mathcal{F}}(x).$$

In the sequel, unless stated otherwise, we will always assume that trace functions are associated with sheaves which are mixed of weights $\leqslant 0$.

[6]via the fixed embedding $\overline{\mathbf{Q}_\ell} \hookrightarrow \mathbf{C}$.

REMARK 3.12. Deligne proved ([**Del80**, Lemme (1.8.1)]) that for a sheaf punctually pure of weight w, for any closed point $x \in \mathbf{P}^1_{\mathbf{F}_q}$, the eigenvalues of $(\mathrm{Frob}_x \, | V_{\mathcal{F}}^{I_x})$ have modulus $\leqslant q_x^{w/2}$. In particular

$$|\operatorname{tr}(\mathrm{Frob}_x \, | V_{\mathcal{F}}^{I_x})| \leqslant \mathrm{rk}(\mathcal{F}) q_x^{w/2}.$$

In particular (assuming that $w = 0$) ℓ^∞-norm of the difference between the extension by 0 of $K_{\mathcal{F}}$ from $U(\mathbf{F}_q)$ to $\mathbf{A}^1(\mathbf{F}_q)$ and the middle-extension (described in Remark 3.7) is bounded by

$$\mathrm{rk}(\mathcal{F})|\mathbf{A}^1(\overline{\mathbf{F}_q}) - U(\overline{\mathbf{F}_q})|.$$

As we will see, we will be interested in situations where this quantity is bounded by an absolute constant (independent of q) the consequence being that whatever extension we choose between the two, it won't make much of a difference.

3.6. Other functions. There are other functions on $\mathbf{F}_q$ of great interest which do not qualify as trace functions under our current definition. For instance the Dirac function at some point $a \in \mathbf{F}_q$

$$\delta_a(n) = \begin{cases} 1 & \text{if } n \equiv a \pmod q \\ 0 & \text{otherwise .} \end{cases}$$

which, extended to $\mathbf{Z}$ is the characteristic function of the arithmetic progression $a + q\mathbf{Z}$ (obviously of considerable interest for analytic number theory). It turns out that such functions can be related to trace functions in our sense by very natural transformations and this will allow us to make some progress on problems from "classical" analytic number theory.

REMARK 3.13. In fact this function could be interpreted as the trace function of a *skyscraper sheaf* supported at the closed point a but we will not do this here.

3.7. Local monodromy representations. Given $\mathcal{F}$ some ℓ-adic sheaf, let

$$D_{\mathcal{F}}^{ram} \subset \mathbf{P}^1(\overline{\mathbf{F}_q}) - U(\overline{\mathbf{F}_q})$$

be the set of geometric points where the representation $\varrho_{\mathcal{F}}$ is ramified, that is the inertia group I_x acts non-trivially. The restricted representation

$$\varrho_{\mathcal{F}|I_x} = \varrho_{\mathcal{F},x}$$

is called the local monodromy representation of $\mathcal{F}$ at x (cf. Remark 3.3 for the abuse of notation). Although $D_{\mathcal{F}}^{ram}$ is disjoint from $U(\overline{\mathbf{F}_q})$, this finite set of representations is fundamental to study $\mathcal{F}$ and its trace function. Let us recall the exact sequence [**Kat88**, Chap. 1]

$$1 \to P_x \to I_x \to I_x^{tame} \to 1$$

where I_x^{tame} is the *tame inertia quotient* and is isomorphic to $\prod_{p \neq q} \mathbf{Z}_p$, while P_x is the q-Sylow subgroup of I_x and is called the wild inertia subgroup.

DEFINITION 3.14. The sheaf is tamely ramified at x if P_x acts trivially on $V_{\mathcal{F}}$ (so that $\varrho_{\mathcal{F},x}$ factors through I_x^{tame}) and is called wildly ramified otherwise.

3.7.1. *The Swan conductor.* If the representation is wildly ramified one can measure how deep it is by means of a numerical invariant: the Swan conductor. The wild inertia subgroup I_x is equipped with the decreasing *upper numbering filtration* $I_x^{(\lambda)}$ indexed by non-negative real numbers $\lambda \geqslant 0$, such that

$$P_x = I_x^{(>0)} = \bigcup_{\lambda>0} I_x^{\lambda}.$$

Given $V = V_{\mathcal{F}}$ as above there is a P_x-stable direct sum decomposition

$$V = \bigoplus_{\lambda\in \mathrm{Break}(V)} V(\lambda)$$

indexed by a finite set of rational numbers $\mathrm{Break}(V) \subset \mathbf{Q}_{\geqslant 0}$ (the set of *breaks* of the I_x-module V) such that

$$V(0) = V^{P_x},\ V(\lambda)^{I_x^{(\lambda)}} = 0,\ V(\lambda)^{I_x^{(\lambda')}} = V(\lambda),\ \lambda' > \lambda$$

(see [**Kat88**, Chap. 1]). The *Swan conductor* is defined as

$$\mathrm{Swan}_x(\mathcal{F}) = \sum_{\lambda\in \mathrm{Break}(V)} \lambda \dim V(\lambda)$$

and turns out to be an integer [**Kat88**, Prop. 1.9].

In the decomposition

$$V = V(0) \oplus \bigoplus_{\substack{\lambda\in \mathrm{Break}(V)\\ \lambda>0}} V(\lambda) = V(0) \oplus V(>0) := V^{tame} \oplus V^{wild}$$

the first summand is called the *tame part* and the remaining one the *wild part.*

4. Summing trace functions over $\mathbf{F}_q$

Let $K_{\mathcal{F}}$ be the trace function associated to a sheaf $\mathcal{F}$ lisse on $U_{\mathbf{F}_q}$. It is a function on $U(\mathbf{F}_q)$ which we may extend by zero to $\mathbf{A}^1(\mathbf{F}_q) \simeq \mathbf{F}_q = \mathbf{Z}/q\mathbf{Z}$.

The Grothendieck-Lefschetz trace formula provides an alternative expression for the sum of $K_{\mathcal{F}}$ over the whole $\mathbf{A}^1(\mathbf{F}_q)$.

THEOREM 4.1 (Grothendieck-Lefschetz trace formula). *Let $\mathcal{F}$ be lisse on U; there exists three finite dimensional ℓ-adic representations of* $\mathrm{Gal}(\overline{\mathbf{F}_q}/\mathbf{F}_q)$, $H^i_c(U_{\overline{\mathbf{F}_q}}, \mathcal{F})$ *such that*

$$\sum_{x\in U(\mathbf{F}_q)} K_{\mathcal{F}}(x) = \sum_{x\in U(\mathbf{F}_q)} \mathrm{tr}(\mathrm{Fr}_x\,|\mathcal{F}) = \sum_{i=0}^{2}(-1)^i \,\mathrm{tr}(\mathrm{Frob}_q\,|H^i_c(U_{\overline{\mathbf{F}_q}}, \mathcal{F})). \tag{4.1}$$

More generally, for any $n \geqslant 1$,

$$\sum_{x\in U(\mathbf{F}_{q^n})} K_{\mathcal{F},n}(x) = \sum_{x\in U(\mathbf{F}_{q^n})} \mathrm{tr}(\mathrm{Fr}_x\,|\mathcal{F}) = \sum_{i=0}^{2}(-1)^i \,\mathrm{tr}(\mathrm{Frob}_q^n\,|H^i_c(U_{\overline{\mathbf{F}_q}}, \mathcal{F})).$$

The $\overline{\mathbf{Q}_\ell}$-vector spaces $H^i_c(U_{\overline{\mathbf{F}_q}}, \mathcal{F})$ are the so-called compactly supported étale cohomology groups of $\mathcal{F}$ and can also be considered as ℓ-adic sheaves over the point $\mathrm{Spec}(\mathbf{F}_q)$.

The above formula reduces the evaluation of averages of trace functions to that of the three summands

$$\mathrm{tr}(\mathrm{Frob}_q\,|H^i_c(U_{\overline{\mathbf{F}_q}}, \mathcal{F})),\ i = 0, 1, 2,$$

we need therefore to control the dimension of these spaces as well as the size of the eigenvalues. We start with the former.

4.1. Bounding the dimension of the cohomology groups. The extremal cohomology groups have a simple interpretation. First

$$H_c^0(U_{\overline{\mathbf{F}_q}},\mathcal{F}) = \begin{cases} 0 & \text{if } U \neq \mathbf{P}^1_{\mathbf{F}_q} \\ V_{\mathcal{F}}^{G^{\mathrm{geom}}} & \text{if } U = \mathbf{P}^1_{\mathbf{F}_q}. \end{cases}$$

As a $\mathrm{Gal}(\overline{\mathbf{F}_q}/\mathbf{F}_q)$-representation, one has an isomorphism

$$H_c^2(U_{\overline{\mathbf{F}_q}},\mathcal{F}) \simeq V_{\mathcal{F},G^{\mathrm{geom}}}(-1) \tag{4.2}$$

(ie $H_c^2(U_{\overline{\mathbf{F}_q}},\mathcal{F})$ is isomorphic to the quotient of G^{geom}-coinvariants of $V_{\mathcal{F}}$ twisted by $\overline{\mathbf{Q}_\ell}(-1)$). In particular, if $\mathcal{F}$ is geometrically irreducible (non geometrically trivial) or more generally geometrically isotypic (the underlying geometric irreducible representation being non trivial) one has

$$H_c^2(U_{\overline{\mathbf{F}_q}},\mathcal{F}) = 0.$$

In any case, one has

$$\dim H_c^0(U_{\overline{\mathbf{F}_q}},\mathcal{F}),\ \dim H_c^2(U_{\overline{\mathbf{F}_q}},\mathcal{F}) \leqslant \mathrm{rk}(\mathcal{F}).$$

The dimension of the middle cohomology group is now determined by the

THEOREM 4.2 (The Grothendieck-Ogg-Shafarevich formula). *One has the following equality*

$$\begin{aligned}\chi(U_{\overline{\mathbf{F}_q}},\mathcal{F}) &= \sum_{i=0}^{2}(-1)^i \dim H_c^i(U_{\overline{\mathbf{F}_q}},\mathcal{F}) \\ &= \mathrm{rk}(\mathcal{F})(2-|\mathbf{P}^1(\overline{\mathbf{F}_q})-U(\overline{\mathbf{F}_q})|) - \sum_{x\in D_{\mathcal{F}}^{ram}(\overline{\mathbf{F}_q})} \mathrm{Swan}_x(\mathcal{F}).\end{aligned}$$

Observe that the quantities that occur are local geometric data associated to the sheaf yet this collection of local data provides global informations.

We then define the following ad-hoc numerical invariant which serves as a measure of the complexity of the sheaf $\mathcal{F}$:

DEFINITION 4.3. The conductor of $\mathcal{F}$ is defined via the following formula

$$C(\mathcal{F}) = \mathrm{rk}(\mathcal{F}) + |\mathbf{P}^1(\overline{\mathbf{F}_q})-U(\overline{\mathbf{F}_q})| + \sum_{x\in D_{\mathcal{F}}^{ram}(\overline{\mathbf{F}_q})} \mathrm{Swan}_x(\mathcal{F}).$$

In view of this definition we have

$$\sum_{i=0}^{2} \dim H_c^i(U_{\overline{\mathbf{F}_q}},\mathcal{F}) \ll C(\mathcal{F})^2. \tag{4.3}$$

4.2. Examples.

4.2.1. *The trivial sheaf.* The trivial representation $\overline{\mathbf{Q}_\ell}$ is everywhere lisse, pure of weight 0, of rank 1 and conductor 1 and

$$K_{\overline{\mathbf{Q}_\ell}}(x) = 1.$$

4.2.2. *Kummer sheaf* [**Del77**]. For any non-trivial Dirichlet character $\chi\colon (\mathbf{F}_q^\times,\times)\to\mathbf{C}^\times$ there exists an ℓ-adic sheaf (a Kummer sheaf) denoted $\mathcal{L}_\chi$ which is of rank 1, pure of weight 0, lisse on $\mathbf{G}_{m,\mathbf{F}_q}=\mathbf{P}^1_{\mathbf{F}_q}-\{0,\infty\}$ with trace function

$$K_{\mathcal{L}_\chi}(x)=\chi(x),\ K_{\mathcal{L}_\chi,n}(x)=\chi(\mathrm{Nr}_{\mathbf{F}_{q^n}/\mathbf{F}_q}(x))=:\chi_n(x)$$

and conductor

$$C(\mathcal{L}_\chi)=3;$$

indeed $\mathrm{Swan}_0(\mathcal{L}_\chi)=\mathrm{Swan}_\infty(\mathcal{L}_\chi)=0$.

4.2.3. *Artin-Schreier sheaf* [**Del77**]. For any additive character $\psi\colon(\mathbf{F}_q,+)\to\mathbf{C}^\times$ there exists an ℓ-adic sheaf (an Artin-Schreier sheaf) denoted $\mathcal{L}_\psi$ which is of rank 1, pure of weight 0, lisse on $\mathbf{A}^1_{\mathbf{F}_q}=\mathbf{P}^1_{\mathbf{F}_q}-\{\infty\}$ with trace function

$$K_{\mathcal{L}_\psi}(x)=\psi(x),\ K_{\mathcal{L}_\psi,n}(x)=\psi(\mathrm{tr}_{\mathbf{F}_{q^n}/\mathbf{F}_q}(x))=:\psi_n(x)$$

and conductor (if ψ is non-trivial)

$$C(\mathcal{L}_\psi)=3.$$

(indeed $\mathrm{Swan}_\infty(\mathcal{L}_\psi)=1$). If $f\in\mathbf{F}_q(X)-\mathbf{F}_q$, the pull-back sheaf $\mathcal{L}_{\psi(f)}$ is geometrically irreducible and has conductor

$$1+\text{ number of poles of } f+\text{ sum of multiplicities of the poles of } f.$$

More generally any character ψ of $(\mathbf{F}_{q^n},+)$ is of the shape

$$x\mapsto\psi_1(\mathrm{tr}_{\mathbf{F}_{q^n}/\mathbf{F}_q}(ax))$$

for ψ_1 a non-trivial character of $(\mathbf{F}_q,+)$ and $a\in\mathbf{F}_{q^n}$, and associated to each such character is an Artin-Schreier sheaf $\mathcal{L}_\psi$.

4.2.4. *(hyper)-Kloosterman sheaves* [**Kat88**]. Hyper-Kloosterman sums are formed by multiplicative convolution out of additive characters.

Given $K_1,K_2\colon\mathbf{F}_q^\times\to\mathbf{C}$ one defines their (normalized) multiplicative convolution:

$$K_1\star K_2\colon x\in\mathbf{F}_q^\times\to\frac{1}{q^{1/2}}\sum_{\substack{x_1,x_2\in\mathbf{F}_q^\times\\ x_1.x_2=x}}K_1(x_1)K_2(x_2)=\frac{1}{q^{1/2}}\sum_{x_1\in\mathbf{F}_q^\times}K_1(x_1)K_2(x/x_1).$$

Similarly for any $n\geqslant 1$ one defines the multiplicative convolution of $K_{1,n},K_{2,n}\colon\mathbf{F}_{q^n}^\times\to\mathbf{C}$ as

$$K_{1,n}\star K_{2,n}\colon x\in\mathbf{F}_{q^n}^\times\to\frac{1}{q^{n/2}}\sum_{\substack{x_1,x_2\in\mathbf{F}_{q^n}^\times\\ x_1.x_2=x}}K_{1,n}(x_1)K_{2,n}(x_2).$$

Now, given a non-trivial additive character ψ of $\mathbf{F}_q$ and $k\geqslant 2$, the hyper-Kloosterman sums can be expressed as the k-fold multiplicative convolutions of ψ:

$$\mathrm{Kl}_{k,\psi}(x;q)=\star_{k\text{ times}}\psi(x)=\frac{1}{q^{(k-1)2}}\sum_{\substack{x_1,\dots,x_k\in\mathbf{F}_q^\times\\ x_1.\dots.x_k=x}}\psi(x_1+\ldots+x_k)$$

and more generally, one defines hyper-Kloosterman sums over $\mathbf{F}_{q^n}^\times$

$$\mathrm{Kl}_{k,\psi}(x;q^n)=\star_{k\text{ times}}\psi_n(x)=\frac{1}{q^{n(k-1)2}}\sum_{\substack{x_1,\dots,x_k\in\mathbf{F}_{q^n}^\times\\ x_1.\dots.x_k=x}}\psi_n(x_1+\ldots+x_k).$$

These are in fact trace functions: their underlying sheaves were constructed by Deligne and were subsequently studies in depth by Katz [**Kat88**]:

THEOREM 4.4. *For any $k \geqslant 2$, there exists an ℓ-adic sheaf (the Kloosterman sheaf) denoted $\mathcal{K}\ell_{k,\psi}$, of rank k, pure of weight 0, geometrically irreducible, lisse on $\mathbf{G}_{m,\mathbf{F}_q}$ with trace function*

$$K_{\mathcal{K}\ell_{k,\psi}}(x) = \mathrm{Kl}_{k,\psi}(x; q)$$

and more generally, for any $n \geqslant 1$

$$K_{\mathcal{K}\ell_{k,\psi},n}(x) = \mathrm{Kl}_{k,\psi}(x; q^n).$$

One has $\mathrm{Swan}_0(\mathcal{K}\ell_{k,\psi}) = 0$ *and* $\mathrm{Swan}_\infty(\mathcal{K}\ell_{k,\psi}) = 1$ *so that the conductor of that sheaf equals*

$$C(\mathcal{K}\ell_{k,\psi}) = k + 2 + 1.$$

The Kloosterman sheaves have trivial determinant

$$\det \mathcal{K}\ell_k = \overline{\mathbf{Q}_\ell}$$

and if (and only if) k is even, the Kloosterman sheaf $\mathcal{K}\ell_k$ is self-dual:

$$D(\mathcal{K}\ell_k) \simeq \mathcal{K}\ell_k.$$

REMARK 4.5. When $\psi(\cdot) = \mathrm{e}_q(\cdot)$ we will not mention the additive character e_q in the notation.

4.3. Deligne's Theorem on the weight. Now that we control the dimension of the cohomology groups occurring in the Grothendieck-Lefschetz trace formula, it remains to control the size of their Frobenius eigenvalues. Suppose that $\mathcal{F}$ is pure of weight 0 so that

$$|K_\mathcal{F}(x)| \leqslant \mathrm{rk}(\mathcal{F}).$$

As we have seen, as long as $U \neq \mathbf{P}^1$, $H^0_c(U_{\overline{\mathbf{F}_q}}, \mathcal{F}) = 0$.

By (4.2), the eigenvalues of Frob_q acting on $H^2_c(U_{\overline{\mathbf{F}_q}}, \mathcal{F})$ are of the form

$$q\alpha_i,\ i = 1, \dots, \dim(V_{\mathcal{F},G^{\mathrm{geom}}}) \text{ with } |\alpha_i| = 1.$$

The trace of the Frobenius on the middle cohomology group $\mathrm{tr}(\mathrm{Frob}_q | H^1_c(U_{\overline{\mathbf{F}_q}}, \mathcal{F}))$ is much more mysterious but fortunately we have the following theorem of Deligne [**Del80**].

THEOREM 4.6 (The Generalized Riemann Hypothesis for finite fields). *The eigenvalues of* Frob_q *acting on* $H^1_c(U_{\overline{\mathbf{F}_q}}, \mathcal{F})$ *are complex numbers of modulus* $\leqslant q^{1/2}$.

We deduce from this

COROLLARY 4.7. *Let $\mathcal{F}$ be an ℓ-adic sheaf lisse on some U pure of weight 0; one has*

$$\sum_{x\in\mathbf{F}_q} K_\mathcal{F}(x) - \mathrm{tr}(\mathrm{Frob}_q | H^2_c(U_{\overline{\mathbf{F}_q}}, \mathcal{F})) \ll C(\mathcal{F})^2 q^{1/2}.$$

More generally for any $n \geqslant 1$

$$\sum_{x\in\mathbf{F}_{q^n}} K_{\mathcal{F},n}(x) - \mathrm{tr}(\mathrm{Frob}_q^n | H^2_c(U_{\overline{\mathbf{F}_q}}, \mathcal{F})) \ll C(\mathcal{F})^2 q^{n/2}.$$

In particular if $\mathcal{F}$ is geometrically irreducible or isotypic with no trivial component, one has

$$\sum_{x\in\mathbf{F}_q} K_{\mathcal{F}}(x) \ll C(\mathcal{F})^2 q^{1/2}.$$

Here, the implied constants are absolute.

In practical applications we will be faced with situations where we have a sequence of sheaves $(\mathcal{F}_q)_q$ indexed by an infinite set of primes (with $\mathcal{F}_q$ a sheaf over the field $\mathbf{F}_q$) such that the sequence of conductors $(C(\mathcal{F}_q))_q$ remains uniformly bounded (by C say). In such situation, the above formula represents an asymptotic formula as $q \to \infty$ for the sum of $q - O(1)$ terms

$$\sum_{x\in U(\mathbf{F}_q)} K_{\mathcal{F}}(x)$$

with main term $\mathrm{tr}(\mathrm{Frob}_q \,|H^2_c(U_{\overline{\mathbf{F}_q}}, \mathcal{F}))$ (possibly 0) and an error term of size $\ll C^2q^{1/2}$.

5. Quasi-orthogonality relations

We will often apply the trace formula and Deligne's theorem to the following sheaf: given $\mathcal{F}$ and $\mathcal{G}$ two ℓ-adic sheaves both lisse on some non-empty open set $U \subset \mathbf{A}^1_{\mathbf{F}_q}$ and both pure of weight 0; consider the tensor product $\mathcal{F} \otimes D(\mathcal{G})$. This sheave is also lisse on U and pure of weight 0, moreover from the definition of the conductor (see [**Kat88**, Chap. 1]) one sees that

$$C(\mathcal{F} \otimes D(\mathcal{G})) \leqslant C(\mathcal{F})C(\mathcal{G}). \tag{5.1}$$

The trace functions of $\mathcal{F} \otimes D(\mathcal{G})$ are given for $x \in U(\mathbf{F}_{q^n})$ by

$$x \mapsto K_{\mathcal{F}\otimes D(\mathcal{G}),n}(x) = K_{\mathcal{F},n}(x)\overline{K_{\mathcal{G},n}(x)}.$$

Therefore the trace formula can be used to evaluate the correlation sums between the trace function of $\mathcal{F}$ and $\mathcal{G}$,

$$\mathcal{C}(\mathcal{F},\mathcal{G}) := \frac{1}{q}\sum_{x\in\mathbf{F}_q} K_{\mathcal{F}}(x)\overline{K_{\mathcal{G}}(x)};$$

more generally for any $n \geqslant 1$ we set

$$\mathcal{C}_n(\mathcal{F},\mathcal{G}) := \frac{1}{q^n}\sum_{x\in\mathbf{F}_{q^n}} K_{\mathcal{F},n}(x)\overline{K_{\mathcal{G},n}(x)}.$$

Indeed, by Corollary 4.7, one has

$$\mathcal{C}_n(\mathcal{F},\mathcal{G}) = \mathrm{tr}(\mathrm{Frob}_q^n \,|V_{\mathcal{F}\otimes D(\mathcal{G}),G^{\mathrm{geom}}}) + O(\frac{C(\mathcal{F})C(\mathcal{G})}{q^{n/2}}). \tag{5.2}$$

In particular if $C(\mathcal{F})C(\mathcal{G})$ are bounded while $q^n \to \infty$, one obtains an asymptotic formula whose main term is given by the trace of the powers of Frobenius acting on the coinvariants of $\mathcal{F} \otimes D(\mathcal{G}) \simeq \mathrm{Hom}(\mathcal{G},\mathcal{F})$.

5.1. Decomposition of sheaves and trace functions. Using first a weaker version of the formula (with an error term converging to 0 as $n \to \infty$), Deligne, on his way to the proof of Theorem 4.6, established that any ℓ-adic sheaf pure of weight 0 is geometrically semi-simple (the representation $\varrho_{\mathcal{F}|G^{\text{geom}}}$ decomposes into a direct sum of irreducible representations (of G^{geom})) [**Del80**, Théorème (3.4.1)]; the irreducible components occurring in the decomposition of $\varrho_{\mathcal{F}|G^{\text{geom}}}$ are called the *geometric irreducible components of* $\mathcal{F}$.

This is not exactly valid for the arithmetic representation, but considering its semi-simplification, one obtains a decomposition

$$\varrho_{\mathcal{F}}^{ss} = \bigoplus_{i \in I} \varrho_{\mathcal{F}_i}$$

where the $\varrho_{\mathcal{F}_i}$ are arithmetically irreducible (and pure) and lisse on U. Regarding geometric reducibility, each $\varrho_{\mathcal{F}_i}$ is either geometrically isotypic or is induced from a representation of $\mathrm{Gal}(K^{\text{sep}}/k.K)$ for k some finite extension of $\mathbf{F}_q$. Since semi-simplification does not change the trace function, we obtain a decomposition of the trace function

$$K_{\mathcal{F}} = \sum_i K_{\mathcal{F}_i}.$$

Moreover a computation shows that whenever $\mathcal{F}_i$ is induced one has $K_{\mathcal{F}_i} \equiv 0$ on $U(\mathbf{F}_q)$. Therefore we obtain

PROPOSITION 5.1. *The trace function associated to some punctually pure sheaf $\mathcal{F}$ lisse on U can be decomposed into the sum of $\leqslant C(\mathcal{F})$ trace functions associated to sheaves $\mathcal{F}_i$, that are lisse on U, punctually pure of weight 0, geometrically isotypic with conductors $C(\mathcal{F}_i) \leqslant C(\mathcal{F})$.*

This proposition reduces the study of trace functions to trace functions associated to geometrically isotypic or (most of the time) geometrically irreducible sheaves. From now on (unless stated otherwise) we will assume that the trace functions are associated to sheaves that are punctually pure of weight 0 and geometrically isotypic. To ease notations, we say that such sheaves are "isotypic" or "irreducible" omitting the mention "geometrically" and likewise will speak of isotypic or irreducible trace functions. In such situation, using Schur lemma, the formula for (5.2) specializes to the

THEOREM 5.2 (Quasi-orthogonality relations). *Supppose that $\mathcal{F}$ and $\mathcal{G}$ are both geometrically isotypic with $n_{\mathcal{F}}$ copies of the irreducible component $\overline{\mathcal{F}}_{irr}$ for $\mathcal{F}$ and $n_{\mathcal{G}}$ copies of the irreducible component $\overline{\mathcal{G}}_{irr}$ for $\mathcal{G}$. There exists $n_{\mathcal{F}}.n_{\mathcal{G}}$ complex numbers $\alpha_{i,\mathcal{F},\mathcal{G}}$ of modulus 1 such that*

$$\mathcal{C}_n(\mathcal{F},\mathcal{G}) = \Big(\sum_{i=1}^{n_{\mathcal{F}}n_{\mathcal{G}}} \alpha_{i,\mathcal{F},\mathcal{G}}^n\Big)\delta_{\overline{\mathcal{F}}\sim_{geom}\mathcal{G}} + O(C(\mathcal{F})^2C(\mathcal{G})^2q^{-n/2}). \tag{5.3}$$

In particular if $\mathcal{F}$ and $\mathcal{G}$ are both geometrically irreducible there exist $\alpha_{\mathcal{F},\mathcal{G}} \in \mathbf{S}^1$ such that

$$\mathcal{C}_n(\mathcal{F},\mathcal{G}) = \alpha_{\mathcal{F},\mathcal{G}}^n\delta_{\overline{\mathcal{F}}\sim_{geom}\mathcal{G}} + O(C(\mathcal{F})^2C(\mathcal{G})^2q^{-n/2}). \tag{5.4}$$

In both (5.3) *and* (5.4) *the implicit constants are independent of n.*

REMARK 5.3. Observe that for $\mathcal{F}$ and $\mathcal{G}$ either the Kummer or Artin-Schreier sheaves these equalities correspond to the orthogonality relations of characters.

REMARK 5.4. If two geometrically irreducible sheaves $\mathcal{F}, \mathcal{G}$ are geometrically isomorphic, then their trace functions are proportional: more precisely one has for any n

$$K_{\mathcal{F},n} = \alpha^n_{\mathcal{F},\mathcal{G}} K_{\mathcal{G},n}$$

where $\alpha_{\mathcal{F},\mathcal{G}}$ is the complex number of modulus 1 introduced in the previous statement.

When q^n is large compared to $C(\mathcal{F})^2 C(\mathcal{G})^2$, the above formula gives a useful criterion to detect whether $\mathcal{F}$ and $\mathcal{G}$ have geometric irreducible components in common. While our focus is on the case $n = 1$ and $q \to \infty$ (while $C(\mathcal{F})^2 C(\mathcal{G})^2$ remains bounded), the case $n \to \infty$ will also prove useful. We start with the following easy lemma

LEMMA 5.5. *Given $\alpha_1, \ldots, \alpha_d \in \mathbf{S}^1$, arbitrary complex numbers of modulus 1, one has*

$$\limsup_{n\to\infty} (\alpha_1^n + \ldots + \alpha_d^n) = d.$$

Using this lemma together with the decomposition into irreducible representations, one obtains the following

COROLLARY 5.6 (Katz's Diophantine criterion for irreducibility). *Let $\mathcal{F}$ be an ℓ-adic sheaf lisse on U pure of weight 0 with decomposition into geometrically irreducible subsheaves denoted*

$$\mathcal{F}^{geom} = \bigoplus_i \overline{\mathcal{F}}_i^{\oplus n_i}.$$

Then

$$\limsup_{n\to\infty} \mathcal{C}_n(\mathcal{F}, \mathcal{F}) = \sum_{\overline{\mathcal{F}}_i} n_i^2.$$

In particular, $\mathcal{F}$ is geometrically irreducible if and only if

$$\limsup_{n\to\infty} \mathcal{C}_n(\mathcal{F}, \mathcal{F}) = 1.$$

5.2. Counting trace functions. The above orthogonality relations lead to upper bounds for the number of geometric isomorphism classes of ℓ-adic sheaves of bounded conductor (see [**FKM13**] for the proof):

THEOREM 5.7. *Let $C \geqslant 1$, the number of geometric isomorphism classes of geometrically irreducible ℓ-adic sheaves of conductor $\leqslant C$ is finite and bounded by*

$$q^{O(C^6)}$$

where the implied constant is absolute.

PROOF. The principle of the proof is as follows: the sheaf-to-trace-function map $\mathcal{F} \to t_{\mathcal{F}}$ associates to the geometric isomorphism class of some sheaf a line in the q-dimensional Hermitian space $\mathbf{C}^{\mathbf{F}_q}$ of complex-valued functions on $\mathbf{F}_q$ with inner product

$$\langle K, K' \rangle = \frac{1}{q} \sum_{x \in \mathbf{F}_q} K(x) \overline{K'}(x).$$

The quasi-orthogonality relations show that these different lines are almost orthogonal to one another and so one obtains a number of almost orthogonal (circles of) unit vectors in the corresponding unit sphere. A sphere-packing argument for high-dimensional hermitian spaces (see [KL78]) implies that the number of such vectors cannot be too large. □

6. Trace functions over short intervals

In the next few sections, we discuss the correlations between trace functions and other classical arithmetic functions. Indeed given a trace function

$$K_{\mathcal{F}}\colon \mathbf{A}^1(\mathbf{F}_q) = \mathbf{F}_q \to \mathbf{C}$$

(extended from $U(\mathbf{F}_q)$ to $\mathbf{A}^1(\mathbf{F}_q)$ either by zero or by the middle-extension) we obtain a q-periodic function on $\mathbf{Z}$ (which we also denote by $K_{\mathcal{F}}$) via the $(\operatorname{mod} q)$-map

$$K = K_{\mathcal{F}}\colon \mathbf{Z} \to \mathbf{Z}/q\mathbf{Z} = \mathbf{A}^1(\mathbf{F}_q) \to \mathbf{C}.$$

Given some other arithmetic function $\lambda\colon \mathbf{N} \to \mathbf{C}$ it is natural to compare them by evaluating their correlation sums

$$\sum_{n\leqslant N} K(n)\overline{\lambda(n)}$$

as $N \to \infty$ (in suitable ranges of interest depending on $C(\mathcal{F})$ and λ).

6.1. The Pólya-Vinogradov method. We start with the basic case where $\lambda = 1_I$ is the characteristic function of an interval I of $\mathbf{Z}$ (which we may assume is contained in $[0, q-1]$). We want to evaluate non-trivially the sum

$$S(K; I) := \sum_{n\in I} K(n).$$

Remember that we may and do assume that $\mathcal{F}$ is geometrically isotypic and that if $I = [0, q-1]$ such sum can be dealt with by Deligne's theorem.

By Parseval, one has

$$S(K; I) = \sum_{y\in\mathbf{F}_q} \widehat{K}(y)\overline{\widehat{1_I}(y)}$$

where

$$\widehat{K}(y) = \frac{1}{q^{1/2}} \sum_{x\in\mathbf{F}_q} K(x)\mathrm{e}_q(xy) \tag{6.1}$$

and

$$\widehat{1_I}(y) = \frac{1}{q^{1/2}} \sum_{x\in I} \mathrm{e}_q(xy)$$

are the (normalized) Fourier transforms of K and 1_I (for the abelian group $(\mathbf{F}_q, +)$). One has

$$|\widehat{1_I}(y)| \ll \frac{1}{q^{1/2}} \min(|I|, \|\frac{y}{q}\|^{-1}) \ll \frac{1}{q^{1/2}} \min(|I|, \frac{q}{|y|})$$

(here $\|y/q\|$ denote the distance to the nearest integer) which implies that

$$\|\widehat{1_I}\|_1 \ll \frac{|I|}{q^{1/2}} + q^{1/2}\log q.$$

Therefore one has

$$\sum_{n\in I} K(n) \ll \|\widehat{K}\|_\infty q^{1/2}\log q.$$

This leads us to look at the size of the Fourier transform $y \mapsto \widehat{K}(y)$. If K is of the shape $e_q(ax)$ for some $a \in \mathbf{F}_q$, its Fourier transform is a Dirac function

$$\widehat{K}(y) = q^{1/2}\delta_{y=a \,(\mathrm{mod}\, q)}$$

and is therefore highly concentrated. To avoid this we make the following

Definition 6.1. An isotypic sheaf $\mathcal{F}$ is Fourier if its geometric irreducible component is not (geometrically) isomorphic to any Artin-Schreier sheaf $\mathcal{L}_\psi$.

In particular, if K is Fourier of conductor $C(\mathcal{F})$, it follows from Theorem 5.2 that for any $y \in \mathbf{F}_q$

$$\widehat{K}(y) \ll C(\mathcal{F})^2.$$

In that way we obtain the

Theorem 6.2 (Pólya-Vinogradov bound). *Let $\mathcal{F}$ be a Fourier sheaf of conductor $C(\mathcal{F})$ and K its associated trace function. For any interval I of length $\leqslant q$, one has*

$$\sum_{x \in I} K(x) \ll C(\mathcal{F})^2 q^{1/2} \log q;$$

here the implicit constant is absolute.

Remark 6.3. This statement was obtained for the first time by Pólya and Vinogradov, independently, in the case of Dirichlet characters χ. In that case the Fourier transform is the normalized Gauss sum

$$\widehat{\chi}(y) = \varepsilon_\chi(y) = \frac{1}{q^{1/2}} \sum_{x \in \mathbf{F}_q} \chi(x) e_q(xy)$$

which is bounded in absolute value by 1.

Observe that this bound is better than the trivial bound

$$|\sum_{x \in I} K(x)| \leqslant C(\mathcal{F})|I|$$

as long as

$$|I| \gg_{C(\mathcal{F})} q^{1/2} \log q.$$

This range is called the *Pólya-Vinogradov range* and the question of bounding non-trivially for as many trace functions as possible over shorter intervals is a fundamental problem in analytic number theory with many striking applications. At this moment, the problem is solved only in a very limited number of cases. One important example is the celebrated work of Burgess on Dirichlet characters [**Bur62**] which we discuss in §16.1. A lot of the forthcoming lectures will indeed be concerned with breaking this barrier in specific cases or in different contexts, and to give some applications.

6.1.1. *Bridging the Pólya-Vinogradov range.* The following argument of Fouvry, Kowalski, Michel, Rivat, Soundararajan and Raju improves slightly the Pólya-Vinogradov range:

Theorem 6.4. [**FKM$^+$17**] *Let $\mathcal{F}$ be a Fourier sheaf of conductor $C(\mathcal{F})$ and K its associated trace function. For any interval I of length $\sqrt{q} < |I| \leqslant q$, we have*

$$\sum_{x \in I} K(x) \ll C(\mathcal{F})^2 q^{1/2}(1 + \log(|I|/q^{1/2})).$$

PROOF. Given $r \in \mathbf{Z}$, let $I_r = r + I$; this is again an interval and $S(K; I)$ and $S(K; I_r)$ differ only by $O(\|K\|_\infty r)$, which is a useful bound when r is not too large. Moreover

$$\widehat{1_{I_r}}(y) = \mathrm{e}_q(ry)\widehat{1_I}(y).$$

We have therefore

$$S(K; I) = \sum_{|y| \leqslant q/2} \widehat{K}(y)\overline{\widehat{1_I}(y)}\frac{1}{R} \sum_{0 \leqslant r \leqslant R-1} \mathrm{e}_q(-ry).$$

We choose $R = [q^{1/2}] + 1$; using the bounds

$$|\widehat{1_I}(y)| \ll q^{-1/2} \min(|I|, q/|y|), \quad \sum_{0 \leqslant r \leqslant R-1} \mathrm{e}_q(-ry) \ll \min(R, q/|r|)$$

and

$$\|K\|_\infty + \|\widehat{K}\|_\infty \ll C(\mathcal{F})^2$$

we obtain the result. □

6.2. A smoothed version of the Pólya-Vinogradov method. Often in analytic number theory one is not faced with summing a trace function over an interval but instead against some smooth compactly supported function, for instance one has to evaluate sums of the shape

$$\sum_{n \in \mathbf{Z}} K(n)V(\frac{n}{N}), \ V \in C_c^\infty(\mathbf{R}) \text{ fixed.}$$

By the Poisson summation formula one has the identity

$$\sum_{n \in \mathbf{Z}} K(n)V(\frac{n}{N}) = \frac{N}{q^{1/2}} \sum_{n \in \mathbf{Z}} \widehat{K}(n)\widehat{V}(\frac{nN}{q}) \tag{6.2}$$

where

$$\widehat{V}(y) = \int_{\mathbf{R}} V(x)e(xy)dx$$

is the Fourier transform of $V(x)$ (over $\mathbf{R}$).

Observe that $\widehat{V}(y)$ is not compactly supported but at least is of rapid decay:

$$\forall A \geqslant 0, \ \widehat{V}(y) \ll_{V,A} (1 + |y|)^{-A}.$$

Therefore the dual sum in (6.2) decays rapidly for $n \gg q/N$ and we obtain

PROPOSITION 6.5. *We have*

$$\sum_{n \in \mathbf{Z}} K(n)V(\frac{n}{N}) \ll_V q^{1/2}\|\widehat{K}\|_\infty \ll_{V,C(\mathcal{F})} q^{1/2}. \tag{6.3}$$

6.3. The Deligne-Laumon Fourier transform. The Fourier transform

$$K \mapsto \widehat{K} : y \to \frac{1}{q^{1/2}} \sum_{x \in \mathbf{F}_q} K(x)\mathrm{e}_q(-xy)$$

is a well-known and very useful operation on the space of function on $(\mathbf{Z}/q\mathbf{Z}, +)$. It serves to realize the spectral decomposition of the functions on $\mathbf{Z}/q\mathbf{Z}$ in terms of eigenvectors of the irreducible representations (characters) of $\mathbf{Z}/q\mathbf{Z}$. Let us recall that

– The Fourier transform is essentially involutive:

$$\widehat{\widehat{K}}(x) = K(-x);$$

stated otherwise, one has the Fourier inversion formula:

$$K(x) = \sum_{y \in \mathbf{F}_q} \widehat{K}(y) \mathrm{e}_q(yx).$$

– The Fourier transform is an isometry on $L^2(\mathbf{Z}/q\mathbf{Z})$; stated otherwise, one has the Plancherel formula

$$\sum_{x \in \mathbf{F}_q} K(x)\overline{K'(x)} = \sum_{y \in \mathbf{F}_q} \widehat{K}(y)\overline{\widehat{K'}(y)}.$$

– The Fourier transform behaves well with respect to to additive and multiplicative shifts: for $a \in \mathbf{F}_q$, $z \in \mathbf{F}_q^\times$,

$$\widehat{[+a]K}(y) = \mathrm{e}_q(ay)\widehat{K}(y),\ \widehat{[\times z]K}(y) = [\times z^{-1}]\widehat{K}(y) = \widehat{K}(y/z).$$

A remarkable fact, due to Deligne is that, to the Fourier transform for trace functions corresponds a "geometric Fourier transform" for sheaves. The following theorem is due to G. Laumon [**Lau87**]:

THEOREM 6.6. *Let $\mathcal{F}$ be a Fourier sheaf, lisse on U and pure of weight 0. There exists a Fourier sheaf $\widehat{\mathcal{F}}$, lisse on some open set $\widehat{U}$, pure of weight 0, such that if $K_{\mathcal{F},n}$ denotes the (middle-extension of the) trace function of $\mathcal{F}$, the (middle extension of the) trace function of $\widehat{\mathcal{F}}$ is given by the Fourier transform $\widehat{K_{\mathcal{F},n}}$ where*

$$\widehat{K_{\mathcal{F},n}}(x) = \frac{1}{q^{n/2}} \sum_y K_{\mathcal{F},n}(y)\mathrm{e}_q(\mathrm{tr}_{\mathbf{F}_{q^n}/\mathbf{F}_q}(xy)).$$

The map[7] *$\mathcal{F} \mapsto \widehat{\mathcal{F}}$ is called the geometric Fourier transform. The geometric Fourier transform satisfies (for $a \in \mathbf{F}_q$, $z \in \mathbf{F}_q^\times$)*

$$\widehat{\widehat{\mathcal{F}}} = [\times -1]^*\mathcal{F},\ \widehat{[+a]^*\mathcal{F}} = \mathcal{L}_{\mathrm{e}_q(a).} \otimes \widehat{\mathcal{F}},\ \widehat{[\times z]^*\mathcal{F}} = [\times z^{-1}]^*\widehat{\mathcal{F}}.$$

In addition, Laumon also defined local versions of the geometric Fourier transform making possible the computation of the local monodromy representations of $\widehat{\mathcal{F}}$ in terms of those of $\mathcal{F}$; using these results one deduces

PROPOSITION 6.7. *Given $\mathcal{F}$ as above, one has*

$$C(\widehat{\mathcal{F}}) \leqslant 10C(\mathcal{F})^2.$$

Also the Fourier transform preserves irreducibility:

PROPOSITION 6.8. *The Fourier transform maps irreducible (resp. isotypic) sheaves to irreducible (resp. isotypic) sheaves.*

PROOF. Given $\mathcal{F}$ a geometrically irreducible sheaf pure of weight 0, to prove that $\widehat{\mathcal{F}}$ is irreducible, it is enough to show (by Katz's irreducibility criterion) that

$$\limsup_n \mathcal{C}_n(\widehat{\mathcal{F}}, \widehat{\mathcal{F}}) = \limsup_n \frac{1}{q^n} \sum_{x \in \mathbf{F}_{q^n}} |\widehat{K_{\mathcal{F},n}}(x)|^2 = 1$$

[7] This is in fact a functor in the derived category of constructible ℓ-adic sheaves.

but by the Plancherel formula

$$\frac{1}{q^n}\sum_{x\in\mathbf{F}_{q^n}}|\widehat{K_{\mathcal{F},n}}(x)|^2=\frac{1}{q^n}\sum_{y\in\mathbf{F}_{q^n}}|K_{\mathcal{F},n}(y)|^2$$

and

$$\limsup_n\frac{1}{q^n}\sum_{y\in\mathbf{F}_{q^n}}|K_{\mathcal{F},n}(y)|^2=1$$

by Katz's irreducibility criterion applied in the reverse direction. □

EXERCISE 6.9. Prove that the hyper-Kloosterman sheaves are geometrically irreducible (hint: observe that the hyper-Kloosterman sums Kl_{k+1} can be expressed in terms of the Fourier transform of Kl_k).

7. Autocorrelation of trace functions; the automorphism group of a sheaf

The next couple of appplications we are going to discuss involve a special type of correlation sums between a trace function and its transform by an automorphism of the projective line.

Let $\mathcal{F}$ be an ℓ-adic sheaf lisse on $U\subset\mathbf{P}^1_{\mathbf{F}_q}$, pure of weight 0, geometrically irreducible but non trivial, with conductor $C(\mathcal{F})$. Let γ be an automorphism of $\mathbf{P}^1_{\mathbf{F}_q}$: γ is a fractional linear transformation:

$$\gamma\colon z\to\gamma\cdot z=\frac{az+b}{cz+d},\ \begin{pmatrix}a&b\\c&d\end{pmatrix}\in\mathrm{PGL}_2(\mathbf{F}_q).$$

Let $\gamma^*\mathcal{F}$ be the associated pull-back sheaf; it is lisse on $\gamma^{-1}\cdot U$ and its trace function is

$$\gamma^*K(z)=K(\gamma\cdot z)=K(\frac{az+b}{cz+d}).$$

Moreover since γ is an automorphism of $\mathbf{P}^1_{\mathbf{F}_q}$, one has $C(\gamma^*\mathcal{F})=C(\mathcal{F})$.

The correlations sums we will consider are those of K and $\gamma^*K(z)$

$$\mathcal{C}(\mathcal{F},\gamma):=\mathcal{C}(K,\gamma^*K)=\frac{1}{q}\sum_z K(z)\overline{K(\gamma\cdot z)}$$

and

$$\mathcal{C}_n(\mathcal{F},\gamma):=\mathcal{C}_n(K,\gamma^*K)=\frac{1}{q^n}\sum_{z\in\mathbf{F}_{q^n}}K_n(z)\overline{K_n(\gamma\cdot z)}$$

which are associated to the tensor product sheaf

$$\mathcal{F}\otimes\gamma^*D(\mathcal{F})$$

which is lisse on $U_\gamma=U\cap\gamma^{-1}\cdot U$.

7.1. The automorphism group. The question of the size of the sums $\mathcal{C}(\mathcal{F},\gamma)$ is largely determined by the following invariant of $\mathcal{F}$ (see [**FKM15**, **FKM14**])

DEFINITION 7.1. Given $\mathcal{F}$ as above, the group of automorphisms of $\mathcal{F}$, denoted $\mathrm{Aut}_{\mathcal{F}}(\mathbf{F}_q)\subset\mathrm{PGL}_2(\mathbf{F}_q)$, is the group of $\gamma\in\mathrm{PGL}_2(\mathbf{F}_q)$ such that

$$\gamma^*\mathcal{F}\simeq_{\mathrm{geom}}\mathcal{F}.$$

The group $\mathrm{Aut}_{\mathcal{F}}(\mathbf{F}_q)$ is the group of $\mathbf{F}_q$-points of an algebraic subgroup, $\mathrm{Aut}_{\mathcal{F}} \hookrightarrow \mathrm{PGL}_2$ defined over $\mathbf{F}_q$. Let $B \subset \mathrm{PGL}_2$ the subgroup generated by upper-triangular matrices; we define

$$B_{\mathcal{F}} := \mathrm{Aut}_{\mathcal{F}} \cap B$$

the subgroup of upper-triangular matrices of $\mathrm{Aut}_{\mathcal{F}}$ and $B_{\mathcal{F}}(\mathbf{F}_q)$ the group of $\mathbf{F}_q$-points.

The relevance of this notion for the above correlations sums is the following

PROPOSITION 7.2. *For $\gamma \notin \mathrm{Aut}_{\mathcal{F}}(\mathbf{F}_q)$, one has*

$$\mathcal{C}(K, \gamma) = O_{C(\mathcal{F})}(q^{-1/2}).$$

In view of this proposition it is important to determine $\mathrm{Aut}_{\mathcal{F}}(\mathbf{F}_q)$ and $B_{\mathcal{F}}(\mathbf{F}_q)$.

EXAMPLE 7.3. Obviously any element of $\mathrm{Aut}_{\mathcal{F}}$ has to leave $\mathbf{P}^1(\overline{\mathbf{F}_q}) - U(\overline{\mathbf{F}_q})$ invariant and all the points in the same orbit have isomorphic local monodromies. This may impose very strong constraints on $\mathrm{Aut}_{\mathcal{F}}$.

- If $\mathcal{F}$ is geometrically trivial then $\mathrm{Aut}_{\mathcal{F}} = \mathrm{PGL}_2$.
- If $\psi\colon (\mathbf{F}_q, +) \to \mathbf{S}^1$ is non trivial then $G_{\mathcal{L}_\psi} = N = \{\begin{pmatrix} 1 & x \\ & 1 \end{pmatrix} \subset \mathrm{PGL}_2\}$.
- If $\chi\colon (\mathbf{F}_q, +) \to \mathbf{S}^1$ is non trivial, then
$$G_{\mathcal{L}_\chi} = T^{0,\infty} = \{\begin{pmatrix} a & 0 \\ 0 & d \end{pmatrix} \subset \mathrm{PGL}_2\}$$
is the diagonal torus, unless χ is quadratic in which case $G_{\mathcal{L}_\chi} = N(T^{0,\infty})$ is the normalizer of the diagonal torus.
- For the Kloosterman sheaves, one can show that $\mathcal{G}_{\mathcal{K}\ell_k}$ is trivial: since $\mathcal{K}\ell_k$ is not lisse at 0 and ∞, with Swan conductor 0 at 0 and 1 at ∞, one has $\mathcal{G}_{\mathcal{K}\ell_k} \subset T^{0,\infty}$. One can then show (see [**Mic98**]) that $[\times a]^*\mathcal{K}\ell_k \simeq_{\mathrm{geom}} \mathcal{K}\ell_k$ iff $a = 1$.

Given $x \neq y \in \mathbf{P}^1(\overline{\mathbf{F}_q})$, we denote by $T^{x,y}$ the pointwise stabilizer of the pair (x, y) (this is a maximal torus defined over some finite extension of $\mathbf{F}_q$) and $N(T^{x,y})$ its normalizer. The torus $T^{x,y}$ is defined over $\mathbf{F}_q$ if x, y belong to $\mathbf{P}^1(\mathbf{F}_q)$ or if x, y belong to $\mathbf{P}^1(\mathbf{F}_{q^2})$ and are Galois conjugates.

PROPOSITION 7.4. *Suppose $q \geqslant 7$. Given $\mathcal{F}$ as above, at least one of the following holds:*

- *$C(\mathcal{F}) > q$.*
- *q does not divide $|\mathrm{Aut}_{\mathcal{F}}(\mathbf{F}_q)|$ and either $\mathrm{Aut}_{\mathcal{F}}(\mathbf{F}_q)$ is of order $\leqslant 60$ or is a subgroup of the normalizer of some maximal torus $N(T^{x,y})$ defined over $\mathbf{F}_q$.*
- *q divides $|\mathrm{Aut}_{\mathcal{F}}(\mathbf{F}_q)|$ and then $\mathcal{F} \simeq \sigma^*\mathcal{L}_\psi$ for some ψ and $K(x) = \alpha\psi(\sigma.x)$ for for some $\sigma \in \mathrm{PGL}_2(\mathbf{F}_q)$ and $\mathrm{Aut}_{\mathcal{F}}(\mathbf{F}_q) = \sigma N \sigma^{-1}$.*

REMARK 7.5. Observe that in the last case

$$\mathcal{C}(K, \gamma) = |K(0)|^2 \mathcal{C}(\psi(\sigma.x), \gamma)$$

Concerning the size of the group $B_{\mathcal{F}}(\mathbf{F}_q)$, one can show that

THEOREM 7.6. *Let $\mathcal{F}$ be an isotypic sheaf whose geometric components are not isomorphic to $[+x]^*\mathcal{L}_\chi$ for some $x \in \mathbf{F}_q$ and some multiplicative character χ and such that*

$$C(\mathcal{F}) < q.$$

Then

$$|B_{\mathcal{F}}(\mathbf{F}_q)| \leqslant C(\mathcal{F}).$$

The proof of this theorem involves the following rigidity statements [**Kat96**, Lemma 2.6.13]:

PROPOSITION 7.7. *Let $\mathcal{L}$ be geometrically irreducible.*

- *If for some $x \in \mathbf{F}_q^\times$, $[+x]^*\mathcal{L} \simeq \mathcal{L}$, then either*

$$C(\mathcal{L}) > q \text{ or } \mathcal{L} \simeq \mathcal{L}_\psi \text{ for some } \psi.$$

- *If $\mathrm{Aut}_{\mathcal{L}}(\mathbf{F}_q)$ contains a subgroup of order m of diagonal matrices then either*

$$c(\mathcal{L}) > m \text{ or } \mathcal{L} \simeq \mathcal{L}_\chi \text{ for some } \chi.$$

8. Trace functions vs. primes

Another possible question to consider (natural from the viewpoint of analytic number theory at least) is how trace functions correlate with the characteristic function of the primes. In this section, we discuss the structure of the proof of the following result:

THEOREM 8.1 (Trace function vs. primes, [**FKM14**]). *Let $\mathcal{F}$ be a geometrically isotypic sheaf of conductor $C(\mathcal{F})$ whose geometric components are not of the shape $\mathcal{L}_\psi \otimes \mathcal{L}_\chi$ and let K its associated trace function. For any $V \in C_c^\infty(\mathbf{R}_{>0})$, one has*

$$\sum_{\substack{p \text{ prime} \\ p \leqslant X}} K(p) \ll X(1 + q/X)^{1/12} p^{-\eta/2}, \tag{8.1}$$

$$\sum_{p \text{ prime}} K(p) V\Big(\frac{p}{X}\Big) \ll X(1 + q/X)^{1/6} q^{-\eta}, \tag{8.2}$$

for $X \ll q$ and $\eta < 1/24$. The implicit constants depend only on η, $C(\mathcal{F})$ and V. Moreover, the dependency on $C(\mathcal{F})$ is at most polynomial.

REMARK 8.2. This result exhibits cancellations when summing trace functions along the primes in intervals of length larger than $q^{3/4}$. It is really a pity that Dirichlet characters are excluded by our hypotheses: such a bound in that case would amount to a quasi generalized Riemann hypothesis for the corresponding Dirichlet character L-function !

We discuss the proof for $X = q$.

8.1. Combinatorial decomposition of the characteristic function of the primes. As is well-known, the problem is equivalent to bounding the sum

$$\sum_n \Lambda(n) K(n) V\Big(\frac{n}{q}\Big)$$

where

$$\Lambda(n) = \begin{cases} \log p & \text{if } n = p^\alpha \ \alpha \geqslant 1 \\ 0 & \text{otherwise,} \end{cases}$$

is the von Mangoldt function. A standard method in analytic number theory is a combinatorial decomposition of this function as a sum of Dirichlet convolutions; one way to achieve this is to use the celebrated Heath-Brown identity:

LEMMA 8.3 (Heath-Brown). *For any integer $J \geqslant 1$ and $n < 2X$, we have*

$$\Lambda(n) = -\sum_{j=1}^{J}(-1)^j\binom{J}{j}\sum_{m_1,\ldots,m_j\leqslant Z}\mu(m_1)\ldots\mu(m_j)\sum_{m_1\ldots m_j n_1\ldots n_j=n}\log n_1,$$

where $Z = X^{1/J}$.

Hence splitting the range of summation of the various variables appearing (using partition of unity) and separating these variables, our preferred sum decomposes (essentially) into $O((\log X)^{2J})$ sums of the shape

$$\Sigma(M_1,\ldots,M_{2j}) = \sum_{m_1,\ldots m_{2j}}\sum \mu(m_1)\ldots\mu(m_j)K(m_1.\ldots.m_{2j})V_1\Big(\frac{m_1}{M_1}\Big)\ldots V_{2j}\Big(\frac{m_{2j}}{M_{2j}}\Big)$$

for $j \leqslant J$; here V_i, $i = 1,\ldots 2j$ are smooth functions compactly supported in $]1, 2[$, and $(M_1,\ldots,M_{2j})$ is a tuple satisfying

$$M_i =: q^{\mu_i},\ \forall i \leqslant j,\ \mu_i \leqslant 1/J,\ \sum_{i\leqslant 2j}\mu_i = 1 + o(1);$$

The objective is to show that

$$\Sigma(M_1,\ldots,M_{2j}) \ll q^{1-\eta}$$

for some fixed $\eta > 0$. We will take $J = 3$ so that $Z = q^{1/3}$. We may assume that

$$\mu_1 \leqslant \ldots \leqslant \mu_j \leqslant 1/3,\ \mu_{j+1} \leqslant \ldots \leqslant \mu_{2j}.$$

We will bound these sums differently depending on the vector $(\mu_1,\ldots,\mu_{2j})$.

Let $0 < \delta < 1/6$ be some small but fixed parameter to be chosen optimally later.

(1) Suppose that $\mu_{2j} \geqslant 1/2+\delta$. Then m_{2j} is a long "smooth variable" (because the weight attached to it is smooth); therefore using (6.3) to sum over m_{2j} while fixing the other variables, we get

$$\Sigma(M_1,\ldots,M_{2j}) \ll q^{\mu_1+\ldots\mu_{2j-1}}q^{1/2+o(1)} = q^{1-\delta+o(1)}.$$

(In the literature, sum of that shape are called "type I" sums).

(2) We may therefore assume that

$$m_{j+1} \leqslant \ldots \leqslant \mu_{2j} \leqslant 1/2 + \delta;$$

in other words, there is no long smooth variable. What one can then do is to group variables together to form longer ones: for this one partitions the indexing set into two blocks

$$\{1,\ldots,2j\} = \mathcal{I} \sqcup \mathcal{I}',$$

and form the variables

$$m = \prod_{i\in\mathcal{I}} m_i,\ n = \prod_{i'\in\mathcal{I}'} m_{i'}$$

so that denoting by α_m the Dirichlet convolutions of either $\mu(\cdot)V(\frac{\cdot}{M_i})$ or $V(\frac{\cdot}{M_i})$ for $i \in \mathcal{I}$ and similarly for β_n for $i' \in \mathcal{I}'$, we are led to bound bilinear sums of the shape

$$B(K;\alpha,\beta) = \sum\sum_{m\ll M, n\ll N} \alpha_m\beta_n K(mn). \tag{8.3}$$

where

$$M = q^{\mu},\ \mu = \sum_{i\in\mathcal{I}} \mu_i,\ N = q^{\nu},\ \nu = \sum_{i'\in\mathcal{I}'} \mu_{i'}.$$

The weights α_m, β_n are rather irregular and it is difficult to exploit their structure (such sums are called "type II").

Assuming that the irreducible component of $\mathcal{F}$ is not of the shape $\mathcal{L}_\chi \otimes \mathcal{L}_\psi$, we will prove in Theorem 9.1 below the following bound

$$\Sigma(M_1,\ldots,M_{2j}) = B(K;\alpha,\beta) \ll_{C(\mathcal{F})} \|\alpha_M\|_2\|\beta_N\|_2(MN)^{1/2}(\frac{1}{M} + \frac{q^{1/2}\log q}{N})^{1/2}.$$

Assuming that

$$\mu \geqslant \delta \text{ and } \nu \geqslant 1/2+\delta$$

we obtain that

$$B(K;\alpha,\beta) \ll q^{1-\delta/2+o(1)}.$$

(3) It remains to treat the sums for which neither $\mu_{2j} \leqslant 1/2+\delta$ nor a decomposition as in (2) exist. This necessarily implies that $\sum_{i\leqslant j}\mu_i \leqslant 1/3$, $j \geqslant 2$ and $\mu_{2j-1}+\mu_{2j} \geqslant 1-\delta$. Setting $M = M_{2j-1}$ and $N = M_{2j}$, denoting

$$a = m_1\ldots m_{2j-2} \ll q^{\delta},$$

it will be sufficient to obtain a bound of the shape

$$\sum_{m,n\geqslant 1} K(amn)V(\frac{m}{M})W(\frac{n}{N}) \ll_{V,W} (MN)^{1-\eta}$$

for some $\eta > 0$ whenever MN is sufficiently close to q. What we have are is a sum involving two smooth variables which are however too short for the Pólya-Vinogradov method to work, but whose product is rather long. We call these sums "type I1/2". We will then use Theorem 8.4 below whose proof is discussed in §10. Observe that this theorem provides a bound which is non trivial as long as $MN \geqslant q^{3/4}$.

(4) Optimizing parameters in these three approaches leads to Theorem 8.1.

THEOREM 8.4. *Let $\mathcal{F}$ be a geometrically isotypic Fourier sheaf of conductor $C(\mathcal{F})$ and K its associated trace function. For any $V, W \in C_c^\infty(\mathbf{R}_{>0})$, any $M, N \geqslant 1$ and any $\eta < 1/8$, one has*

$$\sum_{m,n\geqslant 1} K(mn)V(\frac{m}{M})W(\frac{n}{N}) \ll_{V,W,C(\mathcal{F})} MN(1+\frac{q}{MN})^{1/2}q^{-\eta/2}.$$

9. Bilinear sums of trace functions

Let K be a trace function associated to some isotypic sheaf $\mathcal{F}$, pure of weight 0 and let $(\alpha_m)_{m\leqslant M}$, $(\beta_n)_{n\leqslant N}$ be arbitrary complex numbers. In this section, we bound the "type II" bilinear sums encountered in the previous section :

$$B(K;\alpha,\beta) = \sum\sum_{m\leqslant M, n\leqslant N} \alpha_m\beta_n K(mn).$$

Using the Cauchy-Schwarz inequality, the trivial bound is

$$|B(K;\alpha,\beta)| \ll_{C(\mathcal{F})} \|\alpha_M\|_2\|\beta_N\|_2(MN)^{1/2}.$$

We wish to improve over this bound.

THEOREM 9.1 (Bilinear sums of trace functions). *Notations as above; assume that $1 \leqslant M, N < q$ and that the irreducible component of $\mathcal{F}$ is not of the shape $\mathcal{L}_\chi \otimes \mathcal{L}_\psi$. Then*

$$B(K;\alpha,\beta) \ll_{C(\mathcal{F})} \|\alpha_M\|_2\|\beta_N\|_2(MN)^{1/2}\Big(\frac{1}{M}+\frac{q^{1/2}\log q}{N}\Big)^{1/2}.$$

REMARK 9.2. This bound is non-trivial as soon as $M \gg 1$ and $N \gg q^{1/2}\log q$.

PROOF. By Cauchy-Schwarz, we have

$$|B(K;\alpha,\beta)|^2 \leqslant \|\beta_N\|_2^2 \sum_{m_1,m_2\leqslant M} \alpha_{m_1}\overline{\alpha_{m_2}} \sum_{n\leqslant N} K(m_1n)\overline{K}(m_2n). \tag{9.1}$$

We do not expect to gain anything from the diagonal terms $m_1 \equiv m_2 \,(\mathrm{mod}\, q)$ (equivalently, $m_1 = m_2$ since $M < q$) and the contribution of such terms is bounded trivially by

$$\ll_{C(\mathcal{F})} \|\alpha_M\|_2^2\|\beta_N\|_2^2 N. \tag{9.2}$$

As for the non-diagonal terms, their contribution is

$$\|\beta_N\|_2^2 \sum_{m_1\not\equiv m_2\,(\mathrm{mod}\,q)} \alpha_{m_1}\overline{\alpha_{m_2}} \sum_{n\leqslant N} K(m_1n)\overline{K}(m_2n).$$

Using the Pólya-Vinogradov method, we are led to evaluate the Fourier transform of

$$n \mapsto K(m_1n)\overline{K}(m_2n).$$

By the Plancherel formula, this Fourier transform equals

$$\begin{aligned} y \mapsto \frac{1}{q^{1/2}}\sum_{x\in\mathbf{F}_q} K(m_1x)\overline{K}(m_2x)\mathrm{e}_q(-yx) &= \frac{1}{q^{1/2}}\sum_{z\in\mathbf{F}_q} \widehat{K}((z-y)/m_1)\overline{\widehat{K}}(z/m_2) \\ &= \frac{1}{q^{1/2}}\sum_{z\in\mathbf{F}_q} \widehat{K}((m_2z-y)/m_1)\overline{\widehat{K}}(z) \\ &= \frac{1}{q^{1/2}}\sum_{z\in\mathbf{F}_q} \widehat{K}(\gamma z)\overline{\widehat{K}}(z) \end{aligned}$$

with

$$\gamma = \begin{pmatrix} m_2/m_1 & -y/m_1 \\ 0 & 1 \end{pmatrix} \in B(\mathbf{F}_q).$$

This sum is $q^{1/2}$ times $\mathcal{C}(\widehat{\mathcal{F}},\gamma)$, the correlation sum associated to the isotypic sheaves $\widehat{\mathcal{F}}$ and $\gamma^*\widehat{\mathcal{F}}$, whose conductors are controlled in terms of $C(\mathcal{F})$.

If $\gamma \notin B_{\mathcal{F}}(\mathbf{F}_q)$ we have

$$\mathcal{C}(\widehat{\mathcal{F}},\gamma) \ll_{C(\mathcal{F})} \frac{1}{q^{1/2}}. \tag{9.3}$$

The condition that the irreducible component of $\mathcal{F}$ is not of the shape $\mathcal{L}_\chi \otimes \mathcal{L}_\psi$ translates into the irreducible component of $\widehat{\mathcal{F}}$ not being of the shape $[+x]^*\mathcal{L}_{\overline{\chi}}$. In that case, by Theorem 7.6, there is a set $S_{\mathcal{F}} \subset \mathbf{F}_q^\times$ such that for any $(m_1, m_2, y) \in \mathbf{F}_q^\times \times \mathbf{F}_q^\times \times \mathbf{F}_q$ for which $m_2/m_1 \notin S_{\mathcal{F}}$ one has

$$\mathcal{C}(\widehat{\mathcal{F}}, \gamma) \ll_{C(\mathcal{F})} q^{-1/2}.$$

Returning to (9.1), we bound trivially (by (9.2)) the contribution of the $O_{\mathcal{F}}(M)$ (m_1, m_2) such that the ratio $m_2/m_1 \pmod q$ is in $S_{\mathcal{F}}$. For the other terms, we may use the Pólya-Vinogradov method and bound these terms by

$$\ll_{C(\mathcal{F})} \|\alpha_M\|_2^2 \|\beta_N\|_2^2 M q^{1/2} \log q.$$

Combining these bounds leads to the final result. □

10. Trace functions vs. modular forms

In this section we discuss the proof of Theorem 8.4. This theorem is a special case of the resolution of another problem: the question of the correlation between trace functions and the Fourier coefficients $(\varrho_f(n))_n$ of some modular Hecke eigenform (cf. [**IK04**, Chap. 14&15] and references herein for a quick introduction to the theory modular forms). Given some trace function, we consider the correlation sum

$$\mathcal{S}(K, f; X) := \sum_{n \leqslant X} \varrho_f(n) K(n)$$

or its smoothed version

$$\mathcal{S}_V(K, f; X) := \sum_{n} \varrho_f(n) K(n) V(\frac{n}{X}).$$

These sums are bounded (using the Rankin-Selberg method) by

$$O_{C(\mathcal{F}), f}(X \log^3 X).$$

It turns out that the problem of bounding $\mathcal{S}(K, f; X)$ and $\mathcal{S}_V(K, f; X)$ non-trivially is most interesting when N is of size q or smaller.

In this section, we sketch the proof of the following

THEOREM 10.1 (Trace function vs. modular forms, [**FKM15**]). *Let $\mathcal{F}$ be an irreducible Fourier sheaf of weight* 0 *and K its associated trace function. Let $(\varrho_f(n))_{n \geqslant 1}$ be the sequence of Fourier coefficients of some modular form f with trivial nebentypus and $V \in C_c^\infty(\mathbf{R}_{>0})$. For $X \geqslant 1$ and any $\eta < 1/8$, we have*

$$\mathcal{S}(K, f; X) \ll X(1 + \frac{q}{X})^{1/2} q^{-\eta/2},$$

and

$$\mathcal{S}_V(K, f; X) \ll X(1 + \frac{q}{X})^{1/2} q^{-\eta}.$$

The implicit constants depend only on η, f, $C(\mathcal{F})$ and V. Moreover, the dependency on $C(\mathcal{F})$ is at most polynomial.

This result shows the absence of correlation when $X \gg q^{1-1/8}$. The proof, which uses the amplification method and the Petersson-Kuznetzov trace formula, will ultimately be a consequence of Theorem 7.4.

We give below an idea of the proof. To simplify matters, we will assume that $X = q$ and we wish to bound non-trivially the sum

$$\mathcal{S}_V(K, f) := \sum_{n \geqslant 1} \varrho_f(n) K(n) V(\frac{n}{q}) \tag{10.1}$$

for V a fixed smooth function. Moreover, to simplify things further, we will assume that f has level 1 and is cuspidal and holomorphic of very large (but fixed) weight.

10.1. Trace functions vs. the divisor function. An important special case of Theorem 10.1 is when f is an Eisenstein series, for instance when

$$f(z) = \frac{\partial}{\partial s}E(z,s)_{|s=1/2} \text{ for } E(z,s) = \frac{1}{2}\sum_{(c,d)=1}\frac{y^s}{|cz+d|^{2s}}$$

is the non-holomorphic Eisenstein series at the central point. In that case we have

$$\varrho_f(n) = d(n)$$

the divisor function, and so one has

$$\sum_{m,n\geqslant 1} K(mn)V(\frac{mn}{X}) \ll_{V,C(\mathcal{F})} X(1+\frac{q}{X})^{1/2}q^{-\eta} \tag{10.2}$$

whenever K is the trace function of a Fourier sheaf. This bound holds similarly for the unitary Eisenstein series $E(z,s)$ at any $s=\frac{1}{2}+it$, where the divisor function is replaced by

$$d_{it}(n) = \sum_{ab=n}(a/b)^{it}.$$

Such general bounds make it possible to separate the variables m,n in (10.2) and eventually to prove Theorem 8.4.

REMARK 10.2. As we will see below, the proof of Theorem 10.1 is not a "modular form by modular form" analysis; instead the proof is global, involving the full automorphic spectrum, and establishes the required bound "for all modular forms f at once", including Eisenstein series and therefore proving Theorem 8.4 on the way.

10.2. Functional equations. Our first objective is to understand why the range $X=q$ is interesting. This come from the functional equations satisfied by modular forms as a consequence of their automorphic properties. These equations present themselves in various shapes. One is the Voronoi summation formula, which in its simplest form is the following:

PROPOSITION 10.3 (Voronoi summation formula). *Let f be a holomorphic modular form of weight k and level 1 with Fourier coefficients $(\varrho_f(n))_n$. Let V be a smooth compactly supported function, $q\geqslant 1$ and $(a,q)=1$. We have for $X>0$*

$$\sum_{n\geqslant 1}\varrho_f(n)V\Big(\frac{n}{X}\Big)e\Big(\frac{an}{q}\Big) = \varepsilon(f)\frac{X}{q}\sum_{n\geqslant 1}\varrho_f(n)e\Big(-\frac{\bar{a}n}{q}\Big)\widetilde{V}\Big(\frac{Xn}{q^2}\Big)$$

where $\varepsilon(f)=\pm 1$ denotes the sign of the functional equation of $L(f,s)$, and

$$\widetilde{V}(y) = \int_0^\infty V(u)\mathcal{J}_k(4\pi\sqrt{uy})du,$$

with

$$\mathcal{J}_k(u) = 2\pi i^k J_{k-1}(u),$$

where

$$J_{k-1}(x) = \sum_{l=0}^{\infty}\frac{(-1)^l}{l!(l+k-1)!}(\frac{x}{2})^{2l+k-1}$$

is the Bessel function of order $k-1$.

There are several possible proofs of this proposition: one can proceed classically from the Fourier expansion of the modular form f using automorphy relations (see [**KMV02**, Theorem A.4]). Another more conceptual approach is to use the Whittaker model of the underlying automorphic representation; this approach extends naturally to higher rank automorphic forms (see [**IT13**]). One could also point out other related works like [**MS06**] as well as the recent paper [**KZ16**]. We can extend this formula to general functions modulo q. Given $K\colon \mathbf{Z} \to \mathbf{C}$ a q-periodic function, we define its *Voronoi transform* $\check{K}$ of K as

$$\check{K}(n) = \frac{1}{\sqrt{q}} \sum_{\substack{h \bmod q \\ (h,q)=1}} \widehat{K}(h)\mathrm{e}_q(\overline{h}n) = \frac{1}{\sqrt{q}} \sum_{\substack{h \bmod q \\ (h,q)=1}} \widehat{K}(h^{-1})\mathrm{e}_q(hn).$$

Combining the above formula with the Fourier decomposition

$$K(n) = \frac{1}{q^{1/2}} \sum_{a \,(\mathrm{mod}\, q)} \widehat{K}(a)\mathrm{e}_q(-an),$$

we get

COROLLARY 10.4. *Notations are above, given K a q-periodic arithmetic function, we have for $X > 0$*

$$\begin{aligned}\sum_{n\geqslant 1} \varrho_f(n)K(n)V\Big(\frac{n}{X}\Big) &= \frac{\widehat{K}(0)}{q^{1/2}} \sum_{n\geqslant 1} \varrho_f(n)V\Big(\frac{n}{X}\Big) + \\ &\quad \varepsilon(f)\frac{X}{q} \sum_{n\geqslant 1} \varrho_f(n)\check{K}(-n)\widetilde{V}\Big(\frac{nX}{q^2}\Big).\end{aligned}$$

REMARK 10.5. Another way to obtain such result is to consider the Mellin transform of (the restriction to $\mathbf{F}_q^\times$ of) K:

$$\tilde{K}(\chi) = \frac{1}{(q-1)^{1/2}} \sum_{x\in\mathbf{F}_q^\times} K(x)\chi(x)$$

so that for $x \in \mathbf{F}_q^\times$

$$K(x) = \frac{1}{(q-1)^{1/2}} \sum_{\chi} \tilde{K}(\chi)\chi^{-1}(x).$$

One can then use the (archimedean) inverse-Mellin transform and the functional equation satisfied by the Hecke L-function

$$L(f\otimes\chi, s) = \sum_{n\geqslant 1} \frac{\varrho_f(n)\chi(n)}{n^s}$$

to obtain the formula. For this, one observes that the Mellin transform of $\check{K}_{|\mathbf{F}_q^\times}$ is proportional to

$$\chi \mapsto \varepsilon(\chi)\tilde{K}(\chi^{-1})$$

where $\varepsilon(\chi)$ is the normalized Gauss sum. This method extends easily to automorphic forms of higher rank but uses the fact that q is prime (so that $\mathbf{F}_q^\times$ is not much smaller that $\mathbf{F}_q$).

The identity of Corollary 10.4 is formal and has nothing to do whether K is a trace function or not. In particular applying it to the Dirac function $\delta_a(n) = \delta_{n\equiv a\,(\mathrm{mod}\,q)}$, for some $a \in \mathbf{F}_q^\times$ we obtain

$$\widehat{\delta_a}(h) = \frac{1}{q^{1/2}}\mathrm{e}_q(ah),\ \check{\delta}_a(n) = \frac{1}{q^{1/2}}\,\mathrm{Kl}_2(an;q)$$

so that

$$q^{1/2}\sum_{n\equiv a\,(\mathrm{mod}\,q)} \varrho_f(n)V\Big(\frac{n}{X}\Big) = \frac{1}{q^{1/2}}\sum_{n\geqslant 1}\varrho_f(n)V\Big(\frac{n}{X}\Big) + \varepsilon(f)\frac{X}{q}\sum_{n\geqslant 1}\varrho_f(n)\,\mathrm{Kl}_2(-an;q)\widetilde{V}\Big(\frac{nX}{q^2}\Big). \tag{10.3}$$

This is an example of a natural transformation which, starting from the elementary function δ_a produces a genuine trace function (Kl_2).

Besides this case we would like to use the formula for K a trace function. We observe that the Voronoi transform $\check{K}$ is "essentially" the Fourier transform of the function

$$h \in \mathbf{F}_q^\times \mapsto \widehat{K}(h^{-1}) = \widehat{K}(w\cdot h)$$

with $w = \begin{pmatrix} 0 & 1\\ 1 & 0\end{pmatrix}$; it is therefore essentially involutive. It would be useful to know that $\check{K}$ is a trace function. Suppose that K is associated to some isotypic Fourier sheaf $\mathcal{F}$, then $\check{K}$ is a (isotypic) trace function as long as $w^*\widehat{\mathcal{F}}$ is a Fourier sheaf. This means that $\widehat{\mathcal{F}}$ has no irreducible constituent of the shape $w^*\mathcal{L}_\psi$ which (by involutivity of the Fourier transform means that $\mathcal{F}$ has no irreducible constituent isomorphic to some Kloosterman sheaf $\mathcal{K}\ell_2$. This reasoning[8] is essentially the reverse of the one leading to (10.3).

Let us assume that $\check{K}$ is also a trace function. Then, integration by parts show that for V smooth and compactly supported, $\widetilde{V}(x)$ has rapid decay for $x \gg 1$. Hence Corollary 10.4 is an equality between a sum of length X and a sum of length about q^2/X (up to the term $\frac{\widehat{K}(0)}{q^{1/2}}\sum_{n\geqslant 1}\varrho_f(n)V\big(\frac{n}{X}\big)$ which is easy to understand). The two lengths are the same when $X = q$.

10.3. The amplification method. As mentioned above Theorem 10.1 is proven "for all modular forms at one" as a consequence of the amplification method.

The principle of the amplification method (invented by H. Iwaniec and which in the special case $K = \chi$ was used first by Bykovskii) consist, in the following. For $L \geqslant 1$ and $(x_l)_{l\leqslant L}$ real numbers we consider the following average over orthogonal bases of modular forms (holomorphic or general) of level q:

$$M_k(K) := \sum_{g\in\mathcal{B}_k(q)} |A(g)|^2|\mathcal{S}_V(g,K)|^2 \tag{10.4}$$

[8] by involutivity of the Voronoi transform

(cf. (10.1) for the definition of $\mathcal{S}_V(g,K)$) and

$$(10.5)\quad M(K) := \sum_{k\equiv 0\,(\mathrm{mod}\,2),\ k>0} \dot{\phi}(k)(k-1) \sum_{g\in\mathcal{B}_k(q)} |A(g)|^2|\mathcal{S}_V(g,K)|^2 + \sum_{g\in\mathcal{B}(q)} \tilde{\phi}(t_g)\frac{4\pi}{\cosh(\pi t_g)}|A(g)|^2|\mathcal{S}_V(g,K)|^2 + \sum_{g\in\mathcal{B}_E(q)} \int_{-\infty}^{\infty} \tilde{\phi}(t)\frac{1}{\cosh(\pi t)}|A(g,t)|^2|\mathcal{S}_V(E_g(t),K)|^2\,dt,$$

where $\mathcal{B}_k(q)$, $\mathcal{B}(q)$, $\mathcal{B}_E(q)$ denote orthonormal bases of Hecke eigenforms of level q (either holomorphic of weight k or Maass or Eisenstein series), $\dot{\phi}$, $\tilde{\phi}$ are weights constructed from some smooth function, ϕ, rapidly decreasing at 0 and ∞, which depend only on the spectral parameters of the forms and for each form g, $A(g)$ ("A" is for amplifier) is the linear form in the Hecke eigenvalues $(\lambda_g(n))_{(n,q)=1}$ given by

$$A(g) = \sum_{l\leqslant L} x_l\lambda_g(l).$$

The weights $\tilde{\phi}$ are positive while the weight $\dot{\phi}(k)$ is positive at least for k large enough; one can then add to this quantity a finite linear combination of the $M_k(K)$, $k \ll 1$ from which one can bound

$$(10.6)\quad |M|(K) := \sum_{k\equiv 0\,(\mathrm{mod}\,2),\ k>0} |\dot{\phi}(k)|(k-1) \sum_{g\in\mathcal{B}_k(q)} |A(g)|^2|\mathcal{S}_V(g,K)|^2 + \sum_{g\in\mathcal{B}(q)} \tilde{\phi}(t_g)\frac{4\pi}{\cosh(\pi t_g)}|A(g)|^2|\mathcal{S}_V(g,K)|^2 + \sum\sum_{g\in\mathcal{B}_E(q)} \int_{-\infty}^{\infty} \tilde{\phi}(t)\frac{1}{\cosh(\pi t)}|A(g,t)|^2|\mathcal{S}_V(E_g(t),K)|^2\,dt.$$

As we explain below one will be able to prove the following bound

$$(10.7)\qquad M(K), M_k(K) \ll_{C(\mathcal{F})} q^{o(1)}(q\sum_{l\leqslant L}|x_l|^2 + q^{1/2}L(\sum_{l\leqslant L}|x_l|)^2).$$

Now if f is a Hecke-eigenform of level 1 (of L^2 norm 1 for the usual inner product on the level one modular curve) then $f/(q+1)^{1/2}$ embeds in an orthonormal basis of forms of level q.

Since all the terms in $|M|(K)$ are non-negative, this sums bounds any of its terms occurring discretely (i.e. when f is a cusp form). Therefore we obtain

$$\frac{1}{q+1}|A(f)|^2|\mathcal{S}_V(f,K)|^2 \ll_{C(\mathcal{F}),f} q^{o(1)}(q\sum_{l\leqslant L}|x_l|^2 + q^{1/2}L(\sum_{l\leqslant L}|x_l|)^2).$$

Now we perform amplification by choosing some bounded sequence $(x_l)_{l\leqslant L}$ tailor made for f such that $A(f)$ is "large". Specifically, choosing

$$x_l = \mathrm{sign}(\lambda_f(l)),$$

we obtain

$$|A(f)| \gg L^{1+o(1)}.$$

Dividing by L we obtain

$$|\mathcal{S}_V(f,K)|^2 \ll q^{o(1)}(q^2/L + q^{3/2}L^2)$$

and the optimal choice is $L = q^{1/6}$ giving us

$$\mathcal{S}_V(f,K) \ll q^{1-1/12+o(1)}.$$

10.4. Computing the moments. We now bound $M(K)$. Opening squares and using the multiplicative properties of Hecke eigenvalues, we are essentially reduced to bounding sums of the shape

$$\sum\sum_{m,n} V(\frac{m}{q})V(\frac{n}{q})K(m)\overline{K(n)}\Delta_{q,\phi}(lm,n) \tag{10.8}$$

and

$$\sum\sum_{m,n} V(\frac{m}{q})V(\frac{n}{q})K(m)\overline{K(n)}\Delta_{q,k}(lm,n) \tag{10.9}$$

where $1 \leqslant l \leqslant L^2$,

$$\Delta_{q,k}(lm,n) = \sum_{g\in\mathcal{B}_k(q)} \varrho_g(lm)\overline{\varrho_g(n)}$$

and

$$\begin{aligned}\Delta_{q,\phi}(lm,n) \quad &= \sum_{k\equiv 0\,(\mathrm{mod}\,2),\ k>0} \dot{\phi}(k)(k-1)\sum_{g\in\mathcal{B}_k(q)} \varrho_g(lm)\overline{\varrho_g(n)}\\ &+ \sum_{g\in\mathcal{B}(q)} \tilde{\phi}(t_g)\frac{4\pi}{\cosh(\pi t_g)}\varrho_g(lm)\overline{\varrho_g(n)}\\ &+ \sum_{g\in\mathcal{B}_E(q)} \int_{-\infty}^{\infty} \tilde{\phi}(t)\frac{1}{\cosh(\pi t)}\varrho_g(lm,t)\overline{\varrho_g(n,t)}\,dt.\end{aligned}$$

The Petersson-Kuznetzov formula expresses $\Delta_{q,k}(m,n)$ $\Delta_{q,\phi}(m,n)$ as sums of Kloosterman sums:

$$\Delta_{q,k}(m,n) = \delta_{m=n} + 2\pi i^{-k}\sum_c \frac{1}{cq}S(m,n;cq)J_{k-1}\left(\frac{4\pi\sqrt{mn}}{cq}\right). \tag{10.10}$$

and

$$\Delta_{q,\phi}(m,n) = \sum_c \frac{1}{cq}S(m,n;cq)\phi\left(\frac{4\pi\sqrt{mn}}{cq}\right), \tag{10.11}$$

where

$$S(m,n;cq) = \sum_{(x,cq)=1} e\left(\frac{mx+n\overline{x}}{cq}\right)$$

is the non-normalized Kloosterman sum of modulus cq (where $x.\overline{x} \equiv 1\,(\mathrm{mod}\,cq)$). In (10.9), because m and n are of size q and ϕ is rapidly decreasing at 0, the contribution of the $c \gg l^{1/2}$ is small. We will simplify further by evaluating only the contribution of $c = 1$, that is

$$\frac{1}{q}\sum\sum_{m,n} V(\frac{m}{q})V(\frac{n}{q})K(m)\overline{K(n)}S(lm,n;q)\phi(\frac{4\pi\sqrt{lmn}}{q}).$$

Our next step is to open the Kloosterman sum and apply the Poisson summation formula on the m and n variables. We obtain

$$\frac{1}{q}\frac{q^2}{(q^{1/2})^2}\sum\sum_{m^*,n^*}\widehat{W}(m^*,n^*)\sum_{x\in\mathbf{F}_q^\times}\widehat{K}(lx+m^*)\overline{\widehat{K}(x^{-1}+n^*)}$$

where

$$W(x,y)=V(x)V(y)\phi(4\pi\sqrt{lxy}).$$

In particular, the Fourier transform $\widehat{W}(m^*,n^*)$ is very small unless $m^*+n^*\ll l$ so the above sum is over $m^*,n^*\ll l$. Setting

$$\gamma_1=\begin{pmatrix} l & m^*\\ & 1\end{pmatrix},\ \gamma_2=\begin{pmatrix} n^* & 1\\ 1 & 0\end{pmatrix}$$

we see that the x-sum is the correlation sum $q\mathcal{C}(K,\gamma_2.\gamma_1^{-1})$ which is $\ll q^{1/2}$ if $\gamma_2.\gamma_1^{-1}$ does not belong to the group of automorphism of $\widehat{\mathcal{F}}$. Using Theorem 7.4 one show that if l is a sufficiently small fixed (positive) power of q, the bound

$$\sum_{x\in\mathbf{F}_q^\times}\widehat{K}(lx+m^*)\overline{\widehat{K}(x^{-1}+n^*)}\ll_{C(\mathcal{F})}q^{1/2}$$

holds for most pairs (m^*,n^*). From this we deduce (10.7).

11. The ternary divisor function in arithmetic progressions to large moduli

Given some arithmetic function $\lambda=(\lambda(n))_{n\geqslant 1}$, a natural question in analytic number theory is to understand how well λ is distributed in arithmetic progressions: given $q\geqslant 1$ and $(a,q)=1$ one would like to evaluate the sum

$$\sum_{\substack{n\leqslant X\\ n\equiv a\ (\mathrm{mod}\ q)}}\lambda(n)$$

as $X\to\infty$ and for q as large as possible with respect to X. It is natural to evaluate the difference

$$E(\lambda;q,a):=\sum_{\substack{n\leqslant X\\ n\equiv a\ (\mathrm{mod}\ q)}}\lambda(n)-\frac{1}{\varphi(q)}\sum_{\substack{n\leqslant X\\ (n,q)=1}}\lambda(n)$$

and assuming that λ is "essentially" bounded the target would be to obtain a bound of the shape

$$E(\lambda;q,a)\ll_A\frac{X}{q}(\log X)^{-A} \tag{11.1}$$

for any $A\geqslant 0$, as $X\to+\infty$ and for q as large as possible compared to X.

The emblematic case is when $\lambda=1_{\mathcal{P}}$ is the characteristic function of the primes. In that case the problem can be approached through the analytic properties of Dirichlet L-functions and in particular the localization of their zeros. The method of Hadamard-de la Vallee-Poussin (adapted to this setting by Landau) and the Landau-Siegel theorem show that (11.1) is satisfied for $q\leqslant(\log X)^B$ for any given B, while the validity of the generalized Riemann hypothesis would give (11.1) for $q\ll X^{1/2-\delta}$ for any fixed $\delta>0$. Considering averages over q, it is possible to reach the GRH range and this is the content of the Bombieri-Vinogradov theorem

THEOREM 11.1 (Bombieri-Vinogradov). *For any $A \geqslant 0$ there exists $B = B(A)$ such that for $Q \leqslant X^{1/2}/\log^B X$*

$$\sum_{q \leqslant Q} \max_{(a,q)=1} |E(1_{\mathcal{P}}; q, a)| \ll X/\log^A X.$$

Passing the GRH/Bombieri-Vinogradov range and reaching the inequality $Q \leqslant x^{1/2+\eta}$ for some $\eta > 0$ is a fundamental problem in analytic number theory with many major applications. For instance, Y. Zhang's breakthrough on the existence of bounded gaps between primes proceeded by establishing a version of the Bombieri-Vinogradov theorem going beyond the $Q = X^{1/2}$ range on average over smooth moduli. [**Zha14**]; we will discuss some of the techniques entering his proof below.

Several arithmetic functions are of interest besides the characteristic function of the primes or other sequences. One of the simplest are the divisor functions

$$d_k(n) = \sum_{n_1 \ldots . n_k = n} 1.$$

For $k = 2$, Selberg and others established the following (still unsurpassed).

THEOREM 11.2 (The divisor function in arithmetic progressions to large moduli). *For every non-zero integer a, every $\varepsilon, A > 0$, every $X \geqslant 2$ and every prime q, coprime with a, satisfying*

$$q \leqslant X^{2/3-\varepsilon},$$

we have

$$E(d_2; q, a) \ll \frac{X}{q}(\log X)^{-A},$$

where the implied constant only depends on ε and A (and not on a).

PROOF. (Sketch) To simplify matters we consider the problem of evaluating the model sum

$$\sum_{n_1 n_2 \equiv a \,(\mathrm{mod}\, q)} V(\frac{n_1}{N_1})V(\frac{n_2}{N_2})$$

for $N_1 N_2 = X$ and $V \in \mathcal{C}_c^\infty(]1,2[)$. We apply the Poisson summation formula to the n_1 variable and to the n_2 variable. The condition $n_1 n_2 \equiv a \,(\mathrm{mod}\, q)$ get transformed into

$$\delta_{n_1 n_2 \equiv a \,(\mathrm{mod}\, q)} \to q^{-1/2} \mathrm{e}_q(an_1/n_2) \to q^{-1/2}\, \mathrm{Kl}_2(an_1 n_2; q).$$

The ranges the ranges N_1, N_2 are transformed into

$$N_1^* = q/N_1, N_2^* = q/N_2$$

and the whole model sum is transformed into a sum of the shape

$$MT(a;q) + ET(a;q)$$

where $MT(a;q)$ is a main term which we will not specify (but is of the right order of magnitude), and $ET(a;q)$ is an error term of the shape

$$ET(a;q) = \frac{1}{q^{1/2}} \frac{N_1}{q^{1/2}} \frac{N_2}{q^{1/2}} \sum_{n_1, n_2} \mathrm{Kl}_2(an_1 n_2; q) \tilde{V}(\frac{n_1}{N_1^*}) \tilde{V}(\frac{n_2}{N_2^*})$$

where $\tilde{V}$ is a rapidly decreasing function. By Weil's bound for Kloosterman sums, the error term is bounded by $q^{1/2+\epsilon}$ which is smaller that $X(\log X)^{-A}/q$ as long as $X \leqslant q^{2/3-2\varepsilon}$. □

REMARK 11.3. Improving the exponent 2/3 is tantamount to detect cancellation in the sum of Kloosterman sums above. We have given such an improvment in (10.2); unfortunately in the present case the range of the variable n_1n_2 is $N_1^*N_2^* = q^2/X \leqslant q^{1/2}$ which is too short with current technology. See however the [**FI92**] for an improvement beyond the $q = x^{2/3}$ limit on average over a family of moduli q admitting a specific factorisation.

We now show how to go beyond the Bombieri-Vinogradov range for the specific case of the ternary divisor function

$$d_3(n) = \sum_{n_1n_2n_3=n} 1$$

(in fact in a stronger form because it is not even necessary to average over the modulus q !). The very first result of that kind is due to Friedlander-Iwaniec [**FI85**] (with $\frac{1}{2}+\eta = \frac{1}{2}+\frac{1}{231}$) and was later improved by Heath-Brown (with $\frac{1}{2}+\eta = \frac{1}{2}+\frac{1}{81}$) [**HB86**]. When the modulus q is prime, the best result to date is to be found in [**FKM15**]:

THEOREM 11.4 (The ternary divisor function in arithmetic progressions to large moduli). *For every non-zero integer a, every $A > 0$, every $X \geqslant 2$ and every prime q, coprime with a, satisfying*

$$q \leqslant X^{\frac{1}{2}+\frac{1}{47}},$$

we have

$$E(d_3; q, a) \ll \frac{X}{q}(\log X)^{-A},$$

where the implied constant only depends on A (and not on a).

REMARK 11.5. One may wonder why these higher order divisor functions are so interesting: one reason is that these problems can be considered as approximations for the case of the von Mangoldt function. Indeed, the Heath-Brown identity (Lemma 8.3) expresses the von Mangoldt function as a linear combination of arithmetic functions involving higher divisor functions, therefore studying higher divisor functions in arithmetic progressions to large moduli will enable to progress on the von Mangoldt function.[9]

PROOF. We consider again a model sum of the shape

$$\sum_{n_1n_2n_3\equiv a\,(\mathrm{mod}\,q)} V(\frac{n_1}{N_1})V(\frac{n_2}{N_2})V(\frac{n_3}{N_3})$$

for $N_1N_2N_3 = X$ and $V \in \mathcal{C}_c^\infty(]1,2[)$. We apply the Poisson summation formula to the variables n_1 n_2 and n_3. The condition $n_1n_2n_3 \equiv a\,(\mathrm{mod}\,q)$ is this time transformed into the hyper-Kloosterman sum

$$\frac{1}{q^{1/2}}\,\mathrm{Kl}_3(an_1n_2n_3; q).$$

The model sum is transformed into a main term (of the correct order of magnitude) and an error term

$$ET_3(a;q) = \frac{1}{q^{1/2}}\frac{N_1}{q^{1/2}}\frac{N_2}{q^{1/2}}\frac{N_3}{q^{1/2}}\sum_{n_1,n_2,n_3}\mathrm{Kl}_2(an_1n_2n_3;q)\tilde{V}(\frac{n_1}{N_1^*})\tilde{V}(\frac{n_2}{N_2^*})\tilde{V}(\frac{n_3}{N_3^*})$$

[9]This was formalised by Fouvry [**Fou85**].

with

$$N_i^* = q/N_i, \ i = 1, 2, 3.$$

The objective is to obtain a bound of the shape

$$\Sigma_3 := \sum_{n_1,n_2,n_3} \mathrm{Kl}_3(an_1n_2n_3; q)\tilde{V}(\frac{n_1}{N_1^*})\tilde{V}(\frac{n_2}{N_2^*})\tilde{V}(\frac{n_3}{N_3^*}) \ll \frac{q}{\log^A q} \tag{11.2}$$

for $X = q^{2-\eta}$ for some fixed $\eta > 0$ (small), or equivalently for

$$N_1^* N_2^* N_3^* = q^{1+\eta}.$$

We will show that when $\eta = 0$, (11.2) holds with the stronger bound $\ll q^{1-\delta}$ for some $\delta > 0$. A variation of this argument will show (11.2) for some positive η. Write

$$N_i^* = q^{\nu_i}, \ i = 1, 2, 3, \ \nu_1 + \nu_2 + \nu_3 = 1;$$

we assume that

$$0 \leqslant \nu_1 \leqslant \nu_2 \leqslant \nu_3.$$

Suppose that $\nu_3 \geqslant 1/2 + \delta$. Then the Pólya-Vinogradov method, applied to the n_3 variable, leads to a bound of the shape

$$\Sigma_3 \ll q^{1-\nu_3+1/2} \log q \ll q^{1-\delta} \log q.$$

Otherwise we have $\nu_3 \leqslant 1/2 + \delta$. We assume now that $\nu_1 \geqslant 2\delta$; then $\nu_1 \leqslant 1/3$, so that grouping the variables n_2n_3 into a single variable n of size $\geqslant q^{2/3}$ (weighted by a divisor like function) and applying Theorem 9.1, we obtain the bound

$$\Sigma_3 \ll q^{1-\delta} \log^3 q.$$

We may therefore assume that

$$\nu_1 \leqslant 2\delta, \ \nu_2 + \nu_3 \geqslant 1 - 2\delta.$$

The n_2n_3-sum is similar to the sum in (10.2) (for $K(n) = \mathrm{Kl}_3(an_1n; q)$) and indeed the same bound holds, so that for any $\varepsilon > 0$, we have

$$\Sigma_3 \ll_\varepsilon q^{\nu_1 + \frac{\nu_2+\nu_3}{2} + \frac{1}{2} - \frac{1}{8} + \epsilon} \ll_\varepsilon q^{2\delta + 1 - \frac{1}{8} + \epsilon}$$

which gives the required bounds if δ is chosen $< 1/24$. □

12. The geometric monodromy group and Sato-Tate laws

In this section we discuss an important invariant attached an ℓ-adic sheaf: its geometric monodromy group. This will be crucial in the next section to study more advanced sums of trace functions (multicorrelation sums). Another rather appealing outcome of this notion are the *Sato-Tate* type laws which describe the distribution of the set of values of trace functions as q^n grows.

12.1. Sato-Tate laws for elliptic curves. The term "Sato-Tate law" comes from the celebrated *Sato-Tate Conjecture* for elliptic curves over $\mathbf{Q}$ which is now a theorem established in a series of papers principally by Clozel, Harris, Shepherd-Barron and Taylor [**CHT08, HSBT10, Tay08, BLGHT11**]. Let $E/\mathbf{Q}$ be an elliptic curve defined over $\mathbf{Q}$ with a model over $\mathbf{Z}$ –for instance given by the Weierstrass equation

$$E\colon zy^2 = x^3 - azx^2 - bz^3,\ a, b \in \mathbf{Z},\ \Delta(a, b) = 4a^3 - 27b^2 \neq 0.$$

For any prime q, we denote by $E(\mathbf{F}_q)$ the reduction modulo q of E; we have (Hasse bound)

$$a_q(E) := q + 1 - |E(\mathbf{F}_q)| \in [-2q^{1/2}, 2q^{1/2}];$$

we can then define the angle $\theta_{E,q} \in [0, \pi]$ of E at the prime q by the formula

$$a_q(E)/q^{1/2} = 2\cos(\theta_{E,q}).$$

THEOREM 12.1 (Sato-Tate law for an elliptic curve). *Let $E/\mathbf{Q}$ be a non-CM elliptic curve. As $X \to \infty$, the multiset of angles $\{\theta_{E,q},\ q \leqslant X,\ q \text{ prime}\}$ becomes equidistributed on $[0, \pi]$ with respect to the so-called Sato-Tate measure μ_{ST} whose density is given by*

$$d\mu_{ST} = \frac{2}{\pi}\sin^2(\theta)d\theta.$$

In other words, for any interval $I \subset [0, \pi]$, we have

$$\frac{|\{q \leqslant X,\ q \text{ prime},\ \theta_{E,q} \in I\}|}{\pi(X)} \to \mu_{ST}(I) = \frac{2}{\pi}\int_I \sin^2(\theta)d\theta$$

as $X \to \infty$.

The Sato-Tate measure μ_{ST} introduced in this statement has a more conceptual description: let $\mathrm{SU}_2(\mathbf{C})$ be the special unitary group in two variables and let $\mathrm{SU}_2(\mathbf{C})^\natural$ be its space of conjugacy classes, that space is identified with $[0, \pi]$ via the map

$$\begin{pmatrix} e^{i\theta} & 0 \\ 0 & e^{-i\theta} \end{pmatrix}^\natural \mapsto \theta \,(\mathrm{mod}\,\pi).$$

The Sato-Tate measure μ_{ST} then corresponds to the direct image of the Haar measure on $\mathrm{SU}_2(\mathbf{C})$ under the natural projection $\mathrm{SU}_2(\mathbf{C}) \mapsto \mathrm{SU}_2(\mathbf{C})^\natural$: this follows from the Weyl integration formula. Now let us recall that attached to the elliptic curve E is a Galois representation on its ℓ-adic Tate module[10]

$$\varrho_E\colon \mathrm{Gal}(\overline{\mathbf{Q}}/\mathbf{Q}) \to \mathrm{GL}(V_\ell(E))$$

which is unramified at every prime q not dividing the discriminant (of the integral model) of E and for such a prime, the Frobenius conjugacy class satisfies

$$\mathrm{tr}(\mathrm{Frob}_q\,|V_\ell(E)) = a_q(E) = 2q^{1/2}\cos(\theta_{E,q})$$

hence defines a complex conjugacy class

$$\begin{pmatrix} e^{i\theta_{E,q}} & 0 \\ 0 & e^{-i\theta_{E,q}} \end{pmatrix}^\natural.$$

The Sato-Tate law for non-CM elliptic curves then states that this collection of Frobenius conjugacy classes becomes equidistributed relative to this measure.

[10]which is an ℓ-adic sheaf over $\mathrm{Spec}(\mathbf{Z})$

REMARK 12.2. For CM-elliptic curves there is also a (different) Sato-Tate law which was established by Hecke much earlier: the angles $\theta_{E,q}$ are equidistributed with respect to the uniform measure.

The proof of the Sato-Tate conjecture in the non-CM case is one of the crowning achievements of the Langlands program; several decades before its proof, several variants of this conjecture have been established for *families* of elliptic curves over finite fields: given $a, b \in \mathbf{F}_q$ such that $\Delta(a, b) := 4a^3 - 27b^2 \neq 0$ the Weierstrass equation

$$E_{a,b}\colon y^2 = x^3 - ax^2 - b$$

defines an elliptic curve over $\mathbf{F}_q$ and let

$$a_q(a, b) = q + 1 - |E_{a,b}(\mathbf{F}_q)| = 2q^{1/2}\cos(\theta_{a,b,q}).$$

Using the Selberg trace formula, Birch [**Bir68**], established the following variant of the Sato-Tate law for elliptic curves

THEOREM 12.3. *As $q \to \infty$ the multiset of angles $\{\theta_{a,b,q},\ (a, b) \in \mathbf{F}_q^2,\ \Delta(a, b) \neq 0\}$ becomes equidistributed on $[0, \pi]$ with respect to μ_{ST}: for any interval $I \subset [0, \pi]$, we have*

$$\frac{|\{(a, b) \in \mathbf{F}_q^2,\ \Delta(a, b) \neq 0, \theta_{a,b,q} \in I\}|}{|\{(a, b) \in \mathbf{F}_q^2,\ \Delta(a, b) \neq 0\}|} \to \mu_{ST}(I),\ q \to \infty.$$

There is another variant, spelled out by Katz and which is consequence of Deligne's work [**Del80**]; it concerns one parameter families of elliptic curves: let $a(T), b(T) \in \mathbf{Z}[T]$ be polynomials such that $\Delta(T) := 4a(T)^3 + 27b(T)^2 \neq 0$; for q a sufficiently large prime, the equation over $\mathbf{F}_q$,

$$E_t\colon y^2 = x^3 - a(t)x^2 - b(t)$$

defines a family of elliptic curves indexed by the set $U(\mathbf{F}_q) := \{t \in \mathbf{F}_q,\ \Delta(t) \neq 0\}$. For any $t \in U(\mathbf{F}_q)$ we set

$$\theta_{t,q} := \theta_{a(t),b(t),q} \in [0, \pi].$$

THEOREM 12.4. *Assume that the j-invariant $j(T) = -1728\frac{4a(T)^3}{\Delta(T)}$ is not constant, then the multiset $\{\theta_{t,q},\ t \in U(\mathbf{F}_q)\}$ becomes equidistributed on $[0, \pi]$ with respect to μ_{ST} as $q \to \infty$. In other words, for any interval $I \subset [0, \pi]$, we have*

$$\frac{|\{t \in U(\mathbf{F}_q),\ \theta_{t,q} \in I\}|}{|U(\mathbf{F}_q)|} \to \mu_{ST}(I),\ q \to \infty.$$

REMARK 12.5. Deligne [**Del80**, Proposition 3.5.7] proved another variant of the Sato-Tate law when the parameter set is $U(\mathbf{F}_{q^n})$ with q fixed (large enough) and $n \to \infty$; this is in fact a special case of "Deligne's equidistribution theorem" [**Del80**, Theorem 3.5.3].

Theorem 12.4 is a special case of very general Sato-Tate laws for ℓ-adic sheaves: indeed the function

$$t \in U(\mathbf{F}_q) \mapsto \frac{a_q(t)}{q^{1/2}}$$

is the trace function of some geometrically irreducible ℓ-adic sheaf $\mathcal{E}_{a,b}$ whose associated trace function is given by

$$t \mapsto -\frac{1}{q^{1/2}}\sum_{x \in \mathbf{F}_q}\Big(\frac{x^3 + a(t)x + b(t)}{q}\Big), \tag{12.1}$$

where $\left(\frac{\cdot}{q}\right)$ is the Legendre symbol. A key player for such Sato-Tate law is the

12.2. The geometric monodromy group of a sheaf.

DEFINITION 12.6 ([**Kat88**] Chap. 3). Let $\mathcal{F}$ be a sheaf pure of weight 0 and let $\varrho_{\mathcal{F}}$ be the associated Galois representation. The geometric (resp. arithmetic) monodromy group $G_{\mathcal{F},\mathrm{geom}}$ (resp. $G_{\mathcal{F},\mathrm{arith}}$) is the Zariski closure of $\varrho_{\mathcal{F}}(G^{\mathrm{geom}})$ (resp. $\varrho_{\mathcal{F}}(G^{\mathrm{arith}})$) inside $\mathrm{GL}(V_{\mathcal{F}})$; in particular

$$G_{\mathcal{F},\mathrm{geom}} \subset G_{\mathcal{F},\mathrm{arith}}.$$

It follows from [**Del80**, Théorème (3.4.1)] that the connected component $G^0_{\mathcal{F},\mathrm{geom}}$ of $G_{\mathcal{F},\mathrm{geom}}$ is semisimple.

EXAMPLE 12.7.
- In the case of the trace function (12.1), Deligne showed [**Del80**, Lemme 3.5.5], that if $q > 2$ and the j-invariant $j(T) \pmod q$ is not constant, one has
$$G_{\mathcal{E}_{a,b},\mathrm{geom}} = G_{\mathcal{E}_{a,b},\mathrm{arith}} = \mathrm{SL}_2\,.$$
- In his numerous books [**Kat88, Kat90a, Kat90b, Kat05a, Kat05b, Kat12**] Katz computed the monodromy groups of various classes of sheaves: for instance, he proved in [**Kat88**, Theorem 11.1] that for Kloosterman sheaves one has (for $q > 2$)
$$G_{\mathcal{K}\ell_k,\mathrm{geom}} = G_{\mathcal{K}\ell_k,\mathrm{arith}} = \begin{cases} \mathrm{SL}_k & \text{if } k \text{ is odd} \\ \mathrm{Sp}_k & \text{if } k \text{ is even.} \end{cases}$$

12.3. Sato-Tate laws. In the sequel we make the simplifying hypothesis that

$$G_{\mathcal{F},\mathrm{geom}} = G_{\mathcal{F},\mathrm{arith}}. \tag{12.2}$$

12.3.1. *Moments of trace functions.* Before presenting the Sato-Tate laws in general, let us consider the very specific concrete problem of evaluating the *moments* of a trace function K. For $l \geqslant 0$ an integer, the $2l$-th moment of K is the average

$$\mathcal{M}_{2l}(K) = \frac{1}{q}\sum_{x\in\mathbf{F}_q} |K(x)|^{2l}.$$

The possibility of evaluating these comes from the fact that $x \mapsto |K(x)|^{2l}$ is indeed a trace function (not necessarily and in fact almost never irreducible). Indeed let $\mathrm{Std}\colon G_{\mathcal{F},\mathrm{geom}} \hookrightarrow \mathrm{GL}(V_{\mathcal{F}})$ be the standard representation of the group $G_{\mathcal{F},\mathrm{geom}}$ and let $\varrho_{l,l}$ be the representation

$$\varrho_{l,l} = (\mathrm{Std}\otimes\mathrm{Std}^*)^{\otimes l}.$$

Because of our assumption (12.2) , the composition

$$\varrho_{l,l}(\mathcal{F})\text{''} = \text{''}\varrho_{l,l}\circ\varrho_{\mathcal{F}}$$

is a representation of $G_{\mathcal{F},\mathrm{arith}}$ hence defines an ℓ-adic sheaf pure of weight 0 whose trace function is[11] $x \mapsto |K(x)|^{2l}$.

The decomposition of this representation into irreducible representations of $G_{\mathcal{F},\mathrm{geom}}$

$$\varrho_{l,l} = m_1(\varrho_{l,l}).1 \oplus \bigoplus_{1\neq r\in\mathrm{Irr}(G_{\mathcal{F},\mathrm{geom}})} m_r(\varrho_{l,l}).r$$

[11] at least at the x where it is lisse

yields a decomposition of $\varrho_{l,l}(\mathcal{F})$ into a sum of geometrically irreducible sheaves

$$\varrho_{l,l} \circ \mathcal{F} = m_1(\varrho_{l,l})\overline{\mathbf{Q}_\ell} \oplus \bigoplus_{1 \neq r \in \mathrm{Irr}(G_{\mathcal{F},\mathrm{geom}})} m_r(\varrho_{l,l}) r \circ \mathcal{F}$$

and a decomposition of $|K(x)|^{2l}$ as a sum of trace functions

$$|K(x)|^{2l} = m_1(\varrho_{l,l}) + \sum_{1 \neq r} m_r(\varrho_{l,l}) K_{r \circ \mathcal{F}}(x).$$

From Deligne's Theorem (Cor. 4.7) one deduce that

$$\frac{1}{q} \sum_x |K(x)|^{2l} = m_1(\varrho_{l,l}) + O_{C(\mathcal{F}),l}(q^{-1/2})$$

where $m_1(\varrho_{l,l})$ is the multiplicity of the trivial representation in the representation $(\mathrm{Std} \otimes \mathrm{Std}^*)^{\otimes l}$ of $G_{\mathcal{F},\mathrm{geom}}$. In the same way, we could evaluate (in terms of the representation theory of the group $G_{\mathcal{F},\mathrm{geom}}$) more general moments like

$$\frac{1}{q} \sum_{x \in \mathbf{F}_q} |K(x)|^{2l} K(x)^{l'}$$

for integers $l, l' \geqslant 0$.

12.3.2. *Equidistribution of Frobenius conjugacy classes.* There is a more conceptual interpretation of these moments. For any $x \in U(\mathbf{F}_q)$, the Frobenius at x acting on $V_{\mathcal{F}}$ produces a $\varrho_{\mathcal{F}}(G^{\mathrm{arith}})$-conjugacy class

$$\varrho_{\mathcal{F}}(\mathrm{Frob}_x) \subset G_{\mathcal{F},\mathrm{arith}}(\mathbf{C}) = G_{\mathcal{F},\mathrm{geom}}(\mathbf{C}).$$

The *Frobenius conjugacy class* of $\mathcal{F}$ at x is by definition the $G_{\mathcal{F},\mathrm{geom}}(\mathbf{C})$-conjugacy class of its semisimple part (in the sense of Jordan decomposition) and is denoted $\theta_{x,\mathcal{F}}$. Let K be any maximal compact subgroup of $G_{\mathcal{F},\mathrm{geom}}(\mathbf{C})$ and $K^\natural$ its space of conjugacy classes. As explained in [**Kat88**](Chap. 3), the conjugacy class $\theta_{x,\mathcal{F}}$ defines a unique conjugacy class in K, also denoted $\theta_{x,\mathcal{F}} \in K^\natural$. The Sato-tate laws describe the distribution of the set $\{\theta_{x,\mathcal{F}},\ x \in U(\mathbf{F}_q)\}$ inside $K^\natural$ as $q \to \infty$.

More precisely, let G be a connected semisimple algebraic group over $\overline{\mathbf{Q}_\ell}$ and $K \subset G(\mathbf{C})$ a maximal compact subgroup. Let $\mu^\natural$ be the direct image of the Haar probability measure on K under the projection $K \mapsto K^\natural$.

THEOREM 12.8 (Sato-Tate law). *Let G and $K \subset G(\mathbf{C})$ as above. Suppose we are given a sequence of primes $q \to \infty$ and for each such prime some ℓ-adic sheaf $\mathcal{F}$ over $\mathbf{F}_q$, satisfying* (12.2)*, whose conductor $C(\mathcal{F})$ is bounded independently of q, such that*

$$G_{\mathcal{F},\mathrm{geom}} = G_{\mathcal{F},\mathrm{arith}} = G.$$

For any such q and $x \in U(\mathbf{F}_q)$ let $\theta_{x,\mathcal{F}} \in K^\natural$ be the conjugacy class of $\mathcal{F}$ at x relative to K.

As $q \to \infty$ the sets of conjugacy classes

$$\{\theta_{x,\mathcal{F}},\ x \in U(\mathbf{F}_q)\}$$

become equidistributed with respect to the measure $\mu^\natural$: the probability measure

$$\frac{1}{|U(\mathbf{F}_q)|} \sum_{x \in U(\mathbf{F}_q)} \delta_{\theta_{x,\mathcal{F}}}$$

converges weakly to $\mu^\natural$. *In other words, for any* $f \in \mathcal{C}(K^\natural)$

$$\frac{1}{|U(\mathbf{F}_q)|}\sum_{x\in\mathbf{F}_q} f(\theta_{x,\mathcal{F}}) \to \int_{K^\natural} f(\theta)d\mu^\natural(\theta),\ q\to\infty. \tag{12.3}$$

PROOF. By the Peter-Weyl theorem, the functions

$$\mathrm{tr}(r)\colon \theta\in K^\natural \to \mathrm{tr}(r(\theta))\in\mathbf{C}$$

when r ranges over all the irreducible representations of G, form an orthonormal basis of $L^2(K^\natural,\mu^\natural)$ and generates a dense subspace of the space of continuous functions on $K^\natural$. By Weyl equidistribution criterion it is therefore sufficient to show that for any r irreducible and non-trivial, one has

$$\frac{1}{|U(\mathbf{F}_q)|}\sum_{x\in U(\mathbf{F}_q)} \mathrm{tr}(r(\theta_{x,\mathcal{F}})) \to \mu^\natural(\mathrm{tr}(r)) = 0.$$

The function

$$K_{r,\mathcal{F}}\colon x\in U(\mathbf{F}_q)\to r(\theta_{x,\mathcal{F}})$$

is the trace function associated to the sheaf $r\circ\mathcal{F}$ corresponding to the representation of $G_{\mathcal{F},\mathrm{arith}}$, $r\circ\varrho_{\mathcal{F}}$ (because of (12.2) this composition is well defined). That sheaf is by construction geometrically irreducible, non-trivial and its conductor is bounded in terms of $C(\mathcal{F})$ and r only, so it follows from Deligne's Theorem that

$$\frac{1}{|U(\mathbf{F}_q)|}\sum_{x\in U(\mathbf{F}_q)} \mathrm{tr}(r(\theta_{x,\mathcal{F}})) \ll_{C(\mathcal{F}),r} q^{-1/2}\to 0.$$

□

12.3.3. *The case of Kloosterman sums.* As we have seen above, for the Kloosterman sums $\mathrm{Kl}_2(x;q)$, we have

$$G=\mathrm{Sp}_2=\mathrm{SL}_2,\ K=\mathrm{SU}_2(\mathbf{C})$$

and, via the identification $K^\natural\simeq[0,\pi]$, the measure $\mu^\natural$ is identified with the Sato-Tate measure μ_{ST}.

For $x\in\mathbf{F}_q^\times$, we define the angle $\theta_{q,x}\in[0,\pi]$ of the Kloosterman sum $\mathrm{Kl}_2(x;q)$ as

$$\mathrm{Kl}_2(x;q)=\mathrm{tr}\begin{pmatrix} e^{i\theta_{q,x}} & 0\\ 0 & e^{-i\theta_{q,x}}\end{pmatrix} = 2\cos(\theta_{q,x}).$$

The Sato-Tate law becomes the following explicit statement (due to Katz):

THEOREM 12.9 (Sato-Tate law for Kloosterman sums). *For any interval* $I\subset[0,\pi]$

$$\frac{1}{q-1}|\{x\in\mathbf{F}_q^\times,\ \theta_{q,x}\in I\}| \to \frac{2}{\pi}\int_I \sin^2(\theta)d\theta,\ q\to\infty.$$

The above Sato-Tate law is called "vertical" as it describes the distribution of Kloosterman sums with varying parameters $x\in\mathbf{F}_q^\times$ as $q\to\infty$; such law is analogous to the Sato-Tate law of Theorem 12.4.

In [**Kat80**], Katz in analogy with the original Sato-Tate conjecture (Theorem 12.1) asked for the distribution of the Kloosterman sums for a fixed value of the parameter (say $x=1$) and for a varying prime modulus q. Katz made the following

CONJECTURE 12.10 (Horizontal Sato-Tate law for Kloosterman sums). *As $X \to \infty$, the multiset of Kloosterman angles $\{\theta_{q,1},\ q \leqslant X,\ prime\}$ becomes equidistributed with respect to the Sato-Tate measure: for any $[a, b] \subset [0, \pi]$, we have*

$$\frac{1}{\pi(X)}|\{q \leqslant X,\ q\ prime,\ \theta_{q,1} \in [a, b]\}| \to \frac{2}{\pi}\int_a^b \sin^2(\theta)d\theta$$

as $X \to \infty$.

REMARK 12.11. There are other variants of this vertical equidistribution conjecture that have been established recently:

- Heath-Brown and Patterson [**HBP79**] have proven that the angles of cubic Gauss sums of varying prime moduli are equidistributed with respect to the uniform measure.
- Even closer to the current discussion, Duke, Friedlander and Iwaniec [**DFI95**] have proven the vertical equidistribution of the angles $\theta^S_{q,1}$ of *Salié* sums defined by
$$S(1; q) := \frac{1}{q^{1/2}} \sum_{\substack{x,y\in\mathbf{F}_q^\times \\ xy=1}} (\frac{x}{q})e\left(\frac{x+y}{q}\right) =: 2\cos(\theta^S_{q,1})$$
again with respect to the uniform measure.

12.4. Towards the horizontal Sato-Tate conjecture for almost prime moduli. Unlike the original Sato-Tate conjecture the prospect for a proof of Conjecture 12.10 seem very distant at the moment. Even the following very basic consequences of this conjecture seem today completely out of reach:

- There exist infinitely many primes q such that $|\operatorname{Kl}_2(1; q)| \geqslant 2017^{-2017}$,
- There exist infinitely many primes q such that $\operatorname{Kl}_2(1; q) > 0$ (resp. $\operatorname{Kl}_2(1; q) < 0$)

In this section we will explain how some of the results discussed so far enable to say something non-trivial as the cost of replacing the prime moduli q by *almost prime* moduli (that is squarefree-integers with an absolutely bounded number of prime factors).

Recall that for $c \geqslant 1$ a squarefree integer and $(a, c) = 1$ the normalized Kloosterman sum of modulus c and parameter a is

$$\operatorname{Kl}_2(a; c) = \frac{1}{c^{1/2}} \sum_{x\in(\mathbf{Z}/c\mathbf{Z})^\times} e\left(\frac{\overline{x} + ax}{c}\right).$$

By the Chinese remainder theorem, Kloosterman sums satisfy the *twisted multiplicativity* relation: for $c = c_1c_2$, $(c_1, c_2) = 1$ one has

$$\operatorname{Kl}_2(a; c) = \operatorname{Kl}_2(a\overline{c_2}^2; c_1) \operatorname{Kl}_2(a\overline{c_1}^2; c_2) \tag{12.4}$$

so that by Weil's bound one has

$$|\operatorname{Kl}_2(a; c)| \leqslant 2^{\omega(c)}$$

where $\omega(c)$ is the number of prime factors of c. We can then define the corresponding Kloosterman angle by

$$\cos(\theta_{c,a}) = \frac{\operatorname{Kl}_2(a; c)}{2^{\omega(c)}}.$$

It is then natural to make the following

CONJECTURE 12.12 (Horizontal Sato-Tate law for Kloosterman sums with composite moduli). *Given $k \geqslant 1$ un integer, let $\pi_k(X)$ be the number of squarefree integers $\leqslant X$ with exactly k prime factors and let $\mu_{ST,k}$ be the Sato-Tate measure of order k, defined as the push-forward of the measure $\mu_{ST}^{\otimes k}$ on $[0,\pi]^k$ by the map*

$$(\theta_1,\dots,\theta_k) \in [0,\pi]^k \mapsto \arccos(\cos(\theta_1)\times\dots\times\cos(\theta_k))) \in [0,\pi].$$

for any $k \geqslant 1$, the multiset of Kloosterman angles

$$\{\theta_{c,1},\ c \leqslant X,\ c \text{ is squarefree with } k \text{ prime factors}\}$$

becomes equidistributed with respect to $\mu_{ST,k}$ as $X \to \infty$.

This conjecture for any $k \geqslant 2$ seem as hard as the original one (and is not implies by it). On the other hand it is possible to establish some of its consequences:

THEOREM 12.13. *There exists $k \geqslant 2$ such that*

(1) *for infinitely many square-free integers c with at most k prime factors,*

$$|\,\mathrm{Kl}_2(1;c)| \geqslant 2017^{-2017};$$

(2) *for infinitely many square-free integers c with at most k prime factors,*

$$\mathrm{Kl}_2(1;c) > 0;$$

(3) *for infinitely many square-free integers c with at most k prime factors,*

$$\mathrm{Kl}_2(1;c) < 0.$$

The first statement above was proven in [**Mic95**] for $k = 2$ (with 2017^{-2017} replaced by $4/25$; the second and the third were first proven in [**FM07**] for $k = 23$; this value was subsequently improved by Sivak, Matomäki and Ping who holds the current record with $k = 7$ [**SF09, Mat11, Xi15, Xi16**].

12.4.1. *Kloosterman sums can be large.* We start with the first statement which we prove for $c = pq$ a product of two distinct primes. The main idea is to use the twisted multiplicativity relation

$$\mathrm{Kl}_2(1;pq) = \mathrm{Kl}_2(\overline{p}^2;q)\,\mathrm{Kl}_2(\overline{q}^2;p)$$

and to establish the existence of some κ for which there exist infinitely many pairs of distinct primes (p,q) such that

$$|\,\mathrm{Kl}_2(\overline{p}^2;q)|\;|\,\mathrm{Kl}_2(\overline{q}^2;p)| \geqslant \kappa.$$

Indeed, for such pairs we have

$$|\,\mathrm{Kl}_2(1;pq)| \geqslant \kappa^2.$$

Given X large, we will consider pairs (p,q) such that $p,q \in [X^{1/2},2X^{1/2}[$ and will show that for κ small enough the two sets

$$\{(p,q),\ p+q \in [X^{1/2},2X^{1/2}[,\ p,q \text{ primes }|\,\mathrm{Kl}_2(\overline{p}^2;q)| \geqslant \kappa\}$$

$$\{(p,q),\ p+q \in [X^{1/2},2X^{1/2}[,\ p,q \text{ primes }|\,\mathrm{Kl}_2(\overline{q}^2;p)| \geqslant \kappa\}$$

are large enough to have a non-empty (and in fact large) intersection as $X \to \infty$. This is a consequence of the following equidistribution statement

PROPOSITION 12.14. *Given $X \geqslant 1$, and a prime $q \in [X^{1/2}, 2X^{1/2}]$, the (multi)-set of Kloosterman angles*

$$\{\theta_{q,\overline{p}^2},\ p \in [X^{1/2}, 2X^{1/2}[,\ p \text{ prime},\ p \neq q\}$$

is equidistributed with respect to the Sato-Tate measure: for any interval $[a,b] \subset [0,\pi]$

$$\frac{|\{p \in [X^{1/2}, 2X^{1/2}[,\ p \neq q \text{ prime},\ \theta_{q,\overline{p}^2} \in [a,b]\}|}{|\{p \in [X^{1/2}, 2X^{1/2}[,\ p \neq q \text{ prime}\}|} \to \frac{2}{\pi}\int_a^b \sin^2(\theta)d\theta$$

as $X \to \infty$.

PROOF. We consider the pull-back sheaf $\mathcal{K} := [x \to x^{-2}]^*\mathcal{K}\ell_2$ whose trace function is given by $x \to \mathrm{Kl}_2(\overline{x}^2; q)$. As a representation of the geometric Galois group, it corresponds to restricting the representation $\mathcal{K}\ell_2$ to a subgroup of index 2. Since the geometric monodromy group of $\mathcal{K}\ell_2$ is SL_2, the same is true for the pull-back (the algebriac group SL_2 has no non-trivial finite-index subgroups); therefore

$$G_{\mathcal{K},\mathrm{geom}} = G_{\mathcal{K},\mathrm{arith}} = \mathrm{SL}_2\,.$$

The non-trivial irreducible representations of SL_2 are the symmetric powers of the standard representation, $\mathrm{Sym}_k(\mathrm{Std})$, $k \geqslant 1$. Given $k \geqslant 1$ the composed sheaf

$$\mathcal{K}_k = \mathrm{Sym}_k \circ \mathcal{K}$$

is by construction geometrically irreducible, has rank $k+1$ with conductor bounded in terms of k only and its trace function equals

$$K_k(x) = \mathrm{tr}(\mathrm{Sym}_k\begin{pmatrix} e^{i\theta_{q,\overline{x}^2}} & 0 \\ 0 & e^{-i\theta_{q,\overline{x}^2}}\end{pmatrix}) = \sum_{j=0}^{k} e^{i(k-j)\theta_{q,\overline{x}^2}} e^{-ij\theta_{q,\overline{x}^2}} = \frac{\sin((k+1)\theta_{q,\overline{x}^2})}{\sin(\theta_{q,\overline{x}^2})}.$$

In particular $\mathcal{K}_k$ cannot be geometrically isomorphic to any tensor product of an Artin-Schreier sheaf and a Kummer sheaf (as they have rank 1). Hence by a simple variant of Theorem 8.1 we obtain that

$$\frac{1}{\pi(2X^{1/2}) - \pi(X^{1/2})}\sum_{\substack{p \neq q \\ p \sim X^{1/2}}} K_k(p) \to 0 = \frac{2}{\pi}\int_0^\pi \frac{\sin((k+1)\theta)}{\sin(\theta)}\sin^2(\theta)d\theta$$

∎

Averaging over q, we deduce the existence of some $\kappa > 0$ ($\kappa = 0,4$) such that for X large enough

$$\frac{|\{(p,q),\ p \neq q \in [X^{1/2}, 2X^{1/2}[,\ p,q \text{ primes},\ |\mathrm{Kl}_2(\overline{p}^2;q)| \geqslant \kappa\}|}{|\{(p,q),\ p \neq q \in [X^{1/2}, 2X^{1/2}[,\ p,q \text{ primes}\}|} \geqslant 0,51$$

hence

$$\begin{aligned} (12.5)\qquad &|\{(p,q),\ p \neq q \in [X^{1/2}, 2X^{1/2}[,\ p,q \text{ primes } |\mathrm{Kl}_2(1;pq)| \geqslant \kappa^2\}| \\ &\geqslant (0,01 + o(1))\frac{X}{(\frac{1}{2}\log X)^2}. \end{aligned}$$

12.4.2. *Kloosterman sums change sign.* We now discuss briefly the proof of the remaining two statements: to establish the existence of sign changes, it suffices to prove that given $V \in \mathcal{C}_c^\infty(]1,2[)$ some non-zero non-negative smooth function, there exists $u > 0$ such that, for X large enough

$$(12.6) \qquad \Big|\sum_{\substack{c\geqslant 1\\ p|c\Rightarrow p\geqslant X^{1/u}}} \mathrm{Kl}_2(1;c)V(\frac{c}{X})\Big| < \sum_{\substack{c\geqslant 1\\ p|c\Rightarrow p\geqslant X^{1/u}}} |\mathrm{Kl}_2(1;c)|V(\frac{c}{X}).$$

which will prove the existence of sign changes for Kloosterman sums $\mathrm{Kl}_2(1;c)$ whose modulus has at most $1/u$ prime factors. Using sieve methods and the Petersson-Kuznetzov formulas to express sums of Kloosterman sums in terms of Fourier coefficients of modular forms ((10.10) and (10.11)) and using the theory of automorphic forms, one can show that (see [**FM07**] for a proof)

PROPOSITION 12.15. *For any $\eta > 0$, there exists $u = u(\eta) > 0$ such that*

$$\Big|\sum_{\substack{c\geqslant 1\\ p|c\Rightarrow p\geqslant X^{1/u}}} \mathrm{Kl}_2(1;c)V(\frac{c}{X})\Big| \leqslant \eta\frac{X}{\log X}$$

for X large enough (depending on η and V).

To conclude, it is sufficient to show that for some $u = u_0$, one has

$$(12.7) \qquad \sum_{\substack{c\geqslant 1\\ p|c\Rightarrow p\geqslant X^{1/u}}} |\mu^2(c)\,\mathrm{Kl}_2(1;c)|V(\frac{c}{X}) \gg_V \frac{X}{\log X}$$

(the left-hand side is an increasing function of u so the above inequality remains valid for any $u \geqslant u_0$). The inequality (12.5) points in the right direction (for $u_0 = 2$), however as stated it is off by a factor $\log X \log\log X$. One can however recover this factor $\log X$ entierely and prove the lower bound

$$\sum_{\substack{c\geqslant 1\\ p|c\Rightarrow p\geqslant X^{3/8}}} \mu^2(c)|\,\mathrm{Kl}_2(1;c)|V(\frac{c}{X}) \gg_V \frac{X}{\log X}.$$

The reason is that Theorem 8.1 applies also when p is significantly smaller than q (if $q \simeq X^{1/2+\delta}$ one can obtain a non-trivial bound in (8.2) for p of size $X^{1/2-\delta}$ for $\delta \in [0,1/8[$). The details involve making a partition of unity and we leave it to the interested reader. Another possibility (the one followed originally in [**FM07**]) is to establish the lower bound (12.7) for a suitable u by restricting to moduli c which are products of exactly three prime factors, using the techniques discussed so far.

13. Multicorrelation of trace functions

So far we have mainly discussed the evaluation of correlation sums associated to two trace functions K_1 and K_2 (especially the case $K_1 = K$ and $K_2 = \gamma^*K$), namely

$$\mathcal{C}(K_1,K_2) = \frac{1}{q}\sum_x K_1(x)\overline{K_2(x)}.$$

In many applications, multiple correlation sums occur: sums of the shape

$$\mathcal{C}(K_1,K_2,\dots,K_L) := \frac{1}{q}\sum_x K_1(x)K_2(x)\dots K_L(x)$$

where the K_i, $i = 1, \ldots, L$ are trace functions; of course rewriting the inner term of the sum above as a product of two factors reduces to evaluating a double correlation sum, say associated to the sheaves

$$\mathcal{F} = \mathcal{K}_1 \otimes \ldots \mathcal{K}_l, \ \mathcal{G} = \mathcal{K}_{l+1} \otimes \ldots \mathcal{K}_L$$

but it would remain to determine if $\mathcal{F}$ and $\mathcal{G}$ share a common irreducible component and this may be a hard task. In practice, the multicorrelation sums that occur (due to the application of some Hölder inequality and of the Pólya-Vinogradov method) are often of the shape

$$\mathcal{C}(K, \boldsymbol{\gamma}, h) = \frac{1}{q}\sum_{x} K(\gamma_1 \cdot x) \ldots K(\gamma_l \cdot x)\overline{K(\gamma_1' \cdot x) \ldots K(\gamma_l' \cdot x)}\mathrm{e}_q(xh)$$

for K the trace function of some geometrically irreducible sheaf $\mathcal{F}$, pure of weight 0,

$$\boldsymbol{\gamma} = (\gamma_1, \ldots, \gamma_l, \gamma_1', \ldots, \gamma_l') \in \mathrm{PGL}_2(\mathbf{F}_q)^{2l}$$

and some $h \in \mathbf{F}_q$.

This sum is the correlation associated to the trace functions of the sheaves

$$\gamma_1^*\mathcal{F} \otimes \ldots \otimes \gamma_l^*\mathcal{F} \text{ and } {\gamma'}_1^*\mathcal{F} \otimes \ldots \otimes {\gamma'}_l^*\mathcal{F} \otimes \mathcal{L}_\psi$$

whose conductors are bounded polynomially in terms of $C(\mathcal{F})$. If $\mathcal{F}$ has rank one, the two sheaves above have rank one and it is usually not difficult to determine whether these sheaves are geometrically isomorphic or not.

For $\mathcal{F}$ of higher rank, we describe a method due to Katz which has been axiomatized in [**FKM15**]: this method rests on the notion of geometric monodromy group which we discussed in the previous section.

13.1. A theorem on sums of products of trace functions. In this section we discuss some general result making it possible to evaluate multicorrelations sums of trace functions of interest for analytic number theory. The method is basically due to Katz and was used on several occasions, for instance in [**Mic95, FM98**]. The general result presented here is a special case of the results of [**FKM15**]. For this we need to introduce the following variants of the group of automorphism of a sheaf: one is the group of projective automorphisms

$$\mathrm{Aut}^p_{\mathcal{F}}(\mathbf{F}_q) = \{\gamma \in \mathrm{PGL}_2(\mathbf{F}_q),\ \exists \text{ some rank one sheaf } \mathcal{L} \text{ s.t. } \gamma^*\mathcal{F} \simeq_{geom} \mathcal{F} \otimes \mathcal{L}\},$$

the other is the right-$\mathrm{Aut}^p_{\mathcal{F}}(\mathbf{F}_q)$-orbit

$$\mathrm{Aut}^d_{\mathcal{F}}(\mathbf{F}_q) = \{\gamma \in \mathrm{PGL}_2(\mathbf{F}_q),\ \exists \text{ some rank one sheaf } \mathcal{L} \text{ s.t. } \gamma^*\mathcal{F} \simeq_{geom} D(\mathcal{F}) \otimes \mathcal{L}\}.$$

Let $\mathcal{F}$ be a weight 0, rank k, irreducible sheaf. We assume that

- the geometric monodromy group equals $G_{\mathcal{F},\mathrm{geom}} = \mathrm{SL}_k$ or Sp_k, (we then say that $\mathcal{F}$ is of SL or Sp-*type*),
- the equality (12.2) holds,
- $\mathrm{Aut}^p_{\mathcal{F}}(\mathbf{F}_q) = \{\mathrm{Id}\}$; in particular $\mathrm{Aut}^d_{\mathcal{F}}(\mathbf{F}_q)$ is either empty or is reduced to a single element, $\xi_{\mathcal{F}}$ which is a possibly trivial involution ($\xi_{\mathcal{F}}^2 = \mathrm{Id}$) and is called the *special involution*.

EXAMPLE 13.1. The Kloosterman sheaves $\mathcal{K}\ell_k$ have this property [**Kat88**]. The special involution is either Id if k is even ($\mathcal{K}\ell_k$ is self-dual) or the matrix $\xi = \begin{pmatrix} -1 & \\ & 1 \end{pmatrix}$ for k odd.

Finally we introduce the following ad-hoc definition:

DEFINITION 13.2. Given

$$\boldsymbol{\gamma} = (\gamma_1, \ldots, \gamma_l, \gamma_1', \ldots, \gamma_l') \in \mathrm{PGL}_2(\mathbf{F}_q)^{2l},$$

one says that

- $\boldsymbol{\gamma}$ is normal if there is $\gamma \in \mathrm{PGL}_2(\mathbf{F}_q)$ such that
$$|\{i,\ \gamma_i = \gamma\}| + |\{j,\ \gamma_j' = \gamma\}| \equiv 1 \,(\mathrm{mod}\, 2).$$
- For $k \geqslant 3$, $\boldsymbol{\gamma}$ is k-normal if there exists $\gamma \in \mathrm{PGL}_2(\mathbf{F}_q)$ such that
$$|\{i,\ \gamma_i = \gamma\}| - |\{\gamma_j' = \gamma\}| \not\equiv 0 \,(\mathrm{mod}\, k).$$
- For $k \geqslant 3$, and $\xi \in \mathrm{PGL}_2(\mathbf{F}_q)$ a non-trivial involution, $\boldsymbol{\gamma}$ is k-normal w.r.t. ξ if there exist $\gamma \in \mathrm{PGL}_2(\mathbf{F}_q)$ such that

$$|\{i,\ \gamma_i = \gamma\}| + |\{j,\ \gamma_j' = \xi\gamma\}| - |\{j,\ \gamma_j' = \gamma\}| - |\{i,\ \gamma_i = \xi\gamma\}| \not\equiv 0 \,(\mathrm{mod}\, k).$$

THEOREM 13.3. *Let K be the trace function of a sheaf $\mathcal{F}$ as above, $l \geqslant 1$, $\boldsymbol{\gamma} \in \mathrm{PGL}_2(\mathbf{F}_q)^{2l}$ and $h \in \mathbf{F}_q$. We assume that either*

(1) the sheaf $\mathcal{F}$ is self-dual (so that K is real-valued) and $\boldsymbol{\gamma}$ is normal

(2) the $\mathcal{F}$ is of SL*-type of rank $k \geqslant 3$, $q > r$, and $\boldsymbol{\gamma}$ is k-normal or k-normal w.r.t. the special involution of $\mathcal{F}$, if it exists.*

(3) or $h \neq 0$.

We have

$$\mathcal{C}(K, \boldsymbol{\gamma}, h) = \frac{1}{q} \sum_x K(\gamma_1 \cdot x) \ldots K(\gamma_l \cdot x) \overline{K(\gamma_1' \cdot x) \ldots K(\gamma_l' \cdot x)} \mathrm{e}_q(xh) \ll_{l, C(\mathcal{F})} \frac{1}{q^{1/2}}.$$

PROOF. We discuss the proof only in the self-dual case for simplicity. We group together identical γ_i, γ_j' and the sum becomes

$$\frac{1}{q} \sum_x K(\gamma_1'' \cdot x)^{m_1} \ldots K(\gamma_t'' \cdot x)^{m_t} \mathrm{e}_q(xh)$$

where $t \leqslant 2l$, the γ_i'' are distinct and by hypothesis one of the m_i is odd. The above sum is associated to the trace function of the sheaf

$$\bigotimes_{i=1}^{t} \mathrm{Std}(\gamma_i''^* \mathcal{F})^{\otimes m_i} \otimes \mathcal{L}_\psi$$

where $\psi(\cdot) = \mathrm{e}_q(h \cdot)$ and Std is the tautological representation. We decompose each representation into irreducible

$$\varrho_{m,0} = \mathrm{Std}(G)^{\otimes m} = \sum_r m_r(\varrho_{m,0}) r$$

and are reduced to considering various sheaves of the shape

$$\bigotimes_{i=1}^{t} r_i(\gamma_i''^* \mathcal{F}) \otimes \mathcal{L}_\psi \tag{13.1}$$

where $(r_i)_{i \leqslant t}$ is a tuple of irreducible representations of G; by our hypothesis, we know that either $\mathcal{L}_\psi$ is not trivial or at least one of the r_i is not trivial (and necessarily of dimension > 1).

It is then sufficient to show that, under these assumptions, the sheaves (13.1) are irreducible. For this we consider the direct sum sheaf

$$\bigoplus_i \gamma_i''^{*}\mathcal{F}$$

and let $G_{\oplus,\mathrm{geom}} \subset \prod_i G$ be the Zariski closure of the image of G^{geom} under the sum of representations. The following very useful criterion is due to Katz

THEOREM 13.4 (Goursat-Kolchin-Ribet criterion). *Let $(\mathcal{F}_i)_i$ be a tuple of geometrically irreducible sheaves lisse on $U \subset \mathbf{A}^1_{\mathbf{F}_q}$, pure of weight 0, with geometric monodromy groups G_i. We assume that*

- *For every i, $G_i = \mathrm{Sp}_{k_i}$ or SL_{k_i},*
- *for any rank 1 sheaf $\mathcal{L}$ and any $i \neq j$ there is no geometric isomorphism between $\mathcal{F}_i \otimes \mathcal{L}$ and $\mathcal{F}_j$,*
- *for any rank 1 sheaf $\mathcal{L}$ and any $i \neq j$ there is no geometric isomorphism between $\mathcal{F}_i \otimes \mathcal{L}$ and $D(\mathcal{F}_j)$.*

Then the geometric monodromy group of the sheaf $\bigoplus_i \mathcal{F}_i$ equals $\prod_i G_i$.

Our assumptions (the projective automorphism group of $\mathcal{F}$ is trivial, γ is normal and the geometric monodromy group is either SL or Sp) imply that the above criterion holds and this implies that

$$\bigotimes_i r_i(\gamma_i''^{*}\mathcal{F}) \otimes \mathcal{L}_\psi$$

is always irreducible. □

13.2. Application to non-vanishing of Dirichlet L-functions. We now discuss a beautiful application of bounds for multicorrelation sums due to R. Khan and H. Ngo [**KN16**]. It concerns the proportion of non-vanishing of Dirichlet L-functions at the central point 1/2. The interest in this kind of problems from analytic number theory was renewed with the work of Iwaniec and Sarnak in their celebrated attempt to prove the non-existence of a Landau-Siegel zero [**IS00**]. Their approach was based on the following general problem: *given a family of L-functions*

$$\{L(f,s) = \sum_{n\geqslant 1} \frac{\lambda_f(n)}{n^s},\ f \in \mathcal{F}\}$$

indexed by a "reasonable" family of automorphic forms $\mathcal{F}$[12]*, show that for many $f \in \mathcal{F}$, one has*

$$L(f,1/2) \neq 0.$$

In their work [**IS00**], Iwaniec and Sarnak showed specifically that for $\mathcal{F} = \mathcal{S}_2(q)$ the set of holomorphic new-forms of weight 2 and prime level q (with trivial nebentypus), if one could show that for q large enough at least $(25 + 2017^{-2017})\%$ of the central L-values $L(f,1/2)$ do not vanish (more precisely that at least $(25 + 2017^{-2017})\%$ of these central values are larger than $\log^{-2017} q$) then there would be no Landau-Siegel zero. They eventually proved

THEOREM 13.5 ([**IS00**]). *As $q \to \infty$ along the primes one has*

$$\frac{|\{f \in \mathcal{S}_2(q),\ L(f,1/2) \geqslant \log^{-2} q\}|}{|\mathcal{S}_2(q)|} \geqslant 1/4 - o(1).$$

[12]A reasonable definition of the notion of "reasonable" can be found in [**Kow13, SST16**]

This is "just" at the limit.

The possibility of producing a positive proportion of non-vanishing is not limited to this specific family and one of the most powerful and general tools to achieve this is via the *mollification method.* The principle of mollification method is as follows: given the family $\mathcal{F}$, one considers for some parameter $L \geqslant 1$ and some suitable vector $\boldsymbol{x}_L = (x_\ell)_{\ell \leqslant L} \in \mathbf{C}^\ell$ the linear form

$$\mathcal{L}(\mathcal{F}, \boldsymbol{x}_L) := \frac{1}{|\mathcal{F}|} \sum_{f \in \mathcal{F}} L(f, 1/2) M(f, \boldsymbol{x}_L) \tag{13.2}$$

and the quadratic form

$$\mathcal{Q}(\mathcal{F}, \boldsymbol{x}_L) := \frac{1}{|\mathcal{F}|} \sum_{f \in \mathcal{F}} |L(f, 1/2) M(f, \boldsymbol{x}_L)|^2 \tag{13.3}$$

where $M(f, \boldsymbol{x}_L)$ is the linear form (called "mollifier")

$$M(f, \boldsymbol{x}_L) = \sum_{\ell \leqslant L} \frac{\lambda_f(\ell)}{\ell^{1/2}} x_\ell$$

and the x_ℓ are coefficients to be chosen in an optimal way with the idea of approximating the inverse $L(f, 1/2)^{-1}$. Such coefficients are almost bounded, i.e. satisfy:

$$x_\ell = |\mathcal{F}|^{o(1)}.$$

By Cauchy's inequality one has

$$\frac{|\{f \in \mathcal{F},\ L(f, 1/2) \neq 0\}|}{|\mathcal{F}|} \geqslant \frac{|\mathcal{L}(\mathcal{F}, \boldsymbol{x}_L)|^2}{\mathcal{Q}(\mathcal{F}, \boldsymbol{x}_L)}.$$

For suitable families one can evaluate asymptotically $\mathcal{L}(\mathcal{F}, \boldsymbol{x}_L)$ and $\mathcal{Q}(\mathcal{F}, \boldsymbol{x}_L)$ (the hard case being $\mathcal{Q}$) when $L = |\mathcal{F}|^\lambda$ for $\lambda > 0$ some fixed constant and (upon minimizing $\mathcal{Q}(\mathcal{F}, \boldsymbol{x}_L)$ with respect to $\mathcal{L}(\mathcal{F}, \boldsymbol{x}_L)$) one usually shows that

$$\frac{|\mathcal{L}(\mathcal{F}, \boldsymbol{x}_L)|^2}{\mathcal{Q}(\mathcal{F}, \boldsymbol{x}_L)} = F(\lambda) + o(1) \tag{13.4}$$

for F some increasing rational fraction with $F(0) = 0$. In [**IS00**], Iwaniec and Sarnak have also implemented this strategy for the (simpler) family of Dirichlet L-functions of modulus q

$$\{L(\chi, s) = \sum_{n \geqslant 1} \frac{\chi(n)}{n^s},\ \chi \in (\widehat{\mathbf{Z}/q\mathbf{Z}})^\times\}$$

and were able to evaluate (13.2) and (13.3) for any $\lambda < 1/2$ and to prove (13.4) with

$$F(\lambda) = \frac{\lambda}{\lambda + 1}$$

hence:

THEOREM 13.6 ([**IS99**]). *As $q \to \infty$ along the primes one has*

$$\frac{|\{\chi \,(\mathrm{mod}\, q),\ L(\chi, 1/2) \neq 0\}|}{|\{\chi \,(\mathrm{mod}\, q)\}|} \geqslant 1/3 - o(1).$$

Thus the proportion of non-vanishing can be arbitrarily close to $33.33\ldots\%$. Shortly after, Michel and Vanderkam [**MV00**] obtained the same proportion by a slightly different method: taking into account the fact that for a complex character, the L-function $L(\chi,s)$ is not self-dual ($L(\chi,s)\neq L(\overline{\chi},s)$) and has root number

$$\varepsilon_\chi = i^{\mathfrak{a}}\frac{\tau(\chi)}{q^{1/2}},\ \mathfrak{a}=\frac{\chi(-1)-1}{2}$$

were $\tau(\chi)$ is the Gauss sum, they introduced a symmetrized mollifier of the shape

$$M^s(\chi,\boldsymbol{x}_L)=M(\chi,\boldsymbol{x}_L)+\overline{\varepsilon_\chi}M(\overline{\chi},\boldsymbol{x}_L)=\sum_{\ell\leqslant L}\frac{\chi(\ell)+\overline{\varepsilon_\chi}.\overline{\chi}(\ell)}{\ell^{1/2}}x_\ell.$$

Because of the oscillation of the root number ε_χ, they could evaluate (13.3) only in the shorter range $\lambda<1/4$. However this weaker range is offset by the fact that the symmetrized mollifier is more effective: indeed the rational fraction $F(\lambda)$ is then replaced by

$$F^s(\lambda)=\frac{2\lambda}{2\lambda+1}$$

which takes value $1/3$ at $\lambda=1/4$.

Recently R. Khan and H. Ngo founds a better method to bound the exponential sums considered in [**MV00**] building on Theorem 13.3 and they increased the allowed range from $\lambda<1/4$ to $\lambda<3/10$:

THEOREM 13.7 ([**KN16**]). *As $q\to\infty$ along the primes one has*

$$\frac{|\{\chi\ (\mathrm{mod}\, q),\ L(\chi,1/2)\neq 0\}|}{|\{\chi\ (\mathrm{mod}\, q)\}|}\geqslant 3/8-o(1).$$

The key step in their proof is the asymptotic evaluation of the second mollified moment

$$\frac{1}{\varphi(q)}\sum_{\chi\ (\mathrm{mod}\, q)}|L(\chi,1/2)|^2|M^s(\chi,\boldsymbol{x}_L)|^2 \tag{13.5}$$

for $L=q^\lambda$, and any fixed $\lambda<3/10$. By (nowadays) standard methods[13] the L-value $L(\chi,1/2)$ can be written as a sum of rapidly converging series (cf. [**IK04**, Theorem 5.3]): for q prime and $\chi\neq 1$

$$|L(\chi,1/2)|^2=2\sum_{n_1,n_2\geqslant 1}\frac{\chi(n_1)\overline{\chi}(n_2)}{(n_1n_2)^{1/2}}V(\frac{n_1n_2}{q})$$

where V is a rapidly decreasing function which depends on χ only through its parity $\chi(-1)=\pm 1$. Plugging this expression in the second moment (13.5) and unfolding, one finds that the key point is to obtain a bound of the following shape[14]

$$\sum\sum_{\substack{\ell_1,\ell_2\leqslant L,n_1,n_2\\(l_1l_2n_1n_2,q)=1}}\frac{x_{l_1}\overline{x_{l_2}}}{(ql_1l_2n_1n_2)^{1/2}}V(\frac{n_1n_2}{q})e\left(\frac{n_2\overline{l_1}l_2n_1}{q}\right)\ll q^{-\delta} \tag{13.6}$$

for some $\delta=\delta(\lambda)>0$ for any fixed $\lambda<3/10$. This sum can be decomposed in various sub-sums in which the variables are localized to specific ranges. The

[13] inappropriately called "approximate functional equation"

[14] for simplicity we ignore the dependency of V in the parity of the χ's

problem becomes essentially that of bounding by $O(q^{-\delta})$ the family of bilinear sums

$$\Sigma(L_1,L_2,N_1,N_2)=\frac{1}{(qL_1L_2N_1N_2)^{1/2}}\sum\sum_{\substack{l_i\sim L_i, i=1,2\\ n_1,n_2}} x_{l_1}\overline{x_{l_2}}W(\frac{n_1}{N_1})W(\frac{n_2}{N_2})e\left(\frac{n_2\overline{l_1}l_2n_1}{q}\right)$$

where $W\in\mathcal{C}_c(]1/2,2[)$, $L_1,L_2\leqslant L$ and $N_1N_2\leqslant q$.

The n_2-sum is essentially a geometric series bounded by

$$\ll\min(N_2,\|\overline{l_1}l_2n_1/q\|^{-1})$$

where $\|\cdot\|$ is the distance to the nearest integer. Hence

$$\begin{aligned}\Sigma(L_1,L_2,N_1,N_2)&\ll\frac{q^{\varepsilon}}{(qL_1L_2N_1N_2)^{1/2}}\sum_{m\approx L_1L_2N_1}\min(N_2,\|\overline{m}/q\|^{-1})\\ &\ll\frac{q^{2\varepsilon}}{(qL_1L_2N_1N_2)^{1/2}}\max_{1\leqslant U\leqslant q/2}\min(N_2,\frac{q}{U})\sum\sum_{\substack{m\approx L_1L_2N_1,\ ,\ u\sim U\\ um\equiv\pm1\,(\mathrm{mod}\,q)}}1\\ &\ll\frac{q^{2\varepsilon}}{(qL_1L_2N_1N_2)^{1/2}}\max_{1\leqslant U\leqslant q/2}\min(N_2,\frac{q}{U})(\frac{L_1L_2N_1U}{q}+1)\\ &\ll q^{2\varepsilon}\frac{L}{q^{1/2}}(\frac{N_1}{N_2})^{1/2}. \qquad (13.7)\end{aligned}$$

(Observe that for $\frac{L_1L_2N_1U}{q}\ll 1$ the equation $um\equiv\pm1\,(\mathrm{mod}\,q)$ has no solution unless $L_1L_2N_1U\ll 1$).

Alternatively, applying the Poisson summation formula to the n_1 variable we obtain a sum of the shape

$$\Sigma(L_1,L_2,N_1,N_2)$$
$$=\frac{1}{(qL_1L_2N_1N_2)^{1/2}}\frac{N_1}{q^{1/2}}\sum\sum_{\substack{l_i\sim L_i, i=1,2\\ n_1,n_2}} x_{l_1}\overline{x_{l_2}}\widetilde{W}(\frac{n_1}{q/N_1})W(\frac{n_2}{N_2})\,\mathrm{Kl}_2(\overline{l_1}\overline{l_2}n_1n_2;q)$$

where $\widetilde{W}$ is bounded and rapidly decreasing. Bounding this sum trivially (using that $|\,\mathrm{Kl}_2(m;q)|\leqslant 2$) yields

$$\Sigma(L_1,L_2,N_1,N_2)\ll q^{\varepsilon}L(\frac{N_2}{N_1})^{1/2}. \qquad (13.8)$$

The expression $\min(\frac{L}{q^{1/2}}(\frac{N_1}{N_2})^{1/2},L(\frac{N_2}{N_1})^{1/2})$ is maximal for $\frac{N_1}{N_2}=q^{1/2}$ and equals $L/q^{1/4}$ which is $O(q^{-\delta})$ if $\lambda<1/4$.

The bound (13.8) did not exploit cancellation from the n_1,n_2,l_1,l_2 averaging and indeed this is not evident because in the limiting case $N_1=q^{3/4}$, $N_2=q/N_1=q^{1/4}$, $L_1=L_2=L=q^{1/4}$, one has

$$n_1\approx n_2\approx l_1\approx l_2\approx q^{1/4}$$

which is pretty short. Nevertheless Khan and Ngo where able to detect further cancellation from summing of these short variables. The idea, which we have met already, is to group some of these variables to form longer variables. One possibility could be to group together n_1, n_2 on the one hand and l_1, l_2 on the other hand with the idea of applying the methods of §9. However, the new variables would have size $q^{1/2}$, which is the Pólya-Vinogradov range at which point the standard completion method just fails. Instead, one can group n_1, n_2 and l_2 together and leave l_1 alone. The variable $r = n_1 n_2 \overline{l_2} \pmod q$ takes essentially $q^{3/4}$ distinct values but over all of $\mathbf{F}_q^\times$ and does not vary along an interval. To counter this defect, one uses the Holder inequality instead of Cauchy-Schwarz.

Proceeding as above, we write

$$\Sigma(L_1, L_2, N_1, N_2) = \frac{1}{(qL_1L_2N_1N_2)^{1/2}} \frac{N_1}{q^{1/2}} \sum\sum_{r\in\mathbf{F}_q^\times, l_1} x_{l_1} \nu(r) \operatorname{Kl}_2(\overline{l_1} r; q)$$

where

$$\nu(r) = \sum\sum_{\substack{l_2, n_1, n_2 \\ r = n_1 n_2 \overline{l_2} (q)}} \overline{x_{l_2}} \widetilde{W}(\frac{n_1}{q/N_1}) W(\frac{n_2}{N_2}).$$

Under the assumption

$$L_2 \frac{q}{N_1} N_2 < q/100 \Longrightarrow L_2 \frac{N_2}{N_1} < 1/100 \tag{13.9}$$

we have

$$\sum_r |\nu(r)| + \sum_r |\nu(r)|^2 \ll q^\varepsilon L_2 \frac{q}{N_1} N_2.$$

Indeed under (13.9) one has

$$\bar{l}_2 n_1 n_2 \equiv \bar{l}'_2 n'_1 n'_2 \pmod q \Longleftrightarrow l'_2 n_1 n_2 \equiv l_2 n'_1 n'_2 \pmod q \Longleftrightarrow l'_2 n_1 n_2 = l_2 n'_1 n'_2$$

and the choice of l'_2, n_1, n_2 determines l_2, n'_1, n'_2 up to $O(q^\varepsilon)$ possibilities. Hence, applying Cauchy's inequality twice, we obtain

$$\Sigma(L_1, L_2, N_1, N_2) = \frac{q^\varepsilon}{(qL_1L_2N_1N_2)^{1/2}} \frac{N_1}{q^{1/2}} (L_2 \frac{q}{N_1} N_2)^{3/4} \times \Bigl(\sum_{r\in\mathbf{F}_q^\times} |\sum_{l\sim L_1} x_l \operatorname{Kl}_2(\bar{l} r; q)|^4 \Bigr)^{1/4}.$$

Now (using that $\operatorname{Kl}_2(n; q) \in \mathbf{R}$)

$$\sum_{r\in\mathbf{F}_q^\times} |\sum_{l\sim L_1} x_l \operatorname{Kl}_2(\bar{l} r; q)|^4 \ll q^\varepsilon \sum_{\boldsymbol{l}} |\sum_{r\in\mathbf{F}_q^\times} \prod_{i=1}^4 \operatorname{Kl}_2(\overline{l_i} r; q)|$$

where $\boldsymbol{l} = (l_1, l_2, l_3, l_4) \in [L_1, 2L_1[^4$.

Theorem 13.3, applied to the Kloosterman sheaf, gives

$$\sum_{r\in\mathbf{F}_q^\times} \prod_{i=1}^4 \operatorname{Kl}_2(\overline{l_i} r; q) \ll q^{1/2}$$

unless there exists a partition $\{1, 2, 3, 4\} = \{i, j\} \sqcup \{k, l\}$ such that

$$l_i = l_j, \; l_k = l_l.$$

In this case, we use the trivial bound

$$\sum_{r\in\mathbf{F}_q^\times} \prod_{i=1}^4 \operatorname{Kl}_2(\overline{l_i} r; q) \ll q.$$

Hence

$$\sum_{l} \Big| \sum_{r\in\mathbf{F}_q^\times} \prod_{i=1}^{4} \mathrm{Kl}_2(\overline{l}_i r; q)\Big| \ll L_1^2 q + L_1^4 q^{1/2}$$

and

$$\Sigma(L_1, L_2, N_1, N_2) \ll \frac{q^\varepsilon}{(qL_1L_2N_1N_2)^{1/2}} \frac{N_1}{q^{1/2}} (L_2 \frac{q}{N_1} N_2)^{3/4} (L_1^{1/2} q^{1/4} + L_1 q^{1/8})$$

$$\ll q^\varepsilon L (\frac{N_2}{N_1})^{1/2} (Lq\frac{N_2}{N_1})^{-1/4} (L^{-1/2} q^{1/4} + q^{1/8}). \quad (13.10)$$

For $L \geqslant q^{1/4}$ (the range one would like to improve) one obtains under (13.9)

$$\Sigma(L_1, L_2, N_1, N_2) \ll q^\varepsilon L (\frac{N_2}{N_1})^{1/2} (Lq^{1/2}\frac{N_2}{N_1})^{-1/4}. \quad (13.11)$$

Suppose now we are in a limiting case for (13.8), namely $L^2 N_2/N_1 = 1$. Then (13.9) holds as long as $L \gg 1$ and (13.11) improves over (13.8) by a factor $(q^{1/2}/L)^{1/4}$, which is < 1 as long as $L < q^{1/2}$.

A more detailed analysis combining (13.7), (13.8) and (13.11) shows that (13.6) holds for any fixed $\lambda < 3/10$, and hence leads to Theorem 13.7.

14. Advanced completion methods: the q-van der Corput method

In this section and the next ones, we discuss general methods to evaluate trace functions along intervals of length smaller than the Pólya-Vinogradov range discussed in §6.

14.1. The q-van der Corput method. One of the most basic techniques encountered in analytic number to estimate sums of (analytic) exponentials is the *van der Corput method* (see [**IK04**, Chap. 8]). The q-Van der Corput method is an arithmetic variant due to Heath-Brown which replace archimedean analysis with q-adic analysis. That method concerns c-periodic functions for c a *composite number*. Suppose (to simplify the presentation) that $c = pq$ for two primes p and q and let

$$K_c = K_p K_q \colon \mathbf{Z}/c\mathbf{Z} \to \mathbf{C}$$

be some function modulo c which is the product of two trace functions modulo p and q (of conductor bounded by some constant C). We consider the sum

$$S_V(K, N) := \sum_n K_c(n) V(\frac{n}{N}) = \sum_n K_p(n \,(\mathrm{mod}\, p)) K_q(n \,(\mathrm{mod}\, q)) V(\frac{n}{N})$$

where $V \in \mathcal{C}^\infty(]1, 2[)$ and $2N < c = pq$. We will explain the proof of the following result

THEOREM 14.1 (q-van der Corput method). *Let $c = pq$ a product of two primes and $K_c = K_p.K_q$ as above; assume that K_q is the trace function associated with a geometrically irreducible sheaf $\mathcal{F}$, which is not geometrically isomorphic to a linear or quadratic phase (i.e. not of the shape $[P]^*\mathcal{L}_\psi$ for P a polynomial of degre $\leqslant 2$). Then for $2N < pq$, we have*

$$S_V(K_c, N) \ll_C N^{1/2}(p + q^{1/2})^{1/2}.$$

REMARK 14.2. This bound is non trivial as long as

$$N \geqslant \max(p, q^{1/2}),$$

which is a weaker condition than $N \geqslant (pq)^{1/2}$ as long as

$$1 < p < q.$$

We have therefore improved over the Pólya-Vinogradov range; moreover the range of non triviality is maximal when $p \approx c^{1/3}$ and $q \approx c^{2/3}$. In that case, one obtains

$$S_V(K, N) \ll_C N^{1/2} c^{1/6} \tag{14.1}$$

which is non-trivial as long as

$$N \geqslant c^{1/3}.$$

PROOF. The proof makes use of the (semi-)invariance of K under translations:

$$K(n + ph) = K_p(n) K_q(n + ph).$$

For $H \leqslant N/100p$ we have

$$\begin{aligned}
S_V(K, N) &= \frac{1}{2H+1} \sum_{|h| \leqslant H} \sum_n K_p(n) K_q(n + ph) V(\frac{n + ph}{N}) \\
&= \frac{1}{2H+1} \sum_{|n| \leqslant 3N} K_p(n) \sum_{|h| \leqslant H} K_q(n + ph) V(\frac{n + ph}{N}) \\
&\ll \frac{1}{2H+1} N^{1/2} \Big(\sum_{|n| \leqslant 3N} \Big| \sum_{|h| \leqslant H} K_q(n + ph) V(\frac{n + ph}{N}) \Big|^2 \Big)^{1/2} \\
&= \frac{N^{1/2}}{H} \Big(\sum\sum_{|h|, |h'| \leqslant H} \sum_n K_q(n + ph) \overline{K_q(n + ph')} W_{p,h,h'}(\frac{n}{N}) \Big)^{1/2}
\end{aligned}$$

where

$$W_{p,h,h'}(\frac{n}{N}) = V(\frac{n + ph}{N}) \overline{V(\frac{n + ph'}{N})}.$$

We split the h, h'-sum into its diagonal and non-diagonal contribution

$$\sum\sum_{|h|, |h'| \leqslant H} \cdots = \sum\sum_{\substack{|h|, |h'| \leqslant H \\ h = h'}} \cdots + \sum\sum_{\substack{|h|, |h'| \leqslant H \\ h \neq h'}} \cdots .$$

The diagonal sum contributes by $O(NH)$ and it remains to consider the correlation sums

$$\mathcal{C}(K_q, h, h') := \sum_n K_q(n + ph) \overline{K_q(n + ph')} W_{p,h,h'}(\frac{n}{N})$$

for $h \neq h'$.

Observe that this is the sum of a trace function of modulus q of length $\approx N$. By comparison with the initial sum, we had a trace function of modulus pq of length $\approx N$ so the relative length of n compared to the modulus has increased ! By the Pólya-Vinogradov method, it is sufficient to determine whether the sheaf

$$[+ph]^* \mathcal{F} \otimes [+ph']^* D(\mathcal{F})$$

has an Artin-Schreier sheaf in its irreducible components. This is equivalent to whether one has an isomorphism

$$[+p(h - h')]^* \mathcal{F} \simeq \mathcal{F} \otimes \mathcal{L}_\psi$$

for some Artin-Schreier sheaf. We will answer this question in a slighly more general form:

DEFINITION 14.3. For d an integer satisfying $1 \leqslant d < q$, a polynomial phase sheaf of degree d is a sheaf of the shape $[P]^*\mathcal{L}_\psi$ for P a polynomial of degree d and ψ a non-trivial additive character. It is lisse on $\mathbf{A}^1_{\mathbf{F}_q}$, ramified at infinity with Swan conductor equal to d and its trace function equals

$$x \mapsto \psi(P(x)).$$

We can now invoke the following

PROPOSITION 14.4 ([**Pol14a**]). *Let d be an integer satisfying $1 \leqslant d < q$. Suppose that $\mathcal{F}$ is geometrically irreducible, not isomorphic to a polynomial phase of degree $\leqslant d$ and that $C(\mathcal{F}) \leqslant q^{1/2}$. Then for any $h \in \mathbf{F}_q - \{0\}$ and any non-constant polynomial P of degree $\leqslant d-1$,*

$$[+h]^*\mathcal{F} \text{ and } \mathcal{F} \otimes [P]^*\mathcal{L}_\psi$$

are not geometrically isomorphic.

PROOF. We will only give the easiest part of it and refer to [**Pol14a**, Thm. 6.15] for the complete argument. Suppose that $\mathcal{F}$ is ramified at some point $x_0 \in \mathbf{A}^1(\overline{\mathbf{F}_q})$, since polynomial phases are ramified only at ∞ the isomorphism

$$[+h]^*\mathcal{F} \simeq \mathcal{F} \otimes [P]^*\mathcal{L}_\psi$$

restricted to the inertia group I_x implies that $\mathcal{F}$ is ramified at $x_0 - h$ and iterating at $x_0 - nh$ for any $n \in \mathbf{Z}$, this would imply that $C(\mathcal{F}) \geqslant q$ which is excluded. It remains to deal with the case where $\mathcal{F}$ is ramified only at ∞. □

Under our assumptions the above proposition implies that for $h \neq h'$

$$\mathcal{C}(K_q, h, h') = O(q^{1/2})$$

and that

$$S_V(K,N) \ll N^{1/2}(\frac{N}{H} + q^{1/2})^{1/2}$$

and we choose $H = N/100p$ to conclude the proof. □

14.2. Iterating the method. Suppose more generally that c is a squarefree number and that

$$K_c = \prod_{q|c} K_q$$

is a product of trace functions associated to sheaves not containing any polynomial phases. One can repeat the above argument after factoring c into a product of squarefree coprime moduli $r.s$ and decompose accordingly

$$K_c = K_r.K_s.$$

Thus, we have to bound sums of the shape

$$\sum_n K_s(n+rh)\overline{K_s(n+rh')}W_{r,h,h'}(\frac{n}{N}). \tag{14.2}$$

This time we need to be a bit more careful and decompose the h, h' sum according to the gcd $(h-h', s)$. After applying the Poisson summation formula (cf. (6.2)) we can factor the resulting Fourier transform modulo s into sums over prime moduli $q|s$:

$$\widehat{K_s}(y) = \prod_{q|s} \widehat{K_q}(\overline{s_q}y \,(\mathrm{mod}\, q)),\ y \in \mathbf{Z}/s\mathbf{Z},\ s_q = s/q.$$

If $q|h-h'$ we use the trivial bound $\widehat{K_q}(\overline{s_q}y\,(\mathrm{mod}\,q)) \ll q^{1/2}$ and if $q \nmid h-h'$ we use the non-trivial bound $\widehat{K_q}(\overline{s_q}y\,(\mathrm{mod}\,q)) \ll 1$. We eventually obtain (see [**Pol14a**])

THEOREM 14.5. *Let $C \geqslant 1$, let c be squarefree and let $K_c\colon \mathbf{Z}/c\mathbf{Z} \to \mathbf{C}$ be a product of trace functions K_q such that for any prime $q|c$ the underlying sheaf $\mathcal{F}_q$ is of conductor $\leqslant C$, is geometrically irreducible and is not geometrically isomorphic to any polynomial phase of degree $\leqslant 2$. Then*

$$S_V(K_c, N) \ll_{C,\varepsilon} c^{\varepsilon} N^{1/2}(r + s^{1/2})^{1/2}$$

for any $\varepsilon > 0$.

If s is not a prime, we could also iterate, factor s into $s = r_2 s_2$ and instead of applying the Pólya-Vinogradov completion method to the sum (14.2), we could also apply the q-van der Corput method with the trace functions

$$n \mapsto K_q(n + rh)\overline{K_q(n + rh')},\ q|s_1.$$

This leads us to the quadruple correlation sum

$$\mathcal{C}(K_q, \gamma, \alpha) = \frac{1}{q}\sum_x K_q(\gamma_1 \cdot x)K_q(\gamma_2 \cdot x)\overline{K_q(\gamma_1' \cdot x)K_q(\gamma_2' \cdot x)}\mathrm{e}_q(\alpha x)$$

where the $\gamma_i, \gamma_j',\ i, j = i, 2$ are unipotent matrices

$$\gamma_i = \begin{pmatrix} 1 & h_i \\ 0 & 1 \end{pmatrix},\ \gamma_i' = \begin{pmatrix} 1 & h_j' \\ 0 & 1 \end{pmatrix}.$$

In suitable situations, we can then apply Theorem 13.3 from the previous section.

An important example is when

$$K_c(n) = \mathrm{Kl}_k(n; c) = \frac{1}{c^{(k-1)/2}} \sum_{\substack{x_1,\dots,x_k \in (\mathbf{Z}/c\mathbf{Z})^\times \\ x_1.\dots.x_k = n}} e\left(\frac{x_1 + \dots + x_k}{c}\right)$$

is a hyper-Kloosterman sum. For any $q|c$, one has

$$K_q(y) = \mathrm{Kl}_k(\overline{c_q}^k y; q) \text{ with } c_q = c/q$$

and the underlying sheaf is the multiplicatively shifted Kloosterman sheaf $\mathcal{F}_q = [\times\overline{c_q}^k]^*\mathcal{K}\ell_k$. In that case Theorem 13.3 applies and we eventually obtain the bound

$$S_V(\mathrm{Kl}_k(\cdot; c), N) \ll_k c^{\varepsilon} N^{1/2} \left(r + (N^{1/2}(s_1 + s_2^{1/2}))^{1/2}\right)^{1/2}$$

for any factorisation $c = rs_1s_2$. In particular, if there exists a factorisation $c = rs_1s_2$ such that

$$r \approx c^{1/4},\ s_1 \approx c^{1/4},\ s_2 \approx c^{1/2}$$

we obtain

$$S_V(\mathrm{Kl}_k(\cdot; c), N) \ll_k N^{1-\eta}$$

for some $\eta = \eta(\delta) > 0$ as long as

$$N \geqslant c^{1/4+\delta}.$$

Iterating once more we see that for any factorisation $c = rs_1s_2s_3$ one has

(14.3) $$S_V(\mathrm{Kl}_k(\cdot;c),N) \ll_{k,\varepsilon} c^{\varepsilon}N^{1/2}\Big(r + (N^{1/2}(s_1 + (N^{1/2}(s_2 + s_3^{1/2}))^{1/2}))^{1/2}\Big)^{1/2}$$

so if there exists a factorisation $c = rs_1s_2s_3$ such that

$$r \approx c^{1/5},\ s_1 \approx c^{1/5},\ s_2 \approx c^{1/5},\ s_3 \approx c^{2/5}$$

then

$$S_V(\mathrm{Kl}_k(\cdot;c),N) \ll_{k,\varepsilon} N^{1-\eta}$$

for some $\eta = \eta(\delta) > 0$ as long as

$$N \geqslant c^{1/5+\delta}.$$

We can continue this way as long as enough factorisation for c are available. Such availability is garanteed by the notion of friability:

Definition 14.6. An integer $c \neq 0$ is Δ-friable if

$$q|c\ (q \text{ prime }) \Rightarrow q \leqslant \Delta.$$

Using the reasoning above, Irving [**Irv15**] proved the following result for $k = 2$ (in a quantitative form):

Theorem 14.7. *For any $L \geqslant 2$ there exists $l = l(L) \geqslant 1$ and $\eta = \eta(L) > 0$ such that for c a squarefree integer which is $c^{1/l}$-friable and any $k \geqslant 2$, one has,*

$$S_V(\mathrm{Kl}_k(\cdot;c),N) \ll_{k,V} N^{1-\eta}$$

whenever $N \geqslant c^{1/L}$.

Therefore one can obtain non-trivial bounds for extremely short sums of hyper-Kloosterman sums as long as their modulus is firable enough. In particular for $k = 2$ we have seen in Remark 11.3 that improving on Selberg's 2/3-exponent for the distribution of the divisor function in arithmetic progressions to large moduli (Theorem 11.2) was essentially equivalent to bounding non-trivially sums of the shape

$$\sum_{n_1,n_2}\sum \mathrm{Kl}_2(an_1n_2;c)V(\frac{n_1}{N_1^*})V(\frac{n_2}{N_2^*})$$

for

$$N_1^*N_2^* \approx c^{1/2}.$$

If $N_1^*N_2^* \approx c^{1/2}$ then $\max(N_1^*,N_2^*) \gg c^{1/4}$ and we can use the (14.3) to bound non-trivially the above sum granted that c is friable enough. This leads to the following theorem (compare with Theorem 11.2 for c a prime):

Theorem 14.8. [**Irv15**] *There exists $L \geqslant 4$ and $\eta > 0$ such that for any $c \geqslant 1$ which is squarefree and $c^{1/L}$-friable and any a coprime with c, one has for $c \leqslant X^{2/3+\eta}$ and any $A \geqslant 0$*

$$E(d_2;c,a) \ll_A \frac{X}{c}(\log X)^{-A}.$$

See [**Irv16**] and [**WX16**] for further applications of these ideas.

15. Around Zhang's theorem on bounded gaps between primes

Some of the arguments of the previous chapter can be found in Yitang Zhang's spectacular proof of the existence of bounded gaps between the primes:

THEOREM 15.1 ([**Zha14**]). *Let $(p_n)_{n\geqslant 1}$ be the sequence of primes in increasing order ($p_1 = 2, p_2 = 3, p_3 = 5, \ldots$). There exists an absolute constant C such that*

$$p_{n+1} - p_n \leqslant C$$

for infinitely many n.

Besides Zhang's original paper, we refer to [**Gra15, Kow15**] for a detailed description of Zhang's proof and the methods involved and historical background. Let us however mention a few important facts:

– The question of the existence of small gaps between primes has occupied analytic number theorists for a very long time and has been the motivations for the invention of many techniques, in particular the *sieve method* to detect primes with additional constraints. A conceptual breakthrough occurred with the work of Goldston, Pintz and Yıldırım [**GPY09**] who proved the weaker result

$$\liminf_{n} \frac{p_{n+1} - p_n}{\log p_n} = 0$$

and who on this occasion invented a technique which is also key to Zhang's approach (see Soundararajan's account of their works [**Sou07**]).
Zhang's theorem can be seen as an approximation to the twin prime conjecture:

There exist infinitely many primes p such that $p + 2$ is prime.

Indeed, Zhang's theorem with $C = 2$ is equivalent to the twin prime conjecture.

– A value for the constant C can be given explicitly : Zhang himself gave

$$C = 70.10^6$$

and mentioned that this could certainly be improved. Improving the value of this constant was the objective of the Polymath8 project: following and optimizing Zhang's method in several aspects (some to be explained below), the value was reduced to

$$C = 4680.$$

However Maynard [**May16**] made independently another conceptual breakthrough, simplifying the whole proof and making it possible to obtain stronger results and improving the constant to

$$C = 600.$$

Eventually the Polymath8 project joined with Maynard ; optimizing his argument, the value

$$C = 246$$

was reached (cf. [**Pol14b**]).

A side-effect of Maynard's approach is that what we are going to describe now plays no role anymore in this specific application. Nevertheless, it adresses another important question in analytic number theory.

15.1. The Bombieri-Vinogradov theorem and beyond. The breakthrough of Goldston, Pintz and Yıldırım that is at the origin of Zhang's work builds on the use of sieve methods to detect the existence of infinitely many pairs of primes at distance $\leqslant C$ from one another. The fuel to be put in this sieve machine are results concerning the distribution of primes in arithmetic progressions to moduli large with respect to the size of the primes which are sought after. In this respect the Bombieri-Vinogradov theorem already discussed in §11 is a powerful substitute to GRH:

THEOREM 15.2 (Bombieri-Vinogradov). *For any $A > 0$ there is $B = B(A) > 0$ such that for $x \geqslant 2$*

$$\sum_{q \leqslant x^{1/2}/\log^B x} \max_{(a,q)=1} \left| \psi(x; q, a) - \frac{\psi(x; q)}{\varphi(q)} \right| \ll \frac{x}{\log^A x}.$$

For the question of the existence of bounded gaps between primes, the exponent $1/2$ appearing in the constraint $q \leqslant x^{1/2}/\log^B x$ turns out to be crucial. In their seminal work [**GPY09**], Goldston-Pintz-Yıldırım had pointed out that the Bombieri-Vinogradov theorem with the exponent $1/2$ replaced by any strictly larger constant would be sufficient to imply Theorem 15.1.

The possibility of going beyond Bombieri-Vinogradov is not unexpected: the Elliott-Halberstam conjecture predicts that any fixed exponent < 1 could replace $1/2$. That this conjecture is not wishful thinking comes from the work of Fouvry, Iwaniec and Bombieri-Friedlander-Iwaniec from the 80's [**FI83, Fou84, BFI86**] who proved versions of the Bombieri-Vinogradov theorem with exponents $> 1/2$ but for "fixed" congruences classes (for instance with the sum involving the difference $|\psi(x; q, 1) - \frac{\psi(x;q)}{\varphi(q)}|$ instead of $\max_{(a,q)=1} |\psi(x; q, a) - \frac{\psi(x;q)}{\varphi(q)}|$). Zhang's groundbreaking insight has been to nail down a beyond-Bombieri-Vinogradov type theorem that could be established unconditionally and would be sufficient to establish the existence of bounded gaps between primes. The following theorem is a variant of Zhang's theorem ([**Pol14a**, Thm 1.1]). Let us recall that an integer $q \geqslant 1$ is Δ-friable if any prime p dividing q is $\leqslant \Delta$.

THEOREM 15.3. *Let $\boldsymbol{a} = (a_p)_{p \in \mathcal{P}}$ be a sequence of integers indexed by the primes such that a_p is coprime with p for all p. For any squarefree integer q, let $a_q \pmod{q}$ be the unique congruence class modulo q such that*

$$\forall p|q, \ a_q \equiv a_p \pmod{p};$$

in particular $a_q \in (\mathbf{Z}/q\mathbf{Z})^\times$. There exist absolute constants $\theta > 1/2$ and $\delta > 0$, independent of $\boldsymbol{a}$, such that for any $A > 0$, $x > 2$ one has

$$\sum_{\substack{q \leqslant x^\theta, \ sqfree \\ q \ x^\delta-friable}} |\psi(x; q, a_q) - \frac{\psi(x; q)}{\varphi(q)}| \ll \frac{x}{\log^A x}.$$

Here the implicit constant depends only on A, but not on $\boldsymbol{a}$.

REMARK 15.4. Zhang essentially proved this theorem for $\theta = 1/2 + 1/585$ and in an effort to improve Zhang's constant, the Polymath8 project improved $1/585$ to $7/301$.

We will now describe some of the principles of the proof of this theorem and especially at the points where algebraic exponential sums occur. We refer to the introduction of [**Pol14a**] and to E. Kowalski's account in the Bourbaki seminar [**Kow15**].

Let us write $c(q)$ for $\mu^2(q)$ times the sign of the difference $\psi(x;q,a_q)-\frac{\psi(x;q)}{\varphi(q)}$. The above sum equals

$$\sum_{\substack{q\leqslant x^\theta\\ q\ x^\delta-\text{friable}}} c(q)\sum_{n\leqslant x}\Lambda(n)\Delta_{\boldsymbol{a}}(n;q).$$

where

$$\Delta_{\boldsymbol{a}}(n):=\delta_{n\equiv a_q\,(\mathrm{mod}\,q)}-\frac{\delta_{(n,q)=1}}{\varphi(q)}$$

As is usual when counting primes numbers, the next step is to decompose the von Mangoldt function $\Lambda(n)$ into a sum of convolution of arithmetic functions (for instance by using Heath-Brown's identity Lemma 8.3 as in §8): we essentially arrive at the problem of bounding $(\log x)^{O_J(1)}$ of the following model sums (for $j\leqslant J$ and J is a fixed and large integer)

$$\Sigma(\mathbf{M};\boldsymbol{a},Q)$$
$$:=\sum_{\substack{q\sim Q\\ q\ x^\delta-\text{friable}}} c(q)\sum\sum_{m_1,\ldots,m_{2j}}\mu(m_1)\ldots\mu(m_j)V_1\Big(\frac{m_1}{M_1}\Big)\ldots V_{2j}\Big(\frac{m_{2j}}{M_{2j}}\Big)\Delta_{a_q}(m_1\ldots m_{2j})$$

where V_i, $i=1,\ldots,2j$ are smooth functions compactly supported in $]1,2[$ and $\mathbf{M}=(M_1,\ldots,M_{2j})$ is a tuple satisfying

$$Q\leqslant x^\theta,\ M_i=:x^{\mu_i},\ \forall i\leqslant j,\ \mu_i\leqslant 1/J,\ \sum_{i\leqslant 2j}\mu_i=1+o(1).$$

Our target is the bound

$$\Sigma(\mathbf{M};\boldsymbol{a},Q)\overset{?}{\ll}\frac{x}{\log^A x}. \tag{15.1}$$

The most important case is when

$$Q=x^\theta=x^{1/2+\varpi}$$

for some fixed sufficiently small $\varpi>0$.

The variables with index $j+1\leqslant\leqslant 2j$ are called *smooth* because they are weighted by smooth functions and this makes it possible to use the Poisson summation formula on them to analyze the congruence condition mod q. This is going to be efficient if the range M_i is sufficiently big relatively to $q\sim Q$. The variables with indices $1\leqslant i\leqslant j$ are weighted by the Möbius function but (at least as long as some strong form of the Generalized Riemann Hypothesis is not available) we cannot exploit this information and we will consider the Möbius functions like arbitrary bounded functions. The tradeoff to non-smoothness is that the range of these variables is pretty short $M_i\leqslant x^{1/J}$, especially if J is choosen large.

As we did before we will aggregate some of the variables m_i, $i=1,\ldots,2j$ so as to form two new variables whose ranges are located adequately (similarly to what we did in §8) and will use different methods to bound the sums depending on the size and the type of these new variables.

More precisely, we define

$$\alpha_i(m) = \begin{cases} \mu(m)V_i\Big(\frac{m}{M_i}\Big) & 1 \leqslant i \leqslant j \\ V_i\Big(\frac{m}{M_i}\Big) & j+1 \leqslant i \leqslant 2j. \end{cases}$$

Given some partition of the set of m-indices

$$\{1,\dots,2j\} = \boldsymbol{I} \sqcup \boldsymbol{J}$$

let

$$M = \prod_{i\in\boldsymbol{I}} M_i,\ N = \prod_{j\in\boldsymbol{J}} M_j$$

and

$$\mu_{\boldsymbol{I}} := \sum_{i\in\boldsymbol{I}} \mu_i,\ \mu_{\boldsymbol{J}} := \sum_{i\in\boldsymbol{J}} \mu_i.$$

We have

$$\mu_{\boldsymbol{I}} + \mu_{\boldsymbol{J}} = 1 + o(1),\ M = x^{\mu_{\boldsymbol{I}}},\ N = x^{\mu_{\boldsymbol{J}}}.$$

In the sequel we will always make the convention that $N \leqslant M$ or equivalently $\mu_{\boldsymbol{I}} \geqslant \mu_{\boldsymbol{J}}$.

Finally we define the Dirichlet convolution functions

$$\alpha(m) := \star_{i\in\boldsymbol{I}}\alpha_i(m),\ \beta(n) := \star_{i\in\boldsymbol{J}}\alpha_i(n).$$

We are reduced to bound sums of the shape

$$\sum_{\substack{q\sim Q\\ x^\delta-\text{friable}}} c(q)\sum_{\substack{m\sim M\\ n\sim N}}\sum \alpha(m)\beta(n)\Delta_{a_q}(mn) \overset{?}{\ll} \frac{x}{\log^A x}. \tag{15.2}$$

Observe that the functions α,β are essentially bounded

$$\forall\varepsilon>0, \alpha(m),\beta(n) \ll x^\varepsilon$$

so we need only to improve slightly over the trivial bound.

15.2. Splitting into types. The sums (15.2) will be subdivided into three different types and their treatment will depend on which type the sum belong.

This subdivision follows from the following simple combinatorial Lemma (cf. [**Pol14a**, Lem. 3.1]):

LEMMA 15.5. *Let $1/10 < \sigma < 1/2$ and let μ_i, $i = 1,\dots 2j$ be some non-negative real numbers such that*

$$\sum_{i=1}^{2j} \mu_i = 1.$$

One of the following holds

- *Type 0: there exists i such that $\mu_i \geqslant 1/2+\sigma$.*
- *Type II: there exists a partition*

$$\{1,\dots,2j\} = \boldsymbol{I} \sqcup \boldsymbol{J}$$

such that

$$1/2-\sigma \leqslant \sum_{i\in\boldsymbol{J}} \mu_i \leqslant \sum_{i\in\boldsymbol{I}} \mu_i < 1/2+\sigma.$$

- *Type III: there exist distincts i_1, i_2, i_3 such that*

$$2\sigma \leqslant \mu_{i_1} \leqslant \mu_{i_2} \leqslant \mu_{i_3} \leqslant 1/2-\sigma \text{ and } \mu_{i_1}+\mu_{i_2} \geqslant 1/2+\sigma.$$

REMARK 15.6. If $\sigma > 1/6$ the Type III situation never occurs since $2\sigma > 1/2-\sigma$.

Given σ such that

$$1/10 < \sigma < 1/2$$

we assume that J is choosen large enough so that

$$1/J \leqslant \min(1/2-\sigma, \sigma). \tag{15.3}$$

We say that a sum (15.2) is of

- Type 0, if there exists some i_0 such that $\mu_{i_0} \geqslant 1/2+\sigma$. We choose

  $$\boldsymbol{I} = \{i_0\} \text{ and } \boldsymbol{J} \text{ the complement.}$$

  Since for any $i \leqslant j$, one has $\mu_i \leqslant 1/J < 1/2+\sigma$, necessarily $i_0 \geqslant j+1$ corresponds to a smooth variable; the corresponding sum therefore equals

  $$\sum_{\substack{q\sim Q\\ x^\delta-\text{friable}}} c(q) \sum\sum_{m\geqslant 1, n\sim N} V(\frac{m}{M_{i_0}})\beta(n)\Delta_{a_q}(mn). \tag{15.4}$$

- Type I/II if one can partition the set of indices

  $$\{1,\ldots,2j\} = \boldsymbol{I} \sqcup \boldsymbol{J}$$

  in a way that the corresponding ranges

  $$M = \prod_{i\in\boldsymbol{I}} M_i = x^{\mu_{\boldsymbol{I}}} \geqslant N = \prod_{i\in\boldsymbol{J}} M_i = x^{\mu_{\boldsymbol{J}}}$$

  satisfy

  $$1/2-\sigma \leqslant \mu_{\boldsymbol{J}} = \sum_{i\in\boldsymbol{J}} \mu_i \leqslant 1/2 \tag{15.5}$$

- Type III if we are neither in the Type 0 or Type I/II situation: there exist distinct indices i_1, i_2, i_3 such that

  $$2\sigma \leqslant \mu_{i_1} \leqslant \mu_{i_2} \leqslant \mu_{i_3} \leqslant 1/2-\sigma \text{ and } \mu_{i_1}+\mu_{i_2} \geqslant 1/2+\sigma.$$

  We choose

  $$\boldsymbol{I} = \{i_1, i_2, i_3\} \text{ and } \boldsymbol{J} \text{ to be the complement.}$$

  Again, since $1/J < 2\sigma$ by (15.3), the indices i_1, i_2, i_3 are associated to smooth variables and the Type III sums are of the shape

  $$\sum_{\substack{q\sim Q\\ x^\delta-\text{friable}}} c(q) \sum\sum_{\substack{m_1,m_2,m_3\\ n\sim N}} V(\frac{m_1}{M_{i_1}})V(\frac{m_2}{M_{i_2}})V(\frac{m_3}{M_{i_3}})\beta(n)\Delta_{a_q}(m_1m_2m_3n).$$

REMARK 15.7. In the paper [**Pol14a**] the "Type II" sums introduced here were split into two further types that were called "Type I" and "Type II". These are the sums for which the N variable satisfies

$$\text{Type I: } x^{1/2-\sigma} \leqslant N < x^{1/2-\varpi-c}$$

$$\text{Type II: } x^{1/2-\varpi-c} \leqslant N \leqslant x^{1/2}$$

for some extra parameter c satisfying

$$1/2-\sigma < 1/2-\varpi-c < 1/2.$$

This distinction was necessary for optimisation purposes and especially to achieve the exponent $1/2+7/301$ in Theorem 15.3.

Zhang's Theorem now essentially follows from

THEOREM 15.8. *There exist $\varpi, \sigma > 0$ with $1/10 < \sigma < 1/2$ such that the bound* (15.2) *holds for the Type 0, II and III sums.*

For the rest of this section we will succinctly describe how each type of sum is handled.

The case of Type 0 sums (15.4) is immediate: one applies the Poisson summation formula to the m variable to decompose the congruence $mn \equiv a_q \,(\mathrm{mod}\, q)$. The zero frequency contribution is cancelled up to an error term by the second term of $\Delta_{a_q}(mn)$ while the non-zero frequencies contribute a negligible error term as long as the range of the m variable is larger than the modulus, i.e.

$$1/2 + \sigma > 1/2 + \varpi$$

which can be assumed.

15.3. Treatment of type II sums.

15.3.1. *The art of applying Cauchy-Schwarz.* The Type II sums are more complicated to deal with because we have essentially no control on the shape of the coefficients $\alpha(m), \beta(n)$ (except that they are being essentially bounded). The basic principle is to consider the largest variable $m \sim M$, to make it smooth using the Cauchy-Schwarz inequality and then resolve the congruence

$$m \equiv \overline{n} a_q \,(\mathrm{mod}\, q)$$

using the Poisson summation formula. This is the essence of the *dispersion method* of Linnik.

When implementing this strategy one has to decide which variables to put "inside" the Cauchy-Schwarz inequality and which to leave "outside". To be more specific, suppose we need to bound a general trilinear sum

$$\sum\sum_{m\sim M, n\sim N} \sum_{q\sim Q} \alpha_m \beta_n \gamma_q K(m,n,q)$$

and wish to smooth the m variable using Cauchy-Schwarz. There are two possibilities, either

$$\sum\sum_{m\sim M, n\sim N} \sum_{q\sim Q} \alpha_m \beta_n \gamma_q K(m,n,q) \ll \|\alpha\|_2 \|\gamma\|_2 \Big(\sum\sum_{m\sim M, q\sim Q} |\sum_{n\sim N} \beta_n K(m,n,q)|^2 \Big)^{1/2}$$

or

$$\sum\sum_{m\sim M, n\sim N} \sum_{q\sim Q} \alpha_m \beta_n \gamma_q K(m,n,q) \ll \|\alpha\|_2 \Big(\sum_{m\sim M} |\sum\sum_{n\sim N, q\sim Q} \beta_n \gamma_q K(m,n,q)|^2 \Big)^{1/2}.$$

In the first case the inner sum of the second factor equals

$$\sum\sum_{n_1,n_2\sim N} \beta_{n_1}\overline{\beta_{n_2}} \sum\sum_{m\sim M, q\sim Q} K(m,n_1,q)\overline{K(m,n_2,q)}$$

and in the second case

$$\sum\sum_{n_1,n_2\sim N} \sum\sum_{q_1,q_2\sim Q} \beta_{n_1}\gamma_{q_1}\overline{\beta_{n_2}\gamma_{q_2}} \sum_{m\sim M} K(m,n_1,q_1)\overline{K(m,n_2,q_2)}.$$

In either case, one expects to be able to detect cancellation from the m-sum, at least when the other variables (n_1, n_2) or (n_1, n_2, q_1, q_2) are not located on the diagonal (i.e. $n_1 = n_2$ or $n_1 = n_2,\ q_1 = q_2$). If the other variables are on the diagonal, no cancellation is possible but the diagonal is small compared to the space of variables.

We are faced with the following trade-off:

- For the first possibility, the m-sum is simpler (it involves three parameters n_1, n_2, q) but the ratio "size of the diagonal "/" size of the set of parameters" is $N/N^2 = N^{-1}$.
- For the second possibility, the m-sum is more complicated as it involves more auxiliary parameters n_1, n_2, q_1, q_2 but the ratio "size of the diagonal"/" size of the set of parameters" $NQ/N^2Q^2 = 1/NQ$ is smaller (hence more saving can be obtained from the diagonal part).

15.3.2. *The Type II sums.* We illustrate this discussion in the case of Type II sums. If we apply Cauchy with the q variable outside the diagonal $n_1 = n_2$ would not provide enough saving. If, on the other hand, we apply Cauchy with q inside, then the diagonal is large but we have to analyze the congruence

$$mn_1 \equiv a \,(\mathrm{mod}\, q_1),\ \ mn_2 \equiv a \,(\mathrm{mod}\, q_2)$$

which is a congruence modulo $[q_1, q_2]$. Assuming we are in the generic case of q_1, q_2 coprime, the resulting modulus is $q_1q_2 \sim Q^2 = x^{1+2\varpi}$ while $m \sim M \leqslant x^{1/2}$, which is too small for the Poisson formula to be efficient.

There is fortunately a middle-ground: we can use the extra flexibility (due to Zhang's wonderful insight) that our problem involves *friable* moduli: by the greedy algorithm, one can factor $q \sim Q$ into a product $q = rs$ where r and $s \sim Q/r$ vary over ranges that we can essentially choose as we wish (up to a small indeterminacy of x^δ for δ small). In other words, we are reduced to bounding sums of the shape

$$\Sigma(M, N; \boldsymbol{a}, R, S) = \sum_{\substack{r\sim R,\ s\sim S\\ rs\ x^\delta-\text{friable}}}\sum c(rs) \sum_{\substack{m\sim M\\ n\sim N}}\sum \alpha(m)\beta(n)\Delta_{a_{rs}}(mn)$$

for any factorisation $RS = Q$ that fits with our needs. Now, when applying Cauchy-Schwarz, we have the extra flexibility of having the r variable "out" and the s variable "in".

We do this and get

$$\begin{aligned}
&\sum_{r\sim R, s\sim S}\sum c(rs) \sum_{\substack{m\sim M\\ n\sim N}}\sum \alpha(m)\beta(n)\Delta_{a_{rs}}(mn)\\
&= \sum_{r\sim R}\sum_{m\sim M} \alpha(m) \sum_{s} c(rs) \sum_{n\sim N} \beta(n)\Delta_{a_{rs}}(mn)\\
&\ll_\varepsilon R^{1/2}M^{1/2+\varepsilon}\Big(\sum_r \sum_{s_1,s_2,n_1,n_2}\sum c(rs_1)\overline{c(rs_2)}\beta(n_1)\overline{\beta(n_2)}\\
&\qquad \times \sum_m V\Big(\frac{m}{M}\Big)\Delta_{a_{rs_1}}(mn_1)\Delta_{a_{rs_2}}(mn_2)\Big)^{1/2}
\end{aligned}$$

for V a smooth function compactly supported in $[M/4, 4M]$. We choose R of the shape

$$R = Nx^{-\varepsilon} \leqslant Mx^{-\varepsilon}$$

for $\varepsilon > 0$ but small.

Expanding the square, we obtain a sum involving four terms. The most important one comes from the product

$$(15.6)\qquad \Delta_{a_{rs_1}}(mn_1)\Delta_{a_{rs_2}}(mn_2) = \Big(\delta_{mn_1\equiv a_{rs_1}\,(\mathrm{mod}\, rs_1)} - \frac{\delta_{(n,rs_1)=1}}{\varphi(rs_1)}\Big)\Big(\delta_{mn_2\equiv a_{rs_2}\,(\mathrm{mod}\, rs_2)} - \frac{\delta_{(n,rs_2)=1}}{\varphi(rs_2)}\Big).$$

We will concentrate on the contribution of this term from now on.

The generic and main case is when $(s_1, s_2) = 1$, so that m satisfies a congruence modulo $rs_1s_2 \sim RS^2 = Mx^{2\varpi+\varepsilon}$ which is not much larger than M if ϖ is small. Observe that

$$mn_i \equiv a_{rs_i}\,(\mathrm{mod}\, rs_i),\ i = 1, 2 \Longrightarrow n_1 \equiv n_2\,(\mathrm{mod}\, r).$$

We can therefore write $n_1 = n$, $n_2 = n + rl$ with $|l| \ll N/R = x^{\varepsilon}$. By the Poisson summation formula, we have

$$\sum_m V\Big(\frac{m}{M}\Big)\delta_{m\equiv b\,(\mathrm{mod}\, rs_1s_2)} = \frac{M}{rs_1s_2}\widehat{V}(0) + \frac{M}{rs_1s_2}\sum_{h\neq 0}\widehat{V}\Big(\frac{h}{rs_1s_2/M}\Big)e\left(\frac{hb}{rs_1s_2}\right)$$

where $b = b(n, l)\,(\mathrm{mod}\, rs_1s_2)$ is such that

$$b \equiv a_{rs_1s_2}\overline{n}\,(\mathrm{mod}\, r),\ b \equiv a_{rs_1s_2}\overline{n}\,(\mathrm{mod}\, s_1), b \equiv a_{rs_1s_2}\overline{n + lr}\,(\mathrm{mod}\, s_2).$$

The $h = 0$ contribution provides a main term which is cancelled up to an admissible error term by the main contributions coming from the other summands of (15.6). The contribution of the frequencies $h \neq 0$ will turn out to be error terms. We have to show that

$$\sum_r \sum_{s_1,s_2,n,l}\sum c(rs_1)\overline{c(rs_2)}\beta(n)\overline{\beta(n + rl)}\frac{M}{rs_1s_2}\sum_{h\neq 0}\widehat{V}\Big(\frac{h}{rs_1s_2/M}\Big)e\left(\frac{hb}{rs_1s_2}\right) \ll \frac{MN^2}{R}x^{-\eta} = x^{1-\eta+\varepsilon}$$

for some fixed $\eta > 0$. The length of the h sum is essentially

$$H = RS^2/M = Q^2N/(xR) = x^{2\varpi+\varepsilon}$$

which is small (if ϖ and ε are). We therefore essentially need to prove that

$$(15.7)\qquad \frac{1}{H}\sum_{r\sim R}\sum_{l\ll N/R}\sum_n \beta(n)\overline{\beta(n + lr)}\sum_{0\neq h\ll H}\left|\sum_{s_1,s_2} c(rs_1)\overline{c(rs_2)}e\left(h\frac{a_{rs_1s_2}\overline{n}}{rs_1} + h\frac{a_{rs_1s_2}\overline{n + lr}}{rs_2}\right)\right| \ll x^{1-\eta+\varepsilon}.$$

We can now exhibit cancellation in the n-sum by smoothing out the n variable using the Cauchy-Schwarz inequality for any fixed r, l: letting the h variable "in" we obtain exponential sums of the shape

$$\sum_{n\sim N} e\left(h\frac{a_{rs_1s_2}\overline{n}}{rs_1} - h'\frac{a_{rs_1's_2'}\overline{n}}{rs_1'} + h\frac{a_{rs_1s_2}\overline{n + lr}}{rs_2} - h'\frac{a_{rs_1's_2'}\overline{n + lr}}{rs_2'}\right).$$

The generic case is when $h - h', s_1, s_2, s_1', s_2'$ are all coprime. In that case the above exponential sum has length

$$N \in [x^{1/2-\sigma}, x^{1/2}]$$

and the moduli involved are of size

$$RS^4 = Q^4/R^3 = x^{O(\varepsilon)}Q^4/N^3 = [x^{1/2+4\varpi+O(\varepsilon)}, x^{1/2++4\varpi+3\sigma+O(\varepsilon)}].$$

Therefore if $\sigma, \varpi, \varepsilon$ are small, the length N is not much smaller than the modulus so we could apply the completion method to improve over the trivial bound $O(N)$ for the n-sum. If we apply the Pólya-Vinogradov method, the trivial bound is replaced by $O((RS^4)^{1/2+o(1)})$ and we find that the left-hand side of (15.7) is bounded by

$$\frac{1}{H}R.\frac{N}{R}N^{1/2}(H^2S^4(RS^4)^{1/2+o(1)})^{1/2}$$
$$= x^{O(\varepsilon)+o(1)}N^{3/2}S^3R^{1/4} = x^{\frac{7}{8}+3\varpi+\frac{5}{4}\sigma+O(\varepsilon)+o(1)}$$

which is $\ll x^{1-\eta}$ for some $\eta > 0$ whenever $\sigma < 1/10$ and ϖ and ε are small enough.

Instead of using the Pólya-Vinogradov bound, we could take advantage of the fact that the modulus $rs_1s_1's_2s_2'$ is x^δ-friable (again we can take $\delta > 0$ as small as we need) and apply the q-van der Corput method from the previous section. Factoring $rs_1s_1's_2s_2'$ into a product $r's'$ such that $r' \sim (rs_1s_1's_2s_2')^{1/3+O(\delta)}$, $s' \sim (rs_1s_1's_2s_2')^{2/3+O(\delta)}$, a suitable variant of (14.1) bounds the n-sum by $O(N^{1/2}(RS^4)^{1/6+O(\delta)+o(1)})$ and the left-hand side of (15.7) is bounded by

$$\frac{R}{H}\frac{N}{R}N^{\frac{1}{2}}(H^2S^4N^{1/2}(RS^4)^{1/6})^{\frac{1}{2}+o(1)+O(\delta)}$$
$$= x^{O(\varepsilon+\delta)+o(1)}N^{7/4}S^{7/3}R^{1/12} = x^{\frac{11}{12}+\frac{7}{3}\varpi+\frac{1}{2}\sigma+O(\varepsilon+\delta)+o(1)}$$

which is $\ll x^{1-\eta}$ for some $\eta > 0$ whenever $\sigma < 1/6$ and ϖ and ε are small enough.

15.4. Treatment of type III sums. Our objective for the Type III sums is the following bound: for some $\eta > 0$, we have

$$\sum_{\substack{q\sim Q\\ x^\delta-\text{friable}}} c(q)\sum_{n\sim N}\beta(n)\sum_{m}\tau_{3,\mathbf{M}}(m)\Delta_{a_q}(m_1m_2m_3n)\ll x^{1-\eta}, \tag{15.8}$$

where $\mathbf{M} = (M_{i_1}, M_{i_2}, M_{i_3})$ and

$$\tau_{3,\mathbf{M}}(m) := \sum_{m_1m_2m_3=m} V(\frac{m_1}{M_{i_1}})V(\frac{m_2}{M_{i_2}})V(\frac{m_3}{M_{i_3}})$$

and $M_{i_1}, M_{i_2}, M_{i_3}$ satisfy

$$M = M_{i_1}M_{i_2}M_{i_3} \geqslant x^{1/2+3\sigma}.$$

The function

$$m \mapsto \tau_{3,\mathbf{M}}(m)$$

is basically a smoothed version of the ternary divisor function $m \mapsto \tau_3(m)$ that we have discussed in §11.

In fact, while describing the proof of Theorem 11.4, we have shown that for $M = x$, and for q a prime satisfying

$$q \sim x^{1/2+\varpi},\ \varpi = 1/47$$

one has

$$\sum_m \tau_{3,\mathbf{M}}(m)\Delta_{a_q}(m_1m_2m_3n) \ll \frac{x^{1-\eta}}{q}$$

for some $\eta > 0$. We have therefore the required bound but for individual moduli instead of having it on average.

As we have observed when discussing Type II sums, the parameter σ can be taken as close to 1/6 as we wish and in particular $M \in [x^{1+3(\sigma-\frac{1}{6})}, x]$ can be made as close as we wish from x and $N \in [1, x^{3(\frac{1}{6}-\sigma)}]$ as we wish from x (in the logarithmic scale). In particular, this establishes (15.8) for prime moduli $q \sim Q$ for some value of σ (close enough to 1/6), and some value of ϖ (close enough to 0) and some $\eta > 0$.

The case of x^δ-friable moduli uses similar methods and (besides some elementary technical issues) is maybe simpler than in the prime modulus case because of the extra flexibility provided by the friable moduli.

REMARK 15.9. By a more elaborate treatment, involving different uses of the Cauchy-Schwarz inequality and iterations of the q-van der Corput method, it is possible to bounds successfully all the Type II sums associated to some explicit parameter $\sigma > 1/6$. As pointed out in Remark 15.6, this makes the section devoted to Type III sums (and in particular the theory of hyper-Kloosterman sums $\mathrm{Kl}_3(x;q)$) unnecessary. The interest of this remark comes from the fact that the trace functions occurring in the treatment of the sums of Type II are exclusively algebraic exponentials:

$$x \mapsto \mathrm{e}_q(f(x)), \text{ for } f(X) \in \mathbf{F}_q(X).$$

For such trace functions, Corollary 4.7 "only" uses Weil's resolution of the Riemann Hypothesis for curves over finite fields [**Wei41**] and not the full proof of the Weil conjectures by Deligne [**Del80**].

16. Advanced completions methods: the $+ab$ shift

In this last section, we describe another method allowing to break the Pólya-Vinogradov barrier for prime moduli. This method has its origins in the celebrated work of Burgess on short sums of Dirichlet characters [**Bur62**].

16.1. Burgess's bound. Let q be a prime and le $\chi\colon \mathbf{F}_q^\times \to \mathbf{C}^\times$ be a non trivial multiplicative character. Consider the sum

$$S_V(\chi, N) := \sum_n \chi(n) V(\frac{n}{N})$$

where $V \in \mathcal{C}^\infty(]1,2[)$.

THEOREM 16.1 (Burgess). *For any $N \geqslant 1$ and $l \geqslant 1$ such that*

$$q^{1/2l} \leqslant N < \frac{1}{2} q^{1/2+1/4l} \tag{16.1}$$

we have

$$S_V(\chi, N) \ll_{V,l} q^{o(1)} N (N/q^{1/4+1/4l})^{-1/l}.$$

REMARK 16.2. Observe that this bound is non-trivial (sharper than $S_V(\chi, N) \ll N$) whenever

$$q^{1/4+1/4l+o(1)} \leqslant N < \frac{1}{2} q^{1/2+1/4l}.$$

Moreover, for $N \geqslant \frac{1}{2} q^{1/2+1/4l}$, the Pólya-Vinogradov bound $S_V(\chi, N) \ll q^{1/2}$ is non trivial, therefore, we see that by taking l large enough, that (16.1) yields a non-trivial bound for $S_V(\chi, N)$ as long as

$$N \geqslant q^{1/4+\delta}$$

for some fixed $\delta > 0$.

PROOF. Burgess's argument exploits two features in a critical way: the first one is that an interval is "essentially" invariant under sufficiently small additive translations and the second is the multiplicativity of the Dirichlet character.

Let $A, B \geqslant 1$ be parameters such that $AB \leqslant N/2$; we will also assume that $2B < q$.

We have

$$S_V(\chi, N) = \frac{1}{AB} \sum_{|n| \leqslant 2N} \sum\sum_{a\sim A, b\sim B} \chi(n+ab)V(\frac{n+ab}{N}).$$

The next step is to invoke the Fourier inversion formula to separate the variables n and ab: one has

$$V(\frac{n+ab}{N}) = \int_{\mathbf{R}} \widehat{V}(t)e(\frac{tn}{N})e(\frac{tab}{N})dt.$$

Plugging this formula in our sum, we obtain

$$\begin{aligned} S_V(\chi, N) &= \frac{1}{AB}\int_{\mathbf{R}} \sum_{|n|\leqslant 2N} e(\frac{tn}{N}) \sum\sum_{a\sim A, b\sim B} \chi(n+ab)e(\frac{tab}{N})\widehat{V}(t)dt \\ &\leqslant \frac{1}{AB}\int_{\mathbf{R}} \sum_{|n|\leqslant 2N} \sum_{a\sim A} \Big|\frac{\chi(a)}{a}\widehat{V}(\frac{t}{a})\Big|\Big|\sum_{b\sim B}\chi(\bar{a}n+b)e(\frac{tb}{N})\Big|dt \\ &\leqslant \frac{1}{AB}\int_{\mathbf{R}} \sum_{|n|\leqslant 2N} \sum_{a\sim A} \Big|\sum_{b\sim B}\chi(\bar{a}n+b)e(\frac{tAb}{N})\Big||W(t)|dt \end{aligned}$$

for W some bounded rapidly decaying function.

REMARK 16.3. Observe that the factor $\chi(a)$ coming from the identity

$$\chi(n+ab) = \chi(a(\bar{a}n+b)) = \chi(a)\chi(\bar{a}n+b) \tag{16.2}$$

has been absorbed in the absolute value of the first inequality above.

The innermost sum can be rewritten

$$\sum_{|n|\leqslant 2N} \sum_{a\sim A} \Big|\sum_{b\sim B} \chi(\bar{a}n+b)e(\frac{tAb}{N})\Big| = \sum_{r\in\mathbf{F}_q^\times} \nu(x)\Big|\sum_{b\sim B}\eta_b\chi(r+b)\Big|$$

where $\eta_b = e(\frac{tAb}{N})$ and

$$\nu(r) := |\{(a,n) \in [A, 2A[\times[-2N, 2N],\ \bar{a}n = r \,(\mathrm{mod}\, q)\}|.$$

Consider the map

$$(a,n) \in [A, 2A[\times[-2N, 2N] \mapsto \bar{a}n \,(\mathrm{mod}\, q) = r \in \mathbf{F}_q.$$

The function $\nu(r)$ is the size of the fiber of that map above r. We will show that this map is "essentially injective" (has small fibers on average). Suppose that A is chosen such that $4AN < q$; then one has

$$\sum_r \nu(r) \ll AN,\ \sum_r \nu^2(r) \ll (AN)^{1+o(1)}$$

where the first bound is obvious while for the second we observe that

$$\sum_r \nu^2(r) = |\{(a, a', n, n'),\ a, a' \in [A, 2A[,\ |n|, |n'| \ll N,\ an' \equiv an \,(\mathrm{mod}\, q)\}|,$$

then use the fact that $AN < q$ and that the integer an' has at most $(an')^{o(1)}$ decomposition of the shape $an' = a'n$.

This map however is not surjective nor even close to being so in general, so that the change of variable $\overline{a}.n \leftrightarrow x$ is not very effective. A way to moderate ineffectiveness is to use Hölder's inequality.

Let $l \geqslant 1$ be some integer parameter. Applying Hölder's inequality with $1/p = 1 - 1/2l$, $1/q = 1/2l$ and the above estimate one obtains

$$\sum_{x \in \mathbf{F}_q^\times} \nu(x) \big| \sum_{b \sim B} \eta_b \chi(x+b) \big| \leqslant \big(\sum_x \nu(x)^{\frac{2l}{2l-1}}\big)^{1-1/2l} \big(\sum_x \big| \sum_{b \sim B} \eta_b \chi(x+b)\big|^{2l}\big)^{1/2l}$$
$$\ll (AN)^{1-1/2l+o(1)} \big(\sum_x \big| \sum_{b \sim B} \eta_b \chi(x+b)\big|^{2l}\big)^{1/2l}.$$

The x-sum in the rightmost factor equals

$$\sum_{\boldsymbol{b}} \eta_{\boldsymbol{b}} \sum_{r \in \mathbf{F}_q} \chi\Big(\frac{\prod_{i=1}^l (r+b_i)}{\prod_{i=i}^l (r+b_{k+i})}\Big)$$

where $\boldsymbol{b} = (b_1, \dots, b_{2l}) \in [B, 2B[^{2l}$ and $\eta_{\boldsymbol{b}} = \prod_{i=1}^{2l} \eta_{b_i}$. Consider the fraction

$$F_{\boldsymbol{b}}(X) := \frac{\prod_{i=1}^l (X+b_i)}{\prod_{i=i}^l (X+b_{k+i})} \in \mathbf{Q}(X)$$

and the function on $\mathbf{F}_q$

$$r \in \mathbf{F}_q \mapsto \chi(F_{\boldsymbol{b}}(r))$$

(extended by 0 for $r = -b_i \,(\mathrm{mod}\, q)$, $i = 1, \dots, 2l$). This function is the trace function of the rank one sheaf $[F_{\boldsymbol{b}}]^* \mathcal{L}_\chi$ whose conductor is bounded in terms of l only and (because it is of rank 1) which is geometrically irreducible if not-geometrically constant. If not geometrically constant one has[15]

$$\sum_{r \in \mathbf{F}_q} \chi(F_{\boldsymbol{b}}(r)) \ll_l q^{1/2}.$$

If $q > \max(l, 2B)$ this occurs precisely when $F_{\boldsymbol{b}}(X)$ is not constant nor a k-th power, where k is the order of χ. Hence this holds for $\boldsymbol{b}$ outside an explicit set $\mathcal{B}^{bad} \subset [B, 2B[^{2l}$ of size bounded by $O(B^l)$. If $\boldsymbol{b} \in \mathcal{B}^{bad}$, we use the triv,ial bound

$$\big| \sum_{r \in \mathbf{F}_q} \chi(F_{\boldsymbol{b}}(r)) \big| \leqslant q.$$

All in all, we eventually obtain

$$\sum_{\boldsymbol{b}} \eta_{\boldsymbol{b}} \sum_x \chi\left(\frac{\prod_{i=1}^l (x+b_i)}{\prod_{i=i}^l (x+b_{k+i})}\right) \ll |\mathcal{B}^{bad}| q + |\mathcal{B} - \mathcal{B}^{bad}| q^{1/2} \ll B^l q + B^{2l} q^{1/2}.$$

Choosing $B = q^{1/2l}$ (so as to equal the two terms in the bound above) and $A \approx Nq^{-1/2l}$ with the condition $4AN < q$, which is equivalent to (16.1), we obtain that

$$S_V(\chi, N) \ll_l \frac{q^{o(1)}}{AB} (AN)^{1-1/2l} (q^{3/2})^{1/2l} \ll q^{o(1)} N^{1-1/l} q^{3/4l - (1-1/2l)/2l}$$
$$= q^{o(1)} N (N/q^{1/4+1/4l})^{-1/l}.$$

□

[15] It is not necessary to invoke Deligne's main theorem here: this follows from A. Weil's proof of the Riemann hypothesis for curves [**Wei41**].

16.2. The $+ab$-shift for type I sums. It is natural to try to extend this method to other trace functions; unfortunately the above argument breaks down because the identity (16.2) is not valid in general. It is however possible to mitigate this problem by introducing an extra average.

This technique goes back to Karatsuba and Vinogradov (for the function $x \mapsto \chi(x+1)$). It has been also used by Friedlander-Iwaniec [**FI85**] (for the function $x \mapsto e\left(\frac{\overline{x}}{q}\right)$), Fouvry-Michel [**FM98**] and Kowalski-Michel-Sawin [**KMS17, KMS18**].

Instead of a single sum $S_V(K,N)$, one considers the following average of multiplicative shifts

$$B_V(K,\boldsymbol{\alpha},N) := \sum_{m\sim M}\alpha_m\sum_n V(\frac{n}{N})K(mn)$$

where $1\leqslant M<q$ and $(\alpha_m)_{m\sim M}$ is a sequence of complex numbers of modulus $\leqslant 1$ (this includes the averaged sum $\sum_{m\sim M}|\sum_n K(mn)V(\frac{n}{N})| = \sum_m |S_V([\times m]^*K,N)|$). The objective here is to improve over the trivial bound

$$B_V(K,\boldsymbol{\alpha},N) \ll \|K\|_\infty MN.$$

Proceeding as above we have

$$\begin{aligned} B_V(K,\boldsymbol{\alpha},N) &= \frac{1}{AB}\sum_m \alpha_m\sum_n\sum_{a\sim A,b\sim B}\!\!\sum K(m(n+ab))V(\frac{n+ab}{N}) \\ &\leqslant \frac{1}{AB}\int_{\mathbf{R}}\sum_{m\sim M}\alpha_m\sum_{|n|\leqslant 2N}\sum_{a\sim A}|\sum_{b\sim B}K(am(\overline{a}n+b))e(\frac{tAb}{N})||W(t)|dt.\end{aligned}$$

We have

$$\sum_{m\sim M}\alpha_m\sum_{|n|\leqslant 2N}\sum_{a\sim A}|\sum_{b\sim B}K(am(\overline{a}n+b))e(\frac{tAb}{N})| = \sum\sum_{r,s\in\mathbf{F}_q}\nu(r,s)|\sum_{b\sim B}\eta_bK(s(r+b))|$$

with

$$\nu(r,s) = \sum_{m\sim M}\sum_{|n|\leqslant 2N}\sum_{a\sim A}\alpha_m\delta_{\overline{a}n=r,am=s\,(\mathrm{mod}\,q)}.$$

Assuming that $4AN<q$ and evaluating the number of solutions to the equations

$$am=a'm',\ a\overline{n}\equiv a'\overline{n}'\,(\mathrm{mod}\,q),\ (a,m,n)\in[A,2A[\times[M,2M[\times[N,2N[$$

one finds that

$$\sum\sum_{r,s\in\mathbf{F}_q}|\nu(r,s)| \ll AMN,\ \sum\sum_{r,s\in\mathbf{F}_q}|\nu(r,s)|^2 \ll q^{o(1)}AMN$$

which we interpret as saying that the map

$$(a,m,n)\in[A,2A[\times[M,2M[\times[N,2N[\to(r,s)=(\overline{a}.n,am)\in\mathbf{F}_q\times[AM,4AM[$$

is essentially injective (i.e. has small fibers on average). As before, this map is far from being surjective but one can dampen this with Hölder's inequality:

$$\begin{aligned}
&\sum_{\substack{r\in\mathbf{F}_q\\ 1\leqslant s\leqslant 4AM}}\sum \nu(r,s)\Big|\sum_{b\sim B}\eta_b K(s(r+b))\Big|\\
&\qquad\ll \Big(\sum_{r,s}\sum|\nu(r,s)|^{\frac{2l}{2l-1}}\Big)^{1-1/2l}\Big(\sum_{r,s}\sum\Big|\sum_{b\sim B}\eta_b K(s(r+b))\Big|^{2l}\Big)^{1/2l}\\
&\qquad\ll q^{o(1)}(AMN)^{1-1/2l}\Big(\sum_{\boldsymbol{b}}\eta_{\boldsymbol{b}}\sum_{r,s}\prod_{i=1}^{l}K(s(r+b_i))\overline{K(s(r+b_{i+l}))}\Big)^{1/2l}.
\end{aligned}$$

We are now reduced to the problem of bounding the two variable sum

$$\sum_{r,s}\prod_{i=1}^{l}K(s(r+b_i))\overline{K(s(r+b_{i+l}))}=\sum_{r}\sum_{s}\mathbf{K}(sr,s\boldsymbol{b})=\sum_{r}\mathbf{R}(r,\boldsymbol{b}) \tag{16.3}$$

(say) where

$$\mathbf{K}(r,\boldsymbol{b}):=\prod_{i=1}^{l}K(r+b_i)\overline{K(r+b_{i+l})},\ \mathbf{R}(r,\boldsymbol{b})=\sum_{s}\mathbf{K}(sr,s\boldsymbol{b}). \tag{16.4}$$

The bound will depend on the vector $\boldsymbol{b}\in[B,2B[^{2l}$. To get a feeling of what is going on, let us consider one of cases treated in [**FM98**]: let

$$K(x)=\mathrm{e}_q(\overline{x}+x).$$

We have

$$\mathbf{R}(sr,s\boldsymbol{b})=\sum_{s\in\mathbf{F}_q^\times}\mathrm{e}_q\Big(\overline{s}\sum_{i=1}^{l}(\overline{r+b_i}-\overline{r+b_{i+l}})+s\sum_{i=1}^{l}(b_i-b_{i+l})\Big).$$

This sum is either

(1) Equal to $q-1$, if and only if the vector $(b_1,\dots,b_l)$ equals the vector $(b_{l+1},\dots,b_{2l})$ up to permutation of the entries.
(2) Equal to -1 if $\boldsymbol{b}$ is not as in (1) but is in the hyperplane with equation $\sum_{i=1}^{l}(b_i-b_{i+l})=0$.
(3) The Kloosterman sum

$$\mathbf{R}(r,\boldsymbol{b})=q^{1/2}\,\mathrm{Kl}_2\left(\frac{\sum_{i=1}^{l}(\overline{r+b_i}-\overline{r+b_{i+l}})}{\sum_{i=1}^{l}(b_i-b_{i+l})};q\right)$$

otherwise.

The last case is the most interesting. Given $\boldsymbol{b}$ as in the last situation, we have to evaluate

$$q^{1/2}\sum_{r}\mathrm{Kl}_2(G_{\boldsymbol{b}}(r);q)$$

where

$$G_{\boldsymbol{b}}(X)=\frac{\sum_{i=1}^{l}(\overline{X+b_i}-\overline{X+b_{i+l}})}{\sum_{i=1}^{l}(b_i-b_{i+l})}. \tag{16.5}$$

LEMMA 16.4. *For* $\boldsymbol{b} = (b_1, \dots, b_{2l}) \in \mathbf{F}_q^{2l}$ *such that*
(16.6)

$$(b_1, \dots, b_l) \text{ is not equal to } (b_{l+1}, \dots, b_{2l}) \text{ up to permutation and } \sum_{i=1}^{l}(b_i - b_{i+l}) \neq 0,$$

one has

$$\sum_r \mathrm{Kl}_2(G_{\boldsymbol{b}}(r); q) \ll_l q^{1/2}.$$

PROOF. The function

$$r \mapsto \mathrm{Kl}_2(G_{\boldsymbol{b}}(r); q)$$

is the trace function of the rank 2 sheaf $[G_{\boldsymbol{b}}]^*\mathcal{K}\ell_2$ obtained by pull-back from the Kloosterman sheaf $\mathcal{K}\ell_2$ of morphism

$$x \mapsto G_{\boldsymbol{b}}(x)$$

which is non-constant by assumption.

Moreover, one can show that he conductor of $[G_{\boldsymbol{b}}]^*\mathcal{K}\ell_2$ is bounded in terms of l only, and moreover the geometric monodromy group of $[G_{\boldsymbol{b}}]^*\mathcal{K}\ell_2$ is obtained as the (closure of the) image of the representation $\varrho_{\mathcal{K}\ell_2}$ restricted to a finite index subgroup of $\mathrm{Gal}(K^{\mathrm{sep}}/\overline{\mathbf{F}}_q.K)$. Since the geometric monodromy group of $\mathcal{K}\ell_2$ is SL_2 which has no finite index subgroup, the geometric monodromy group of $[G_{\boldsymbol{b}}]^*\mathcal{K}\ell_2$ is SL_2 as well. It follows that the sheaf $[G_{\boldsymbol{b}}]^*\mathcal{K}\ell_2$ is geometrically irreducible (and not geometrically trivial because of rank 2) and the estimate follows by Deligne's theorem. □

It follows from this analysis that

$$\sum_{r,s}\sum\Bigl|\sum_{b\sim B}\eta_b K(s(r+b))\Bigr|^{2l} \ll B^l q^2 + B^{2l} q,$$

hence choosing $B = q^{1/l}$, $AB \approx N$ and $A \approx Nq^{-1/l}$ we obtain

$$B_V(K, \boldsymbol{\alpha}, N) \ll \frac{q^{o(1)}}{AB}(AMN)^{1-1/2l} q^{3/2l} = q^{o(1)} MN\Bigl(\frac{N^2 M}{q^{1+1/l}}\Bigr)^{-1/2l}.$$

To resume we have therefore proven the

THEOREM 16.5. *Let* $K(x) = \mathrm{e}_q(\overline{x} + x)$ *and* $M, N, l \geqslant 1$ *and* $(\alpha_m)_{m\sim M}$ *be a sequence of complex numbers of modulus bounded by* 1. *Assuming that*

$$q^{1/l} \leqslant N < \frac{1}{2} q^{1/2+1/2l}$$

we have

$$\sum_{m\sim M} \alpha_m \sum_n V\Bigl(\frac{n}{N}\Bigr) K(mn) \ll q^{o(1)} MN\Bigl(\frac{N^2 M}{q^{1+1/l}}\Bigr)^{-1/2l}.$$

This bound is non trivial (sharper than $\ll MN$) as long as[16]

$$N^2 M \geqslant q^{1+1/l}.$$

For instance, if $M = q^\delta$ for some $\delta > 0$, the above bound is nontrivial for l large enough and $N \geqslant q^{1/2+\delta/3}$. Alternatively if $M = N$, this bound is non trivial as long as

$$N = M \geqslant q^{1/3+\delta}$$

[16] If $N \geqslant \frac{1}{2} q^{1/2+1/2l}$ the Pólya-Vinogradov inequality is non trivial already.

if l is taken large enough. Therefore this method improves the range of non-triviality in Theorem 9.1.

16.3. The $+ab$-shift for type II sums. With this method, it is also possible to deal with the more general (type II) bilinear sums

$$B(K,\boldsymbol{\alpha},\boldsymbol{\beta}) = \sum\sum_{m\sim M, n\sim N} \alpha_m\beta_n K(mn)$$

where $(\alpha_m)_{m\sim M}$, $(\beta_n)_{n\sim N}$ are sequences of complex numbers of modulus bounded by 1.

We leave it to the interested reader to fill in the details (or to look at [**FM98, KMS17**] or [**KMS18**]). The first step is to apply Cauchy-Schwarz to smooth out the n variable: for a suitable smooth function V, compactly supported in $[1/2, 5/2]$ and bounded by 1, one has

$$|\sum\sum_{m\sim M, n\sim N} \alpha_m\beta_n K(mn)| \leqslant N^{1/2}\Big(\sum\sum_{m_1,m_2\sim M} \alpha_{m_1}\overline{\alpha_{m_2}}\sum_n V(\frac{n}{N})K(m_1n)\overline{K(m_2n)}\Big)^{1/2}.$$

The next step is to perform the $+ab$-shift on the n variable and to make the change of variables

$$\begin{aligned}(a,m_1,m_2,n) \in [A,2A[\times[M,2M[^2\times[N,2N[&\longleftrightarrow (\overline{a}n, am_1, am_2) \,(\mathrm{mod}\, q)\\ &= (r,s_1,s_2)\in \mathbf{F}_q^3.\end{aligned}$$

Considering the fiber counting function for that map, namely

$$\nu(r,s_1,s_2) := \sum\sum_{\substack{(a,n,m_1,m_2)\\ a\sim A, |n|\leqslant 2N,\ m_i\simeq M}} \alpha_{m_1}\overline{\alpha_{m_2}}\delta_{\overline{a}n=r,\ am_i=s_i\,(\mathrm{mod}\, q)}$$

one shows that for $AN < q/2$ one has

$$\sum\sum_{(r,s_1,s_2)\in\mathbf{F}_q{}^3} |\nu(r,s_1,s_2)| \ll AM^2N, \qquad \sum\sum_{(r,s_1,s_2)\in\mathbf{F}_q{}^3} |\nu(r,s_1,s_2)|^2 \leqslant q^{o(1)}AM^2N.$$

Applying Hölder's inequality leads us to the problem of bounding the following complete sum indexed by the parameter $\boldsymbol{b}$

$$\sum_{r\in\mathbf{F}_q} |\mathbf{R}(r,\boldsymbol{b})|^2 - q\sum_{r\in\mathbf{F}_q} |\mathbf{K}(r,\boldsymbol{b})|^2. \tag{16.7}$$

We will explain what is expected in general in a short moment but let us see what happens for our previous case $K(x) = \mathrm{e}_q(\overline{x}+x)$: for $\boldsymbol{b} = (b_1,\ldots,b_{2l}) \in \mathbf{F}_q{}^{2l}$ satisfying (16.6) the sum (16.7) equals

$$q\sum_{\substack{r\in\mathbf{F}_q\\ r\neq -b_i}} |\operatorname{Kl}_2(G_{\boldsymbol{b}}(r);q)|^2 - q\sum_{\substack{r\in\mathbf{F}_q\\ r\neq -b_i}} 1 = q\sum_{\substack{r\in\mathbf{F}_q\\ r\neq -b_i}} (|\operatorname{Kl}_2(G_{\boldsymbol{b}}(r);q)|^2 - 1) + O_l(q)$$

where $G_{\boldsymbol{b}}(X)$ is defined in (16.5)

LEMMA 16.6. *For $\boldsymbol{b} = (b_1,\ldots,b_{2l}) \in \mathbf{F}_q{}^{2l}$ satisfying* (16.6), *one has*

$$\sum_r (|\operatorname{Kl}_2(G_{\boldsymbol{b}}(r);q)|^2 - 1) \ll_l q^{1/2}.$$

PROOF. This follows from the fact that $[G_{\boldsymbol{b}}]^*\mathcal{K}\ell_2$ is geometrically irreducible with geometric monodromy group equal to SL_2: since the tensor product of the standard representation of SL_2 with itself equals the trivial representation plus the symmetric square of the standard representation which is non-trivial and irreducible,

$$x \mapsto |\,\mathrm{Kl}_2(G_{\boldsymbol{b}}(r);q)|^2 - 1$$

is the trace function of a geometrically irreducible sheaf. □

Using this bound and trivial estimates for $\boldsymbol{b}$ not satisfying (16.6), one eventually obtains

THEOREM 16.7. *Let $K(x) = \mathrm{e}_q(\overline{x}+x)$, $1 \leqslant M, N < q$ and $l \geqslant 1$ some integer. Assuming that*

$$N < \frac{1}{2}q^{1/2+1/2l},$$

one has

$$B(K,\boldsymbol{\alpha},\boldsymbol{\beta}) = \sum\sum_{m\sim M, n\sim N} \alpha_m\beta_n K(mn) \ll q^{o(1)}MN\Big(\frac{1}{M} + \Big(\frac{MN}{q^{3/4+3/4l}}\Big)^{-1/4l}\Big)^{1/2}.$$

REMARK 16.8. For l large enough, this bound is non-trivial as long as $M \geqslant q^{\delta}$ and $MN \geqslant q^{3/4+\delta}$, again improving on Theorem 9.1 in this specific case.

16.4. The $+ab$-shift for more general trace functions. For applications to analytic number theory, it is highly desirable to extend the method of the previous section to trace functions as general as possible. This method may be axiomatized in the following way. Let q be a prime, $K\colon \mathbf{F}_q \to \mathbf{C}$ a complex valued function bounded by 1 in absolute value, $1 \leqslant M, N < q$ some parameters and $\boldsymbol{\alpha} = (\alpha_m)_{m\sim M}$, $\boldsymbol{\beta} = (\beta_n)_{n\sim N}$ sequences of complex number bounded by 1. We define the type I sum

$$B(K,\boldsymbol{\alpha},1_N) = \sum\sum_{m\sim M, n\sim N} \alpha_m K(mn)$$

and the type II sum

$$B(K,\boldsymbol{\alpha},\boldsymbol{\beta}) = \sum\sum_{m\sim M, n\sim N} \alpha_m\beta_n K(mn).$$

For $l \geqslant 1$ an integer, let $\mathbf{K}(r,\boldsymbol{b})$ and $\mathbf{R}(r,\boldsymbol{b})$ be the functions of the variables $(r,\boldsymbol{b}) \in \mathbf{F}_q \times \mathbf{F}_q{}^{2l}$ given by (16.4). For $B \geqslant 1$ we set

$$\mathcal{B} = \mathbf{Z}^{2l} \cap [B, 2B[^{2l}.$$

An axiomatic treatment of the type I sums $B(K,\boldsymbol{\alpha},1_N)$ is provided by the following:

THEOREM 16.9. *Notations as above, let $B, C \geqslant 1$ and $\gamma \in [0,2]$ be some real numbers.*

– *Let $\mathcal{B}^{\Delta} \subset \mathcal{B}$ be the set of $\boldsymbol{b} \in \mathcal{B}$ for which*

$$\text{there exists } r \in \mathbf{F}_q \text{ satisfying } |\mathbf{R}(r,\boldsymbol{b})| > Cq^{1/2}. \tag{16.8}$$

– *Let $\mathcal{B}_I^{bad} \subset \mathcal{B}$ be the union of $\mathcal{B}^{\Delta}$ and the set of $\boldsymbol{b} \in \mathcal{B}$ such that*

$$\Big|\sum_{r\in\mathbf{F}_q} \mathbf{R}(r,\boldsymbol{b})\Big| > Cq. \tag{16.9}$$

Suppose that for any $1 \leqslant B < q/2$ *one has*

$$|\mathcal{B}^{\Delta}| \leqslant CB^{l},\ |\mathcal{B}_{I}^{bad}| \leqslant B^{(2-\gamma)l}. \tag{16.10}$$

Then, if N *satisfies*

$$q^{1/l} \leqslant N \leqslant \frac{1}{2}q^{1/2+1/2l},$$

one has for any $\varepsilon > 0$

$$B(K,\boldsymbol{\alpha},1_N) \ll_{C,l,\varepsilon} q^{\varepsilon}MN(\frac{q^{1+1/l}}{MN^2}+\frac{q^{3/2-\gamma+1/l}}{MN^2})^{1/2l}. \tag{16.11}$$

An axiomatic treatment of the type II sums $B(K,\boldsymbol{\alpha},\boldsymbol{\beta})$ is provided by the following

THEOREM 16.10. *Notations as above, let* $B, C \geqslant 1$ *and* $\gamma \in [0,2]$ *be some real numbers,*

– *Let* $\mathcal{B}^{\Delta} \subset \mathcal{B}$ *be the set of* $\boldsymbol{b} \in \mathcal{B}$ *for which*

there exists $r \in \mathbf{F}_q$ *satisfying* $|\mathbf{R}(r,\boldsymbol{b})| > Cq^{1/2}$.

– *Let* $\mathcal{B}_{II}^{bad} \subset \mathcal{B}$ *be the union of* $\mathcal{B}^{\Delta}$ *and the set of* $\boldsymbol{b} \in \mathcal{B}$ *such that*

$$|\sum_{r\in\mathbf{F}_q}|\mathbf{R}(r,\boldsymbol{b})|^2 - q\sum_{r\in\mathbf{F}_q}|\mathbf{K}(r,\boldsymbol{b})|^2| > Cq^{3/2}. \tag{16.12}$$

Assume that for any $B \in [1, q/2[$ *one has*

$$|\mathcal{B}^{\Delta}| \leqslant CB^{l},\ |\mathcal{B}_{II}^{bad}| \leqslant CB^{(2-\gamma)l}. \tag{16.13}$$

Then, if N *satisfies*

$$q^{3/2l} \leqslant N \leqslant \frac{1}{2}q^{1/2+3/4l},$$

one has for any $\varepsilon > 0$,

$$B(K,\boldsymbol{\alpha},\boldsymbol{\beta}) \ll_{C,l,\varepsilon} q^{\varepsilon}MN\Big(\frac{1}{M}+(\frac{q^{1-\frac{3}{4}\gamma+\frac{3}{4l}}}{MN}+\frac{q^{\frac{3}{4}+\frac{3}{4l}}}{MN})^{\frac{1}{l}}\Big)^{1/2}. \tag{16.14}$$

We conclude these lectures with a few remarks concerning these two theorems:

(1) In the case $K(x) = \mathrm{e}_q(\overline{x}+x)$, we have just verified that the conditions (16.10) and (16.13) hold with $\gamma = 1$. In [**FM98**], this was shown to hold more generally for the trace functions

$$K(x) = \mathrm{e}_q(x^{-k}+ax),\ a \in \mathbf{F}_q,\ k \geqslant 1.$$

(2) For more general trace functions, the first condition in (16.10) and (16.13) can be verified using some variant of the "sums of products" Theorem 13.3 and does not constitute a big obstacle. One should also notice that Theorem 13.3 implies that for any $\boldsymbol{b} = (b_1,\dots,b_{2l})$ on the "first" diagonal (i.e. $b_1 = b_{l+1},\dots,b_l = b_{2l}$) one has

$$\mathbf{R}(r,\boldsymbol{b}) = \sum_{s}\prod_{i=1}^{l}|K(s(r+b_i))|^2 = |K(0)|^{2l} + \sum_{s\neq 0}\prod_{i=1}^{l}|K(s(r+b_i))|^2 \gg_l q$$

and therefore

$$|\mathcal{B}^{\Delta}| \geqslant B^{l}.$$

It follows that the first bound in (16.10) and (16.13) is sharp and for the second condition one cannot expect γ to be greater than 1.

(3) In order to reach the best available bound by the above method, it is not necessary to aim for $\gamma = 1$: it is sufficient to establish (16.10) with $\gamma \geqslant 1/2$ and (16.13) with $\gamma \geqslant 1/3$. In such a case, the bounds of Theorem 16.9 and Theorem 16.10 are non trivial as long as

$$MN^2 \geqslant q^{1+1/l}, \; MN \geqslant q^{3/4+3/4l},$$

respectively.

(4) Checking the second bound in (16.10) and (16.13) for general trace functions is much more difficult. In [**KMS17**], with specific applications in mind, these bounds have been established for $l = 2$ and $\gamma = 1/2$ for the hyper-Kloosterman sums

$$K(x) = \mathrm{Kl}_k(x; q), \; k \geqslant 2.$$

Because $l = 2$ is too small, this alone is not sufficient to improve over the Pólya-Vinogradov type bound of Theorem 9.1 (one would have needed $l \geqslant 4$). A more refined treatment is necessary: instead of letting (somewhat wastefully) the variables $s = am \,(\mathrm{mod}\, q)$ or $s_1 = am_1, s_2 = am_2 \,(\mathrm{mod}\, q)$ vary freely over the whole interval $[0, q-1] \simeq \mathbf{F}_q$, one uses the fact that s, s_1, s_2 belong to the shorter interval $[AM, 4AM[$. Applying the Pólya-Vinogradov completion method to detect this inclusion with additive characters, this leads to bounds for complete sums analogous to (16.9) and (16.12) but for the additively twisted variant of $\mathbf{R}(r, \boldsymbol{b})$ defined by

$$\mathbf{R}(r, \lambda, \boldsymbol{b}) = \sum_s \mathbf{K}(sr, s\boldsymbol{b}) e\left(\frac{\lambda s}{q}\right), \quad \text{for } \lambda \in \mathbf{F}_q.$$

Specifically, the bounds are: for all $\boldsymbol{b} \in \mathcal{B} - \mathcal{B}^{\Delta}$, we have

$$\forall \lambda \in \mathbf{F}_q, \; |\mathbf{R}(r, \lambda, \boldsymbol{b})| \leqslant Cq^{1/2},$$

and for all $\boldsymbol{b} \in \mathcal{B} - \mathcal{B}_I^{bad}$, we have

$$\forall \lambda \in \mathbf{F}_q, \; |\sum_r \mathbf{R}(r, \lambda, \boldsymbol{b})| \leqslant Cq,$$

and for all $\boldsymbol{b} \in \mathcal{B} - \mathcal{B}_{II}^{bad}$, we have

$$\forall \lambda, \lambda' \in \mathbf{F}_q, \; \Big|\sum_r \mathbf{R}(r, \lambda, \boldsymbol{b})\overline{\mathbf{R}(r, \lambda', \boldsymbol{b})} - q\delta_{\lambda=\lambda'} \sum_s \prod_{i=1}^{l} |K(s(r+b_i))|^2\Big| \leqslant Cq^{3/2}.$$

In [**KMS17**], these bounds were established for $l = 2$ and $\boldsymbol{b}$ outside the sets $\mathcal{B}^{\Delta}$, $\mathcal{B}_I^{bad}$ and $\mathcal{B}_{II}^{bad}$ satisfying

$$|\mathcal{B}^{\Delta}| \leqslant B^2, \; |\mathcal{B}_{I,II}^{bad}| \leqslant CB^3.$$

(5) In the paper [**KMS18**], the bounds (16.10) and (16.13) are established for the hyper-Kloosterman sums and generalized Kloosterman sums for every $l \geqslant 2$ and $\gamma = 1/2$.

16.5. Some applications of the $+ab$-shift bounds. The problem of estimating bilinear sums of trace functions below the critical Pólya-Vinogradov range already had several applications in analytic number theory. We list some of them below with references for the interested remaining reader(s).

– This method was used by Karatsuba and Vinogradov, for the function

$$K(n) = \chi(n+a)$$

where $(a,q) = 1$ and $\chi \,(\mathrm{mod}\, q)$ is a non-trivial Dirichlet character, to bound non-trivially its sum along the primes over short intervals (now a special case of Theorem 8.1). In particular, Karatsuba [**Kar70**] proved for any $\varepsilon > 0$, the bound

$$\sum_{\substack{p \leqslant x \\ p \text{ prime}}} \chi(p+a) \ll x^{1-\varepsilon^2/1024}$$

whenever $x \geqslant q^{1/2+\varepsilon}$. This bound is therefore non-trivial in a range which is wider than that established in Theorem 8.1 for general trace functions.

– The method was used by Friedlander-Iwaniec for the function

$$K(n) = e\left(\frac{\overline{n}}{q}\right), \quad n.\overline{n} \equiv 1 \,(\mathrm{mod}\, q)$$

to show that the ternary divisor function $d_3(n)$ is well distributed in arithmetic progressions to modulus $q \leqslant x^{1/2+1/230}$, passing for the first time the Bombieri-Vinogradov barrier (see Theorem 11.4).

– In the case of the Kloosterman sums

$$K(n) = \mathrm{Kl}_2(n;q),$$

the bound established in [**KMS17**] together with [**BM15**,**BFK$^+$17**] leads to an asymptotic formula for the second moment of character twists of modular L-functions: for f a fixed primitive cusp form, one has

$$\frac{1}{q-1} \sum_{\chi \,(\mathrm{mod}\, q)} |L(f\otimes\chi, 1/2)|^2 = MT_f(\log q) + O_f(q^{-1/145})$$

for q prime, where $MT_f(\log q)$ is a polynomial in $\log q$ (of degree $\leqslant 1$) depending on f. This completes the work of Young for f an Eisenstein series [**You11**] and of Blomer-Milicevic for f cuspidal and q suitably composite [**BM15**].

– Using this method, Nunes [**Nun17**] obtained non-trivial bounds, below the Pólya-Vinogradov range, for the (smooth) bilinear sum

$$\sum_{\substack{m \leqslant M \\ n \leqslant N}}\sum K(mn^2)$$

where K is the Kloosterman-like trace function

$$K(n;q) := \frac{1}{q^{1/2}} \sum_{x \in \mathbf{F}_q^\times} e_q(a\overline{x}^2 + bx)$$

(where a, b are some integral parameters such that $(ab,q) = 1$). He deduced from this bound that the characteristic function of squarefree integers is well-distributed in arithmetic progression to prime modulus

$$q \leqslant x^{2/3+1/57}.$$

The previous best result, due to Prachar [**Pra58**], was $q \leqslant x^{2/3-\varepsilon}$ (similar to Selberg's Theorem 11.2 for the divisor function $d_2(n)$) dated to 1958 !

Acknowledgements

These expository notes are an expanded version of a series of lectures given by Ph.M. and W.S. during the 2016 Arizona Winter School and based on our recent joint works.

We would like to thank the audience for its attention and its numerous questions during the daily lectures, as well as the teams of student, who engaged in the research activities that we proposed during the evening sessions, for their enthusiasm. Big thanks are also due to Alina Bucur, Bryden Cais and David Zureick-Brown for the perfect organization, making this edition of the AWS a memorable experience. We would also like to thank the referees for correcting many mistakes and typosin earlier versions of this text.

References

[BLGHT11] T. Barnet-Lamb, D. Geraghty, M. Harris, and R. Taylor, *A family of Calabi-Yau varieties and potential automorphy II*, Publ. Res. Inst. Math. Sci. **47** (2011), no. 1, 29–98, DOI 10.2977/PRIMS/31. MR2827723

[Bir68] B. J. Birch, *How the number of points of an elliptic curve over a fixed prime field varies*, J. London Math. Soc. **43** (1968), 57–60, DOI 10.1112/jlms/s1-43.1.57. MR0230682

[BM15] V. Blomer and D. Milićević, *The second moment of twisted modular L-functions*, Geom. Funct. Anal. **25** (2015), no. 2, 453–516, DOI 10.1007/s00039-015-0318-7. MR3334233

[BFK+17] V. Blomer, É. Fouvry, E. Kowalski, P. Michel, and D. Milićević, *On moments of twisted L-functions*, Amer. J. Math. **139** (2017), no. 3, 707–768, DOI 10.1353/ajm.2017.0019. MR3650231

[BFI86] E. Bombieri, J. B. Friedlander, and H. Iwaniec, *Primes in arithmetic progressions to large moduli*, Acta Math. **156** (1986), no. 3-4, 203–251, DOI 10.1007/BF02399204. MR834613

[Bur62] D. A. Burgess, *On character sums and primitive roots*, Proc. London Math. Soc. (3) **12** (1962), 179–192, DOI 10.1112/plms/s3-12.1.179. MR0132732

[CHT08] L. Clozel, M. Harris, and R. Taylor, *Automorphy for some l-adic lifts of automorphic mod l Galois representations*, Publ. Math. Inst. Hautes Études Sci. **108** (2008), 1–181, DOI 10.1007/s10240-008-0016-1. With Appendix A, summarizing unpublished work of Russ Mann, and Appendix B by Marie-France Vignéras. MR2470687

[Del80] P. Deligne, *La conjecture de Weil. II* (French), Inst. Hautes Études Sci. Publ. Math. **52** (1980), 137–252. MR601520

[DFI95] W. Duke, J. B. Friedlander, and H. Iwaniec, *Equidistribution of roots of a quadratic congruence to prime moduli*, Ann. of Math. (2) **141** (1995), no. 2, 423–441, DOI 10.2307/2118527. MR1324141

[Fou84] É. Fouvry, *Autour du théorème de Bombieri-Vinogradov* (French), Acta Math. **152** (1984), no. 3-4, 219–244, DOI 10.1007/BF02392198. MR741055

[Fou85] É. Fouvry, *Sur le problème des diviseurs de Titchmarsh* (French), J. Reine Angew. Math. **357** (1985), 51–76, DOI 10.1515/crll.1985.357.51. MR783533

[FI83] E. Fouvry and H. Iwaniec, *Primes in arithmetic progressions*, Acta Arith. **42** (1983), no. 2, 197–218, DOI 10.4064/aa-42-2-197-218. MR719249

[FI92] É. Fouvry and H. Iwaniec, *The divisor function over arithmetic progressions*, Acta Arith. **61** (1992), no. 3, 271–287, DOI 10.4064/aa-61-3-271-287. With an appendix by Nicholas Katz. MR1161479

[FKM15] É. Fouvry, E. Kowalski, and P. Michel, *Algebraic twists of modular forms and Hecke orbits*, Geom. Funct. Anal. **25** (2015), no. 2, 580–657, DOI 10.1007/s00039-015-0310-2. MR3334236

[FKM13] É. Fouvry, E. Kowalski, and P. Michel, *Counting sheaves using spherical codes*, Math. Res. Lett. **20** (2013), no. 2, 305–323, DOI 10.4310/MRL.2013.v20.n2.a8. MR3151649

[FKM15] É. Fouvry, E. Kowalski, and Ph. Michel, *A study in sums of products*, Philos. Trans. A **373** (2015), no. 2040, 20140309, 26pp. arXiv:1304.3199.

[FKM14] É. Fouvry, E. Kowalski, and P. Michel, *Algebraic trace functions over the primes*, Duke Math. J. **163** (2014), no. 9, 1683–1736, DOI 10.1215/00127094-2690587. MR3217765

[FKM15] É. Fouvry, E. Kowalski, and P. Michel, *On the exponent of distribution of the ternary divisor function*, Mathematika **61** (2015), no. 1, 121–144, DOI 10.1112/S0025579314000096. MR3333965

[FM98] É. Fouvry and P. Michel, *Sur certaines sommes d'exponentielles sur les nombres premiers* (French, with English and French summaries), Ann. Sci. École Norm. Sup. (4) **31** (1998), no. 1, 93–130, DOI 10.1016/S0012-9593(98)80019-0. MR1604298

[FM07] É. Fouvry and Ph. Michel, *Sur le changement de signe des sommes de Kloosterman* (French, with English summary), Ann. of Math. (2) **165** (2007), no. 3, 675–715, DOI 10.4007/annals.2007.165.675. MR2335794

[FKM$^+$17] É. Fouvry, E. Kowalski, P. Michel, C. S. Raju, J. Rivat, and K. Soundararajan, *On short sums of trace functions* (English, with English and French summaries), Ann. Inst. Fourier (Grenoble) **67** (2017), no. 1, 423–449. MR3612337

[FI85] J. B. Friedlander and H. Iwaniec, *Incomplete Kloosterman sums and a divisor problem*, Ann. of Math. (2) **121** (1985), no. 2, 319–350, DOI 10.2307/1971175. With an appendix by Bryan J. Birch and Enrico Bombieri. MR786351

[GPY09] D. A. Goldston, J. Pintz, and C. Y. Yıldırım, *Primes in tuples. I*, Ann. of Math. (2) **170** (2009), no. 2, 819–862, DOI 10.4007/annals.2009.170.819. MR2552109

[Gra15] A. Granville, *Primes in intervals of bounded length*, Bull. Amer. Math. Soc. (N.S.) **52** (2015), no. 2, 171–222, DOI 10.1090/S0273-0979-2015-01480-1. MR3312631

[HSBT10] M. Harris, N. Shepherd-Barron, and R. Taylor, *A family of Calabi-Yau varieties and potential automorphy*, Ann. of Math. (2) **171** (2010), no. 2, 779–813, DOI 10.4007/annals.2010.171.779. MR2630056

[HBP79] D. R. Heath-Brown and S. J. Patterson, *The distribution of Kummer sums at prime arguments*, J. Reine Angew. Math. **310** (1979), 111–130. MR546667

[HB86] D. R. Heath-Brown, *The divisor function $d_3(n)$ in arithmetic progressions*, Acta Arith. **47** (1986), no. 1, 29–56, DOI 10.4064/aa-47-1-29-56. MR866901

[IT13] A. Ichino and N. Templier, *On the Voronoĭ formula for* GL(n), Amer. J. Math. **135** (2013), no. 1, 65–101, DOI 10.1353/ajm.2013.0005. MR3022957

[Irv15] A. James Irving, *The divisor function in arithmetic progressions to smooth moduli*, Int. Math. Res. Not. IMRN **15** (2015), 6675–6698, DOI 10.1093/imrn/rnu149. MR3384495

[Irv16] A. J. Irving, *Estimates for character sums and Dirichlet L-functions to smooth moduli*, Int. Math. Res. Not. IMRN **15** (2016), 4602–4633, DOI 10.1093/imrn/rnv285. MR3564622

[Iwa97] H. Iwaniec, *Topics in classical automorphic forms*, Graduate Studies in Mathematics, vol. 17, American Mathematical Society, Providence, RI, 1997. MR1474964

[IK04] H. Iwaniec and E. Kowalski, *Analytic number theory*, American Mathematical Society Colloquium Publications, vol. 53, American Mathematical Society, Providence, RI, 2004. MR2061214

[IS00] H. Iwaniec and P. Sarnak, *The non-vanishing of central values of automorphic L-functions and Landau-Siegel zeros*. part A, Israel J. Math. **120** (2000), no. part A, 155–177, DOI 10.1007/s11856-000-1275-9. MR1815374

[IS99] H. Iwaniec and P. Sarnak, *Dirichlet L-functions at the central point*, Number theory in progress, Vol. 2 (Zakopane-Kościelisko, 1997), de Gruyter, Berlin, 1999, pp. 941–952. MR1689553

[KL78] G. A. Kabatjanskiĭ and V. I. Levenšteĭn, *Bounds for packings on the sphere and in space* (Russian), Problemy Peredači Informacii **14** (1978), no. 1, 3–25. MR0514023

[Kar70] A. A. Karacuba, *Sums of characters with prime numbers* (Russian), Izv. Akad. Nauk SSSR Ser. Mat. **34** (1970), 299–321. MR0271040

[Kat80] N. M. Katz, *Sommes exponentielles* (French), Astérisque, vol. 79, Société Mathématique de France, Paris, 1980. Course taught at the University of Paris, Orsay, Fall 1979; With a preface by Luc Illusie; Notes written by Gérard Laumon; With an English summary. MR617009

[Kat88] N. M. Katz, *Gauss sums, Kloosterman sums, and monodromy groups*, Annals of Mathematics Studies, vol. 116, Princeton University Press, Princeton, NJ, 1988. MR955052

[Kat90a] N. M. Katz, *Exponential sums and differential equations*, Annals of Mathematics Studies, vol. 124, Princeton University Press, Princeton, NJ, 1990. MR1081536

[Kat90b] N. M. Katz, *Exponential sums over finite fields and differential equations over the complex numbers: some interactions*, Bull. Amer. Math. Soc. (N.S.) **23** (1990), no. 2, 269–309, DOI 10.1090/S0273-0979-1990-15922-1. MR1032857

[Kat96] N. M. Katz, *Rigid local systems*, Annals of Mathematics Studies, vol. 139, Princeton University Press, Princeton, NJ, 1996. MR1366651

[Kat05a] N. M. Katz, *Moments, monodromy, and perversity: a Diophantine perspective*, Annals of Mathematics Studies, vol. 159, Princeton University Press, Princeton, NJ, 2005. MR2183396

[Kat05b] N. M. Katz, *Moments, monodromy, and perversity: a Diophantine perspective*, Annals of Mathematics Studies, vol. 159, Princeton University Press, Princeton, NJ, 2005. MR2183396

[Kat12] N. M. Katz, *Convolution and equidistribution: Sato-Tate theorems for finite-field Mellin transforms*, Annals of Mathematics Studies, vol. 180, Princeton University Press, Princeton, NJ, 2012. MR2850079

[KN16] R. Khan and H. T. Ngo, *Nonvanishing of Dirichlet L-functions*, Algebra Number Theory **10** (2016), no. 10, 2081–2091, DOI 10.2140/ant.2016.10.2081. MR3582014

[KZ16] E. M. Kıral and F. Zhou, *The Voronoi formula and double Dirichlet series*, Algebra Number Theory **10** (2016), no. 10, 2267–2286, DOI 10.2140/ant.2016.10.2267. MR3582019

[Klo27] H. D. Kloosterman, *On the representation of numbers in the form* $ax^2 + by^2 + cz^2 + dt^2$, Acta Math. **49** (1927), no. 3-4, 407–464, DOI 10.1007/BF02564120. MR1555249

[Kow13] E. Kowalski, *Families of cusp forms*, Actes de la Conférence "Théorie des Nombres et Applications", Publ. Math. Besançon Algèbre Théorie Nr., vol. 2013, Presses Univ. Franche-Comté, Besançon, 2013, pp. 5–40. MR3220018

[Kow15] E. Kowalski, *Gaps between prime numbers and primes in arithmetic progressions [after Y. Zhang and J. Maynard]*, Astérisque **367-368** (2015), Exp. No. 1084, ix, 327–366. MR3363595

[KMS17] E. Kowalski, P. Michel, and W. Sawin, *Bilinear forms with Kloosterman sums and applications*, Ann. of Math. (2) **186** (2017), no. 2, 413–500, DOI 10.4007/annals.2017.186.2.2. MR3702671

[KMS18] E. Kowalski, Ph. Michel, and W. Sawin, *Stratification and averaging for exponential sums : bilinear forms with generalized Kloosterman sums* (2018). `https://arxiv.org/abs/1802.09849`.

[KMV02] E. Kowalski, P. Michel, and J. VanderKam, *Rankin-Selberg L-functions in the level aspect*, Duke Math. J. **114** (2002), no. 1, 123–191, DOI 10.1215/S0012-7094-02-11416-1. MR1915038

[Lau87] G. Laumon, *Transformation de Fourier, constantes d'équations fonctionnelles et conjecture de Weil* (French), Inst. Hautes Études Sci. Publ. Math. **65** (1987), 131–210. MR908218

[Mat11] K. Matomäki, *A note on signs of Kloosterman sums* (English, with English and French summaries), Bull. Soc. Math. France **139** (2011), no. 3, 287–295, DOI 10.24033/bsmf.2609. MR2869308

[May16] J. Maynard, *Large gaps between primes*, Ann. of Math. (2) **183** (2016), no. 3, 915–933, DOI 10.4007/annals.2016.183.3.3. MR3488739

[Mic95] P. Michel, *Autour de la conjecture de Sato-Tate pour les sommes de Kloosterman. I* (French), Invent. Math. **121** (1995), no. 1, 61–78, DOI 10.1007/BF01884290. MR1345284

[Mic98] P. Michel, *Minorations de sommes d'exponentielles* (French), Duke Math. J. **95** (1998), no. 2, 227–240, DOI 10.1215/S0012-7094-98-09507-2. MR1652005

[MV00] P. Michel and J. VanderKam, *Non-vanishing of high derivatives of Dirichlet L-functions at the central point*, J. Number Theory **81** (2000), no. 1, 130–148, DOI 10.1006/jnth.1999.2460. MR1743500

[MS06] S. D. Miller and W. Schmid, *Automorphic distributions, L-functions, and Voronoi summation for* GL(3), Ann. of Math. (2) **164** (2006), no. 2, 423–488, DOI 10.4007/annals.2006.164.423. MR2247965

[Nun17] R. M. Nunes, *On the least squarefree number in an arithmetic progression*, Mathematika **63** (2017), no. 2, 483–498, DOI 10.1112/S0025579317000043. MR3706592

[Pol14a] D. H. J. Polymath, *New equidistribution estimates of Zhang type*, Algebra Number Theory **8** (2014), no. 9, 2067–2199, DOI 10.2140/ant.2014.8.2067. MR3294387

[Pol14b] D. H. J. Polymath, *Variants of the Selberg sieve, and bounded intervals containing many primes*, Res. Math. Sci. **1** (2014), Art. 12, 83, DOI 10.1186/s40687-014-0012-7. MR3373710

[Pra58] K. Prachar, *Über die kleinste quadratfreie Zahl einer arithmetischen Reihe* (German), Monatsh. Math. **62** (1958), 173–176, DOI 10.1007/BF01301288. MR0092806

[SST16] P. Sarnak, S. W. Shin, and N. Templier, *Families of L-functions and their symmetry*, Families of automorphic forms and the trace formula, Simons Symp., Springer, [Cham], 2016, pp. 531–578. MR3675175

[Ser79] J.-P. Serre, *Local fields*, Graduate Texts in Mathematics, vol. 67, Springer-Verlag, New York-Berlin, 1979. Translated from the French by Marvin Jay Greenberg. MR554237

[SF09] J. Sivak-Fischler, *Crible asymptotique et sommes de Kloosterman* (French, with English and French summaries), Bull. Soc. Math. France **137** (2009), no. 1, 1–62, DOI 10.24033/bsmf.2568. MR2496700

[Sou07] K. Soundararajan, *Small gaps between prime numbers: the work of Goldston-Pintz-Yıldırım*, Bull. Amer. Math. Soc. (N.S.) **44** (2007), no. 1, 1–18, DOI 10.1090/S0273-0979-06-01142-6. MR2265008

[Tay08] R. Taylor, *Automorphy for some l-adic lifts of automorphic mod l Galois representations. II*, Publ. Math. Inst. Hautes Études Sci. **108** (2008), 183–239, DOI 10.1007/s10240-008-0015-2. MR2470688

[Vau97] R. C. Vaughan, *The Hardy-Littlewood method*, 2nd ed., Cambridge Tracts in Mathematics, vol. 125, Cambridge University Press, Cambridge, 1997. MR1435742

[Wei41] A. Weil, *On the Riemann hypothesis in functionfields*, Proc. Nat. Acad. Sci. U. S. A. **27** (1941), 345–347. MR0004242

[WX16] J. Wu and P. Xi, *Arithmetic exponent pairs for algebraic trace functions and applications* (2016). `https://arxiv.org/abs/1603.07060`.

[Xi15] P. Xi, *Sign changes of Kloosterman sums with almost prime moduli*, Monatsh. Math. **177** (2015), no. 1, 141–163, DOI 10.1007/s00605-014-0653-z. MR3336337

[Xi16] P. Xi, *Sign changes of Kloosterman sums with almost prime moduli, II*, IMRN **2016** (2016), no. 00, 1–28.

[You11] M. P. Young, *The fourth moment of Dirichlet L-functions*, Ann. of Math. (2) **173** (2011), no. 1, 1–50, DOI 10.4007/annals.2011.173.1.1. MR2753598

[Zha14] Y. Zhang, *Bounded gaps between primes*, Ann. of Math. (2) **179** (2014), no. 3, 1121–1174, DOI 10.4007/annals.2014.179.3.7. MR3171761

[Del77] P. Deligne, *Cohomologie étale* (French), Lecture Notes in Mathematics, vol. 569, Springer-Verlag, Berlin, 1977. Séminaire de géométrie algébrique du Bois-Marie SGA $4\frac{1}{2}$. MR463174

Laboratoire de Mathématiques d'Orsay, Université Paris-Sud, CNRS, Université Paris-Saclay, 91405 Orsay, France
Email address: `etienne.fouvry@u-psud.fr`

ETH Zürich – D-MATH, Rämistrasse 101, CH-8092 Zürich, Switzerland
Email address: `kowalski@math.ethz.ch`

EPFL/SB/TAN, Station 8, CH-1015 Lausanne, Switzerland
Email address: `philippe.michel@epfl.ch`

Mathematics Department, Rm 411, MC 4439 2990 Broadway New York NY 10027, Columbia University, USA
Email address: `will.sawin@columbia.edu`

Contemporary Mathematics
Volume **740**, 2019
https://doi.org/10.1090/conm/740/14904

Sato-Tate distributions

Andrew V. Sutherland

ABSTRACT. In this expository article we explore the relationship between Galois representations, motivic L-functions, Mumford-Tate groups, and Sato-Tate groups, and we give an explicit formulation of the Sato-Tate conjecture for abelian varieties as an equidistribution statement relative to the Sato-Tate group. We then discuss the classification of Sato-Tate groups of abelian varieties of dimension $g \leq 3$ and compute some of the corresponding trace distributions. This article is based on a series of lectures presented at the 2016 Arizona Winter School held at the Southwest Center for Arithmetic Geometry.

1. An introduction to Sato-Tate distributions

Before discussing the Sato-Tate conjecture and Sato-Tate distributions in the context of abelian varieties, let us first consider the more familiar setting of Artin motives (varieties of dimension zero).

1.1. A first example. Let $f \in \mathbf{Z}[x]$ be a squarefree polynomial of degree d. For each prime p, let $f_p \in (\mathbf{Z}/p\mathbf{Z})[x] \simeq \mathbf{F}_p[x]$ denote the reduction of f modulo p, and define

$$N_f(p) := \#\{x \in \mathbf{F}_p : f_p(x) = 0\},$$

which we note is an integer between 0 and d. We would like to understand how $N_f(p)$ varies with p. The table below shows the values of $N_f(p)$ for the polynomial $f(x) = x^3 - x + 1$ for primes $p \leq 60$:

p:	2	3	5	7	11	13	17	19	23	29	31	37	41	43	47	53	59
$N_f(p)$	0	0	1	1	1	0	1	1	2	0	0	1	0	1	0	1	3

There does not appear to be any obvious pattern (and we should know not to expect one, because the Galois group of f is nonabelian). The prime $p = 23$ is exceptional because it divides $\operatorname{disc}(f) = -23$, which means that $f_{23}(x)$ has a double root. As we are interested in the distribution of $N_f(p)$ as p tends to infinity, we are happy to ignore such primes, which are necessarily finite in number.

This tiny dataset does not tell us much. Let us now consider primes $p \leq B$ for increasing bounds B, and compute the proportions $c_i(B)$ of primes $p \leq B$ with $N_f(p) = i$. We obtain the following statistics:

2010 *Mathematics Subject Classification.* Primary 11M50; Secondary 11G10, 11G20, 14G10, 14K15.

The author was supported by NSF grants DMS-1115455 and DMS-1522526.

B	$c_0(B)$	$c_1(B)$	$c_2(B)$	$c_3(B)$
10^3	0.323353	0.520958	0.005988	0.155689
10^4	0.331433	0.510586	0.000814	0.157980
10^5	0.333646	0.502867	0.000104	0.163487
10^6	0.333185	0.500783	0.000013	0.166032
10^7	0.333360	0.500266	0.000002	0.166373
10^8	0.333337	0.500058	0.000000	0.166605
10^9	0.333328	0.500016	0.000000	0.166656
10^{12}	0.333333	0.500000	0.000000	0.166666

This leads us to conjecture that the limiting values c_i of $c_i(B)$ as $B \to \infty$ are

$$c_0 = 1/3, \quad c_1 = 1/2, \quad c_2 = 0, \quad c_3 = 1/6.$$

There is a natural motivation for this conjecture (which is, in fact, a theorem), one that would allow us to correctly predict the asymptotic ratios c_i without needing to compute any statistics. Let us fix an algebraic closure $\overline{\mathbf{Q}}$ of $\mathbf{Q}$. The absolute Galois group $\operatorname{Gal}(\overline{\mathbf{Q}}/\mathbf{Q})$ acts on the roots of $f(x)$ by permuting them. This allows us to define the *Galois representation* (a continuous homomorphism)

$$\rho_f\colon \operatorname{Gal}(\overline{\mathbf{Q}}/\mathbf{Q}) \to \operatorname{GL}_d(\mathbf{C}),$$

whose image is a subgroup of the permutation matrices in $\operatorname{O}_d(\mathbf{C}) \subseteq \operatorname{GL}_d(\mathbf{C})$; here O_d denotes the orthogonal group (we could replace $\mathbf{C}$ with any field of characteristic zero). Note that $\operatorname{Gal}(\overline{\mathbf{Q}}/\mathbf{Q})$ and $\operatorname{GL}_d(\mathbf{C})$ are topological groups (the former has the Krull topology), and homomorphisms of topological groups are understood to be continuous. In order to associate a permutation of the roots of $f(x)$ to a matrix in $\operatorname{GL}_d(\mathbf{C})$ we need to fix an ordering of the roots; this amounts to choosing a basis for the vector space $\mathbf{C}^d$, which means that our representation ρ_f is really defined only up to conjugacy.

The value ρ_f takes on $\sigma \in \operatorname{Gal}(\overline{\mathbf{Q}}/\mathbf{Q})$ depends only on the restriction of σ to the splitting field L of f, so we could restrict our attention to $\operatorname{Gal}(L/\mathbf{Q})$. This makes ρ_f an *Artin representation*: a continuous representation $\operatorname{Gal}(\overline{\mathbf{Q}}/\mathbf{Q}) \to \operatorname{GL}_d(\mathbf{C})$ that factors through a finite quotient (by an open subgroup). But in the more general settings we wish to consider this may not always be true, and even when it is, we typically will not be given L; it is thus more convenient to work with $\operatorname{Gal}(\overline{\mathbf{Q}}/\mathbf{Q})$.

To facilitate this approach, we associate to each prime p an *absolute Frobenius element*

$$\operatorname{Frob}_p \in \operatorname{Gal}(\overline{\mathbf{Q}}/\mathbf{Q})$$

that may be defined as follows. Fix an embedding $\overline{\mathbf{Q}}$ in $\overline{\mathbf{Q}}_p$ and use the valuation ideal $\mathfrak{P}$ of $\overline{\mathbf{Q}}_p$ (the maximal ideal of its ring of integers) to define a compatible system of primes $\mathfrak{q}_L := \mathfrak{P} \cap L$, where L ranges over all finite extensions of $\mathbf{Q}$. For each prime $\mathfrak{q}_L$, let $D_{\mathfrak{q}_L} \subseteq \operatorname{Gal}(L/\mathbf{Q})$, denote its decomposition group, $I_{\mathfrak{q}_L} \subseteq D_{\mathfrak{q}_L}$ its inertia group, and $\mathbf{F}_{\mathfrak{q}_L} := \mathbf{Z}_L/\mathfrak{q}_L$ its residue field, where $\mathbf{Z}_L$ denotes the ring of integers of L. Taking the inverse limit of the exact sequences

$$1 \to I_{\mathfrak{q}_L} \to D_{\mathfrak{q}_L} \to \operatorname{Gal}(\mathbf{F}_{\mathfrak{q}_L}/\mathbf{F}_p) \to 1$$

over finite extensions $L/\mathbf{Q}$ ordered by inclusion gives an exact sequence of profinite groups

$$1 \to I_p \to D_p \to \operatorname{Gal}(\overline{\mathbf{F}}_p/\mathbf{F}_p) \to 1.$$

We now define $\mathrm{Frob}_p \in D_p \subseteq \mathrm{Gal}(\overline{\mathbf{Q}}/\mathbf{Q})$ by arbitrarily choosing a preimage of the Frobenius automorphism $x \to x^p$ in $\mathrm{Gal}(\overline{\mathbf{F}}_p/\mathbf{F}_p)$ under the map in the exact sequence above. We actually made two arbitrary choices in our definition of Frob_p, since we also chose an embedding of $\overline{\mathbf{Q}}$ into $\overline{\mathbf{Q}}_p$. Our absolute Frobenius element Frob_p is thus far from canonical, but it exists. Its key property is that if $L/\mathbf{Q}$ is a finite Galois extension in which p is unramified, then the conjugacy class $\mathrm{conj}_L(\mathrm{Frob}_p)$ in $\mathrm{Gal}(L/\mathbf{Q})$ of the restriction of $\mathrm{Frob}_p\colon \overline{\mathbf{Q}} \to \overline{\mathbf{Q}}$ to L is uniquely determined, independent of our choices; note that when p is unramified, I_p is trivial and $D_p \simeq \mathrm{Gal}(\overline{\mathbf{F}}_p/\mathbf{F}_p)$. Everything we have said applies *mutatis mutandi* if we replace $\mathbf{Q}$ by a number field K: put $\overline{K} := \overline{\mathbf{Q}}$, replace p by a prime $\mathfrak{p}$ of K (a nonzero prime ideal of $\mathbf{Z}_K$), and replace $\mathbf{F}_p$ by the residue field $\mathbf{F}_\mathfrak{p} := \mathbf{Z}_K/\mathfrak{p}$.

We now make the following observation: for any prime p that does not divide $\mathrm{disc}(f)$ we have

$$N_f(p) = \mathrm{tr}\,\rho_f(\mathrm{Frob}_p). \tag{1.1}$$

This follows from the fact that the trace of a permutation matrix counts its fixed points. Since p is unramified in the splitting field of f, the inertia group $I_p \subseteq \mathrm{Gal}(\overline{\mathbf{Q}}/\mathbf{Q})$ acts trivially on the roots of $f(x)$, and the action of Frob_p on the roots of $f(x)$ coincides (up to conjugation) with the action of the Frobenius automorphism $x \to x^p$ on the roots of $f_p(x)$, both of which are described by the permutation matrix $\rho_f(\mathrm{Frob}_p)$. The Chebotarev density theorem implies that we can compute c_i via (1.1) by counting matrices in $\rho_f(\mathrm{Gal}(\overline{\mathbf{Q}}/\mathbf{Q}))$ with trace i, and it is enough to determine the trace and cardinality of each conjugacy class.

THEOREM 1.1. CHEBOTAREV DENSITY THEOREM *Let L/K be a finite Galois extension of number fields with Galois group $G := \mathrm{Gal}(L/K)$. For every subset C of G stable under conjugation we have*

$$\lim_{B\to\infty} \frac{\#\{N(\mathfrak{p}) \le B : \mathrm{conj}_L(\mathrm{Frob}_\mathfrak{p}) \subseteq C\}}{\#\{N(\mathfrak{p}) \le B\}} = \frac{\#C}{\#G},$$

where $\mathfrak{p}$ ranges over primes of K and $N(\mathfrak{p}) := \#\mathbf{F}_\mathfrak{p}$ is the cardinality of the residue field $\mathbf{F}_\mathfrak{p} := \mathbf{Z}_K/\mathfrak{p}$.

PROOF. See Corollary 2.13 in Section 2. □

REMARK 1.2. In Theorem 1.1 the asymptotic ratio on the left depends only on primes of inertia degree 1 (those with prime residue field), since these make up all but a negligible proportion of the primes $\mathfrak{p}$ for which $N(\mathfrak{p}) \le B$. Taking $C = \{1_G\}$ shows that a constant proportion of the primes of K split completely in L and in particular have prime residue fields; this special case is already implied by the Frobenius density theorem, which was proved much earlier (in terms of Dirichlet density). In our statement of Theorem 1.1 we do not bother to exclude primes of K that are ramified in L because no matter what value $\mathrm{conj}_L(\mathrm{Frob}_\mathfrak{p})$ takes on these primes it will not change the limiting ratio.

In our example with $f(x) = x^3 - x + 1$, one finds that $G_f := \rho_f(\overline{\mathbf{Q}}/\mathbf{Q})$ is isomorphic to S_3, the Galois group of the splitting field of $f(x)$. Its three conjugacy classes are represented by the matrices

$$\begin{bmatrix} 0 & 1 & 0 \\ 0 & 0 & 1 \\ 1 & 0 & 0 \end{bmatrix}, \qquad \begin{bmatrix} 1 & 0 & 0 \\ 0 & 0 & 1 \\ 0 & 1 & 0 \end{bmatrix}, \qquad \begin{bmatrix} 1 & 0 & 0 \\ 0 & 1 & 0 \\ 0 & 0 & 1 \end{bmatrix},$$

with traces 0, 1, 3. The corresponding conjugacy classes have cardinalities 2, 3, 1, respectively, thus

$$c_0 = 1/3, \quad c_1 = 1/2, \quad c_2 = 0, \quad c_3 = 1/6,$$

as we conjectured.

If we endow the group G_f with the discrete topology it becomes a compact group, and therefore has a *Haar measure* μ that is uniquely determined if we normalize it so that $\mu(G_f) = 1$ (which we always do). Recall that the Haar measure of a compact group G is a translation-invariant Radon measure (in particular, $\mu(gS) = \mu(Sg) = \mu(S)$ for any measurable set S and $g \in G$), and is unique up to scaling.[1] For finite groups the Haar measure μ is just the normalized counting measure. We can compute the expected value of trace (and many other statistical quantities of interest) by integrating against the Haar measure, which in this case amounts to summing over the finite group G_f:

$$\mathrm{E}[\mathrm{tr}] = \int_{G_f} \mathrm{tr}\ \mu = \frac{1}{\#G_f} \sum_{g \in G_f} \mathrm{tr}(g) = \sum_{i=0}^{d} c_i i.$$

The Chebotarev density theorem implies that this is also the average value of $N_f(p)$, that is,

$$\lim_{B \to \infty} \frac{\sum_{p \le B} N_f(p)}{\sum_{p \le B} 1} = \mathrm{E}[\mathrm{tr}].$$

This average is 1 in our example, because $f(x)$ is irreducible; see Exercise 1.1.

The quantities c_i define a probability distribution on the set $\{\mathrm{tr}(g) : g \in G_f\}$ that we can also view as a probability distribution on the set $\{N_f(p) : p \text{ prime}\}$. Picking a random prime p in some large interval $[1, B]$ and computing $N_f(p)$ is the same thing as picking a random matrix g in G_f and computing $\mathrm{tr}(g)$. More precisely, the sequence $(N_f(p))_p$ indexed by primes p is *equidistributed* with respect to the pushforward of the Haar measure μ under the trace map. We discuss the notion of equidistribution more generally in Section 2.

1.2. Moment sequences. There is another way to characterize the probability distribution on $\mathrm{tr}(g)$ given by the c_i; we can compute its *moment sequence*:

$$\mathrm{M}[\mathrm{tr}] := (\mathrm{E}[\mathrm{tr}^n])_{n \ge 0},$$

where

$$\mathrm{E}[\mathrm{tr}^n] = \int_{G_f} \mathrm{tr}^n\ \mu.$$

It might seem silly to include the zeroth moment $\mathrm{E}[\mathrm{tr}^0] = \mathrm{E}[1] = 1$, but in Section 4 we will see why this convention is useful. In our example we have the moment sequence

$$\mathrm{M}[\mathrm{tr}] = (1, 1, 2, 5, 14, 41, \ldots, \tfrac{1}{2}(3^{n-1} + 1), \ldots).$$

The sequence M[tr] uniquely determines[2] the distributions of traces and thus captures all the information encoded in the c_i. It may not seem very useful to replace a finite set of rational numbers with an infinite sequence of integers, but when dealing

[1]For locally compact groups G one distinguishes left and right Haar measures, but the two coincide when G is compact; see [**22**] for more background on Haar measures.

[2]Not all moment sequences uniquely determine an underlying probability distribution, but all the moment sequence we shall consider do (because they satisfy *Carleman's condition*, see [**52**, p. 126], for example).

with continuous probability distributions, as we are forced to do as soon as we leave our weight zero setting, moment sequences are a powerful tool.

If we pick another cubic polynomial $f \in \mathbf{Z}[x]$, we will typically obtain the same result as we did in our example; when ordered by height almost all cubic polynomials f have Galois group $G_f \simeq S_3$. But there are exceptions: if f is not irreducible over $\mathbf{Q}$ then G_f will be isomorphic to a proper subgroup of S_3, and this also occurs when the splitting field of f is a cyclic cubic extension (this happens precisely when $\operatorname{disc}(f)$ is a square in $\mathbf{Q}^\times$; the polynomial $f(x) = x^3 - 3x - 1$ is an example). Up to conjugacy there are four subgroups of S_3, each corresponding to a different distribution of $N_f(p)$:

$f(x)$	G_f	c_0	c_1	c_2	c_3	M[tr]
$x^3 - x$	1	0	0	0	1	$(1, 3, 9, 27, 81, \ldots)$
$x^3 + x$	C_2	0	1/2	0	1/2	$(1, 2, 5, 14, 41, \ldots)$
$x^3 - 3x - 1$	C_3	2/3	0	0	1/3	$(1, 1, 3, 19, 27, \ldots)$
$x^3 - x + 1$	S_3	1/3	1/2	0	1/6	$(1, 1, 2, 5, 14, \ldots)$

One can do the same thing with polynomials of degree $d > 3$. For $d \leq 19$ the results are exhaustive: for every transitive subgroup G of S_d the database of Klüners and Malle [**51**] contains at least one polynomial $f \in \mathbf{Z}[x]$ with $G_f \simeq G$ (including all 1954 transitive subgroups of S_{16}). The non-transitive cases can be constructed as products (of groups and of polynomials) of transitive cases of lower degree. It is an open question whether this can be done for all d (even in principle). This amounts to a strong form of the inverse Galois problem over $\mathbf{Q}$; we are asking not only whether every finite group can be realized as a Galois group over $\mathbf{Q}$, but whether every transitive permutation group of degree d can be realized as the Galois group of the splitting field of an irreducible polynomial of degree d.

1.3. Zeta functions. For polynomials f of degree $d = 3$ there is a one-to-one correspondence between subgroups of S_d and distributions of $N_f(p)$. This is not true for $d \geq 4$. For example, the polynomials $f(x) = x^4 - x^3 + x^2 - x + 1$ with $G_f \simeq C_4$ and $g(x) = x^4 - x^2 + 1$ with $G_g \simeq C_2 \times C_2$ both have $c_0 = 3/4$, $c_1 = c_2 = c_3 = 0$, and $c_4 = 1/4$, corresponding to the moment sequence $\mathrm{M}[\mathrm{tr}] = (1, 1, 4, 16, 64, \ldots)$.

We can distinguish these cases if, in addition to considering the distribution of $N_f(p)$, we also consider the distribution of

$$N_f(p^r) := \#\{x \in \mathbf{F}_{p^r} : f_p(x) = 0\}$$

for integers $r \geq 1$. In our quartic example we have $N_g(p^2) = 4$ for almost all p, whereas $N_f(p^2)$ is 4 or 2 depending on whether p is a square modulo 5 or not. In terms of the matrix group G_f we have

$$N_f(p^r) = \operatorname{tr}\big(\rho_f(\mathrm{Frob}_p)^r\big) \tag{1.2}$$

for all primes p that do not divide $\operatorname{disc}(f)$. To see this, note that the permutation matrix $\rho_f(\mathrm{Frob}_p)^r$ corresponds to the permutation of the roots of $f_p(x)$ given by the rth power of the Frobenius automorphism $x \mapsto x^p$. Its fixed points are precisely the roots of $f_p(x)$ that lie in $\mathbf{F}_{p^r}$; taking the trace counts these roots, and this yields $N_f(p^r)$.

This naturally leads to the definition of the local *zeta function* of f at p:

$$Z_{f_p}(T) := \exp\left(\sum_{r=1}^{\infty} N_f(p^r)\frac{T^r}{r}\right), \tag{1.3}$$

which can be viewed as a generating function for the sequence of positive integers $(N_f(p), N_f(p^2), N_f(p^3), \ldots)$. This particular form of generating function may seem strange when first encountered, but it has some very nice properties. For example, if $f, g \in \mathbf{Z}[x]$ are squarefree polynomials with no common factor, then their product fg is also square free, and for all $p \nmid \operatorname{disc}(fg)$ we have

$$Z_{(fg)_p} = Z_{f_p g_p} = Z_{f_p} Z_{g_p}.$$

REMARK 1.3. The identity (1.2) is a special case of the Grothendieck-Lefschetz trace formula. It allows us to express the zeta function $Z_{f_p}(T)$ as a sum over powers of the traces of the image of Frob_p under the Galois representation ρ_f. In general one considers the trace of the Frobenius endomorphism acting on étale cohomology, but in dimension zero the only relevant cohomology is H^0.

While defined as a power series, in fact $Z_{f_p}(T)$ is a rational function of the form

$$Z_{f_p}(T) = \frac{1}{L_p(T)},$$

where $L_p(T)$ is an integer polynomial whose roots lie on the unit circle. This can be viewed as a consequence of the Weil conjectures in dimension zero,[3] but in fact it follows directly from (1.2). Indeed, for any matrix $A \in \mathrm{GL}_d(\mathbf{C})$ we have the identity

$$\exp\left(\sum_{r=1}^{\infty} \operatorname{tr}(A^r)\frac{T^r}{r}\right) = \det(1 - AT)^{-1}, \tag{1.4}$$

which can be proved by expressing the coefficients on both sides as symmetric functions in the eigenvalues of A; see Exercise 1.2. Applying (1.2) and (1.4) to the definition of $Z_{f_p}(T)$ in (1.3) yields

$$Z_{f_p}(T) = \frac{1}{\det(1 - \rho_f(\mathrm{Frob}_p)T)},$$

thus

$$L_p(T) = \det(1 - \rho_f(\mathrm{Frob}_p)T).$$

The polynomial $L_p(T)$ is precisely the polynomial that appears in the Euler factor at p of the (partial) *Artin L-function* $L(\rho_f, s)$ for the representation ρ_f:

$$L(\rho_f, s) := \prod_p L_p(p^{-s})^{-1},$$

at least for primes p that do not divide $\operatorname{disc}(f)$; for the definition of the Euler factors at ramified primes (and the Gamma factors at archimedean places), see [**60**, Ch. 2].[4] The Euler product for $L(\rho_f, s)$ defines a function that is holomorphic and nonvanishing on $\mathrm{Re}(s) > 1$. We shall not be concerned with the Euler factors at ramified primes, other than to note that they are holomorphic and nonvanishing on $\mathrm{Re}(s) > 1$.

[3] Provided one accounts for the fact that $f(x) = 0$ does not define an irreducible variety unless $\deg(f) = 1$; in this case $N_f(p^r) = 1$ and $L_p(T) = 1 - T$, which is consistent with the usual formulation of the Weil conjectures (see Theorem 1.8).

[4] The alert reader will note that primes dividing the discriminant of f need not ramify in its splitting field; we are happy to ignore these primes as well, just as we may ignore primes of bad reduction for a curve that are good primes for its Jacobian.

REMARK 1.4. Every representation $\rho\colon \operatorname{Gal}(\overline{\mathbf{Q}}/\mathbf{Q}) \to \mathrm{GL}_d(\mathbf{C})$ with finite image gives rise to an Artin L-function $L(\rho, s)$, and Artin proved that every decomposition of ρ into sub-representations gives rise to a corresponding factorization of $L(\rho, s)$ into Artin L-functions of lower degree. The representation ρ_f we have defined is determined by the permutation action of $\operatorname{Gal}(\overline{\mathbf{Q}}/\mathbf{Q})$ on the formal $\mathbf{C}$-vector space with basis elements corresponding to roots of f. The linear subspace spanned by the sum of the basis vectors is fixed by $\operatorname{Gal}(\overline{\mathbf{Q}}/\mathbf{Q})$, so for $d > 1$ we can always decompose ρ_f as the sum of the trivial representation and a representation of dimension $d-1$, in which case $L(\rho_f, s)$ is thc product of the Riemann zeta function (the Artin L-function of the trivial representation), and an Artin L-function of degree $d-1$. The Artin L-functions $L(\rho_f, s)$ we have defined are thus imprimitive for $\deg f > 1$.

Returning to the question of equidistribution, the Haar measure μ of the group $G_f = \rho_f(\operatorname{Gal}(\overline{\mathbf{Q}}/\mathbf{Q}))$ allows us to determine the distribution of L-polynomials $L_p(T)$ that we see as p varies. Each polynomial $L_p(T)$ is the *reciprocal polynomial* (obtained by reversing the coefficients) of the characteristic polynomial of $\rho_f(\operatorname{Frob}_p)$. If we fix a polynomial $P(T)$ of degree $d = \deg f$, and pick a prime p at random from some large interval, the probability that $L_p(T) = P(T)$ is equal to the probability that the reciprocal polynomial $T^d P(1/T)$ is the characteristic polynomial of a random element of G_f (this probability will be zero unless $P(T)$ has a particular form; see Exercise 1.3).

REMARK 1.5. For $d \le 5$ the distribution of characteristic polynomials uniquely determines each subgroup of S_d (up to conjugacy). This is not true for $d \ge 6$, and for $d \ge 8$ one can find non-isomorphic subgroups of S_d with the same distribution of characteristic polynomials; the transitive permutation groups `8T10` and `8T11` which arise for $x^8 - 13x^6 + 44x^4 - 17x^2 + 1$ and $x^8 - x^5 - 2x^4 + 4x^2 + x + 1$ (respectively) are an example.

1.4. Computing zeta functions in dimension zero. Let us now briefly address the practical question of efficiently computing the zeta function $Z_{f_p}(T)$, which amounts to computing the polynomial $L_p(T)$. It suffices to compute the integers $N_f(p^r)$ for $r \le d$, which is equivalent to determining the degrees of the irreducible polynomials appearing in the factorization of $f_p(x)$ in $\mathbf{F}_p[x]$. These determine the cycle type, and therefore the conjugacy class, of the permutation of the roots of $f_p(x)$ induced by the action of the Frobenius automorphism $x \mapsto x^p$, which in turn determines the characteristic polynomial of $\rho_f(\operatorname{Frob}_p)$ and the L-polynomial $L_p(T) = \det(1 - \rho_f(\operatorname{Frob}_p)T)$; see Exercise 1.3. To determine the factorization pattern of $f_p(x)$, one can apply the following algorithm.

ALGORITHM 1.6. *Given a squarefree polynomial $f \in \mathbf{F}_p[x]$ of degree $d > 1$, compute the number n_i of irreducible factors of f in $\mathbf{F}_p[x]$ of degree i, for $1 \le i \le d$ as follows:*

1. Let $g_1(x)$ be $f(x)$ made monic and put $r_0(x) := x$.
2. For i from 1 to d:
   a. If $i > \deg(g_i)/2$ then for $i \le j \le d$ put $n_j := 1$ if $j = \deg(g_i)$ and $n_j := 0$ otherwise, and then proceed to step 3.
   b. Using binary exponentiation in the quotient ring $\mathbf{F}_p[x]/(g_i)$, compute $r_i := r_{i-1}^p \bmod g_i$.
   c. Compute $h_i(x) := \gcd(g_i, r_i(x) - x) = \gcd(g_i(x), x^{p^i} - x)$ using the Euclidean algorithm.

d. Compute $n_i := \deg(h_i)/i$ and $g_{i+1} := g_i/h_i$ using exact division.
e. If $\deg(g_{i+1}) = 0$ then put $n_j := 0$ for $i < j \le d$ and proceed to step 3.
3. Output $n_1, \ldots, n_d$.

Algorithm 1.6 makes repeated use of the fact that the polynomial

$$x^{p^i} - x = \prod_{a \in \mathbf{F}_{p^i}} (x - a)$$

is equal to the product of all irreducible monic polynomials of degree dividing i in $\mathbf{F}_p[x]$. By starting with $i = 1$ and removing all factors of degree i as we go, we ensure that each h_i is a product of irreducible polynomials of degree i. Using fast algorithms for integer and polynomial arithmetic and the fast Euclidean algorithm (see [**29**, §8-11], for example), one can show that this algorithm uses $O((d \log p)^{2+o(1)})$ bit operations, a running time that is quasi-quadratic in the $O(d \log p)$ bit-size of its input $f \in \mathbf{F}_p[x]$.[5] In practical terms, it is extremely efficient. For example, the table of $c_i(B)$ values for our example polynomial $f(x) = x^3 - x + 1$ with $B = 10^{12}$ took less than two minutes to create using the `smalljac` software library [**48**, **85**], which includes an efficient implementation of basic finite field arithmetic. The NTL [**80**] and FLINT [**33**, **34**] libraries also incorporate variants of this algorithm, as do the computer algebra systems Sage [**67**] and Magma [**11**].

REMARK 1.7. Note that Algorithm 1.6 does *not* output the factorization of $f(x)$, just the degrees of its irreducible factors. It can be extended to a probabilistic algorithm that outputs the complete factorization of $f(x)$ (see [**29**, Alg. 14.8], for example), with an expected running time that is also quasi-quadratic. But no deterministic polynomial-time algorithm for factoring polynomials over finite fields is known, not even for $d = 2$. This is a famous open problem. One approach to solving it is to first prove the generalized Riemann hypothesis (GRH), which would address the case $d = 2$ and many others, but it is not even known whether the GRH is sufficient to address all cases.[6]

1.5. Arithmetic schemes. We now want to generalize our first example. Let us replace the equation $f(x) = 0$ with an *arithmetic scheme* X, a scheme of finite type over $\mathbf{Z}$; the case we have been considering is $X = \operatorname{Spec} A$, where $A = \mathbf{Z}[x]/(f)$. For each prime p the fiber X_p of $X \to \operatorname{Spec} \mathbf{Z}$ is a scheme of finite type over $\mathbf{F}_p$, and we let $N_X(p) := X_p(\mathbf{F}_p)$ count its $\mathbf{F}_p$-points; equivalently, we may define $N_X(p)$ as the number of closed points (maximal ideals) of X whose residue field has cardinality p, and similarly define $N_X(q)$ for prime powers $q = p^r$. The local zeta function of X at p is then defined as

$$Z_{X_p}(T) := \exp\left(\sum_{r=1}^{\infty} N_X(p^r)\frac{T^r}{r}\right).$$

These local zeta functions can then be packaged into a single *arithmetic zeta-function*

$$\zeta_X(s) := \prod_p Z_{X_p}(p^{-s}).$$

[5]One can improve this to $O\big(d^{1.5+o(1)}(\log p)^{1+o(1)} + d^{1+o(1)}(\log p)^{2+o(1)}\big)$ via [**50**]. In our setting d is fixed and $\log p$ is tending to infinity, so this is not an asymptotic improvement, but it does provide a constant factor improvement for large d.

[6]If you succeed with even a special case of this first step, the Clay institute will help fund the remaining work.

In our example with $X = \operatorname{Spec} \mathbf{Z}[x]/(f)$, the zeta function $\zeta_X(s)$ coincides with the Artin L-function $L(\rho_f, s) = \prod L_p(s)^{-1}$ up to a finite set of factors at primes p that divide $\operatorname{disc}(f)$.

The definitions above generalize to any number field K: replace $\mathbf{Q}$ by K, replace $\mathbf{Z}$ by $\mathbf{Z}_K$, replace p by a prime $\mathfrak{p}$ of K (nonzero prime ideal of $\mathbf{Z}_K$), replace $\mathbf{F}_p \simeq \mathbf{Z}/p\mathbf{Z}$ by the residue field $\mathbf{F}_\mathfrak{p} := \mathbf{Z}_K/\mathfrak{p}$. When considering questions of equidistribution we order primes $\mathfrak{p}$ by their norm $N(\mathfrak{p}) := \mathbf{F}_\mathfrak{p}$ (we may break ties arbitrarily), so that rather that summing over $p \le B$ we sum over $\mathfrak{p}$ for which $N(\mathfrak{p}) \le B$.

1.6. A second example. We now leave the world of Artin motives, which are motives of weight 0, and consider the simplest example in weight 1, an elliptic curve $E/\mathbf{Q}$. This is the setting in which the Sato–Tate conjecture was originally formulated. Every elliptic curve $E/\mathbf{Q}$ can be written in the form

$$E\colon y^2 = x^3 + Ax + B,$$

with $A, B \in \mathbf{Z}$. This equation is understood to define a smooth projective curve in $\mathbf{P}^2$ (homogenize the equation by introducing a third variable z), which has a single projective point $P_\infty := (0 : 1 : 0)$ at infinity that we take as the identity element of the group law on E. Recall that an elliptic curve is not just a curve, it is an abelian variety, and comes equipped with a distinguished rational point corresponding ot the identity; by applying a suitable automorphism of $\mathbf{P}^2$ we can always take this to be the point P_∞.

The group operation on E can be defined via the usual chord-and-tangent law (three points on a line sum to zero), which can be used to derive explicit formulas with coefficients in $\mathbf{Q}$, or in terms of the divisor class group $\operatorname{Pic}^0(E)$ (divisors of degree zero modulo principal divisors), in which every divisor class can be uniquely represented by a divisor of the form $P - P_\infty$, where P is a point on the curve. This latter view is more useful in that it easily generalizes to curves of genus $g > 1$, whereas the chord-and-tangent law does not. The Abel–Jacobi map $P \mapsto P - P_\infty$ gives a bijection between points on E and points on $\operatorname{Jac}(E)$ that commutes with the group operation, so the two approaches are equivalent.

For each prime p that does not divide the discriminant $\Delta := -16(4A^3 + 27B^2)$ we can reduce our equation for E modulo p to obtain an elliptic curve $E_p/\mathbf{F}_p$; in this case we say that p is a *prime of good reduction for* E (or simply a *good prime*). We should note that the discriminant Δ is not necessarily minimal; the curve E may have another model with good reduction at primes that divide Δ (possibly including 2), but we are happy to ignore any finite set of primes, including those that divide Δ.[7]

For every prime p of good reduction for E we have

$$N_E(p) := \#E_p(\mathbf{F}_p) = p + 1 - t_p,$$

where the integer t_p satisfies the *Hasse-bound* $|t_p| \le 2\sqrt{p}$. In contrast to our weight zero examples, the integers $N_E(p)$ now tend to infinity with p: we have

[7]All elliptic curves over $\mathbf{Q}$ have a global minimal model for which the primes of bad reduction are precisely those that divide the discriminant, but this model is not necessarily of the form $y^2 = x^3 + Ax + B$. Over general number fields K global minimal models do not always exist (they do when K has class number one).

$N_E(p) = p + 1 + O(\sqrt{p})$. In order to study how the error term varies with p we want to consider the normalized traces

$$x_p := t_p/\sqrt{p} \in [-2, 2].$$

We now want to conduct the following experiment: given an elliptic curve $E/\mathbf{F}_p$, compute x_p for all good primes $p \leq B$ and see how the x_p are distributed over the real interval $[-2, 2]$.

One can see an example for the elliptic curve $E : y^2 = x^3 + x + 1$ in Figure 1, which shows a histogram whose x-axis spans the interval $[-2, 2]$. This interval is subdivided into approximately $\sqrt{\pi(B)}$ subintervals, each of which contains a bar representing the number of x_p (for $p \leq B$) that lie in the subinterval. The gray line shows the height of the uniform distribution for scale (note that the vertical and horizontal scales are not the same). For $0 \leq n \leq 10$, the moment statistics

$$M_n := \frac{\sum_{p \leq B} x_p^n}{\sum_{p \leq B} 1},$$

are shown below the histogram. They appear to converge to the sequence of integers

$$1, 0, 1, 0, 2, 0, 5, 0, 14, 0, 42, \ldots$$

with label A126120 in the Online Encyclopedia of Integer Sequences (OEIS) [**64**]).

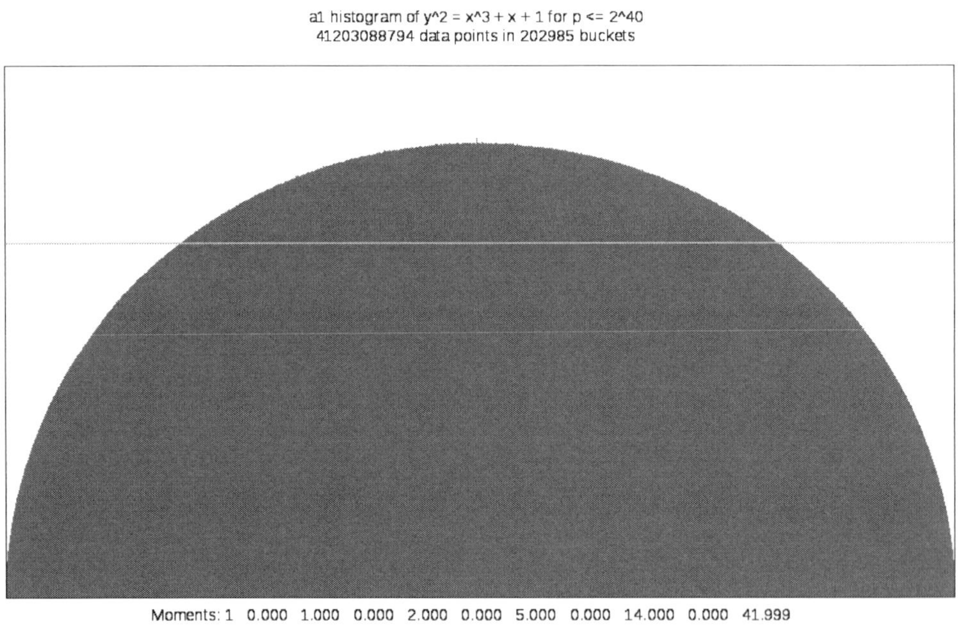

FIGURE 1. Sato-Tate distribution of an elliptic curve over $\mathbf{Q}$ (without CM). Visit `http://math.mit.edu/~drew/g1_D1_a1f.gif` to see an animated version.

The Sato–Tate conjecture for elliptic curves over $\mathbf{Q}$ (now a theorem) implies that for almost all $E/\mathbf{Q}$, whenever we run this experiment we will see the asymptotic distribution of Frobenius traces visible in Figure 1, with moment statistics that converge to the same integer sequence. In order to make this conjecture precise, let us first explain where the conjectured distribution comes from. In our first example we had a compact matrix group G_f associated to the scheme $X = \operatorname{Spec} \mathbf{Z}[x]/(f)$ whose Haar measure governed the distribution of $N_f(p)$. In fact we showed that more is true: there is a direct relationship between characteristic polynomials of elements of G_f and the L-polynomials $L_p(T)$ that appear in the local zeta functions $Z_{f_p}(T)$.

The same is true with our elliptic curve example. In order to identify a candidate group G_E whose Haar measure controls the distribution of normalized Frobenius traces x_p we need to look at the local zeta functions $Z_{E_p}(T)$. Let us recall what the Weil conjectures [**96**] (proved by Deligne [**18, 19**]) tell us about the zeta function of a variety over a finite field. The case of one-dimensional varieties (curves) was proved by Weil [**94**], who also proved an analogous result for abelian varieties [**95**]. This covers all the cases we shall consider, but let us state the general result. Recall that for a compact manifold X over $\mathbf{C}$, the *Betti number* b_i is the rank of the singular homology group $H_i(X, \mathbf{Z})$, and the *Euler characteristic* χ of X is defined by $\chi := \sum(-1)^i b_i$.

THEOREM 1.8 (WEIL CONJECTURES). *Let X be a geometrically irreducible non-singular projective variety of dimension n defined over a finite field $\mathbf{F}_q$ and define the zeta function*

$$Z_X(T) := \exp\left(\sum_{r=1}^{\infty} N_X(q^r)\frac{T^r}{r}\right),$$

where $N_X(q^r) := \#X(\mathbf{F}_{q^r})$. The following hold:

(i) ***Rationality**: $Z_X(T)$ is a rational function of the form*

$$Z_X(T) = \frac{P_1(T)\cdots P_{2n-1}(T)}{P_0(T)\cdots P_{2n}(T)},$$

with $P_i \in 1 + T\mathbf{Z}[T]$.

(i) ***Functional Equation**: the roots of $P_i(T)$ are the same as the roots of $T^{\deg P_{2n-i}} P_{2n-i}(1/(q^n T))$.*[8]

(i) ***Riemann Hypothesis**: the roots of $P_i(T)$ are complex number of absolute value $q^{-i/2}$.*

(i) ***Betti Numbers**: if X is the reduction of a non-singular variety Y defined over a number field $K \subseteq \mathbf{C}$, then the degree of P_i is equal to the Betti number b_i of $Y(\mathbf{C})$.*

The curve E_p is a curve of genus $g = 1$, so we may apply the Weil conjectures in dimension $n = 1$, with Betti numbers $b_0 = b_2 = 1$ and $b_1 = 2g = 2$. This implies that its zeta function can be written as

$$Z_{E_p}(T) = \frac{L_p(T)}{(1-T)(1-pT)}, \tag{1.5}$$

where $L_p \in \mathbf{Z}[T]$ is a polynomial of the form

$$L_p(T) = pT^2 + c_1 T + 1,$$

---

[8]Moreover, one has $Z_X(T) = \pm q^{-n\chi/2} T^{-\chi} Z_X(1/(q^n T))$, where χ is the Euler characteristic of X, which is defined as the intersection number of the diagonal with itself in $X \times X$.

with $|c_1| \le 2\sqrt{p}$ (by the Riemann Hypothesis). If we expand both sides of (1.5) as power series in $\mathbf{Z}[[T]]$ we obtain

$$1 + N_E(p)T + \cdots = 1 + (p + 1 + c_1)T + \cdots ,$$

so we must have $N_E(p) = p + 1 + c_1$, and therefore

$$c_1 = N_E(p) - p - 1 = -t_p.$$

It follows that the single integer $N_E(p)$ completely determines the zeta function $Z_{E_p}(T)$.

Corresponding to our normalization $x_p = t_p/\sqrt{p}$, we define the *normalized L-polynomial*

$$\bar{L}_p(T) := L_p(T/\sqrt{p}) = T^2 + a_1 T + 1,$$

where $a_1 = c_1/\sqrt{p} = -x_p$ is a real number in the interval $[-2, 2]$ and the roots of $\bar{L}_p(T)$ lie on the unit circle. In our first example we obtained the group G_f as a subgroup of permutation matrices in $\mathrm{GL}_d(\mathbf{C})$. Here we want a subgroup of $\mathrm{GL}_2(\mathbf{C})$ whose elements have eigenvalues that

(a) are inverses (by the functional equation);
(b) lie on the unit circle (by the Riemann hypothesis).

Constraint (a) makes it clear that every element of G_E should have determinant 1, so $G_E \subseteq \mathrm{SL}_2(\mathbf{C})$. Constraints (a) and (b) together imply that in fact $G_E \subseteq \mathrm{SU}(2)$. As in the weight zero case, we expect that G_E should in general be as large as possible, that is, $G_E = \mathrm{SU}(2)$.

We now consider what it means for an elliptic curve to be generic.[9] Recall that the endomorphism ring of an elliptic curve E necessarily contains a subring isomorphic to $\mathbf{Z}$, corresponding to the multiplication-by-n maps $P \mapsto nP$. Here

$$nP = P + \cdots + P$$

denotes repeated addition under the group law, and we take the additive inverse if n is negative. For elliptic curves over fields of characteristic zero, this typically accounts for all the endomorphisms, but in special cases the endomorphism ring may be larger, in which case it contains elements that are not multiplication-by-n maps but can be viewed as "multiplication-by-α" maps for some $\alpha \in \mathbf{C}$. One can show that the minimal polynomials of these extra endomorphisms are necessarily quadratic, with negative discriminants, so such an α necessarily lies in an imaginary quadratic field K, and in fact $\mathrm{End}(E) \otimes_{\mathbf{Z}} \mathbf{Q} \simeq K$. When this happens we say that E has *complex multiplication* (CM) by K (or more precisely, by the order in $\mathbf{Z}_K$ isomorphic to $\mathrm{End}(E)$).

We can now state the Sato-Tate conjecture, as independently formulated in the mid 1960's by Mikio Sato (based on numerical data) and John Tate (as an application of what is now known as the Tate conjecture [**88**]), and finally proved in the late 2000's by Richard Taylor et al. [**6, 7, 32**].

THEOREM 1.9 (SATO–TATE CONJECTURE). *Let $E/\mathbf{Q}$ be an elliptic curve without CM. The sequence of normalized Frobenius traces x_p associated to E is equidistributed with respect to the pushforward of the Haar measure on* SU(2) *under the*

[9]The criterion given here in terms of endomorphism rings suffices for elliptic curves (and curves of genus $g \le 3$ or abelian varieties of dimension $g \le 3$), but in general one wants the Galois image to be as large as possible, which is a strictly stronger condition for $g > 3$. This issue is discussed further in Section 3.

trace map. In particular, for every subinterval $[a,b]$ *of* $[-2,2]$ *we have*

$$\lim_{B\to\infty}\frac{\#\{p\le B : x_p\in[a,b]\}}{\#\{p\le B\}}=\frac{1}{2\pi}\int_a^b\sqrt{4-t^2}\,dt.$$

We have not defined x_p for primes of bad reduction, but there is no need to do so; this theorem is purely an asymptotic statement. To see where the expression in the integral comes from, we need to understand the Haar measure on SU(2) and its pushforward onto the set of conjugacy classes conj(SU(2)) (in fact we only care about the latter). Each conjugacy class in SU(2) can be described by an *eigenangle* $\theta\in[0,\pi]$; its eigenvalues are then $e^{\pm i\theta}$ (a conjugate pair on the unit circle, as required). In terms of eigenangles, the pushforward of the Haar measure to conj(SU(2)) is given by

$$\mu=\frac{2}{\pi}\sin^2\theta\,d\theta$$

(see Exercise 2.4), and the trace is $t=2\cos\theta$; from this one can deduce the trace measure $\frac{1}{2\pi}\sqrt{4-t^2}dt$ on $[-2,2]$ that appears in Theorem 1.9. We can also use the Haar measure to compute the nth moment of the trace

$$\mathrm{E}[t^n]=\frac{2}{\pi}\int_0^\pi(2\cos\theta)^n\sin^2\theta d\theta=\begin{cases}0 & \text{if } n \text{ is odd,}\\ \frac{1}{m+1}\binom{2m}{m} & \text{if } n=2m \text{ is even,}\end{cases}\tag{1.6}$$

and find that the $2m$th moment is the mth Catalan number.[10]

1.7. Exercises.

EXERCISE 1.1. Let $f\in\mathbf{Z}[x]$ be a nonconstant squarefree polynomial. Prove that the average value of $N_f(p)$ over $p\le B$ converges to the number of irreducible factors of f in $\mathbf{Z}[x]$ as $B\to\infty$.

EXERCISE 1.2. Prove that the identity in (1.4) holds for all $A\in\mathrm{GL}_d(\mathbf{C})$.

EXERCISE 1.3. Let $f_p\in\mathbf{F}_p[x]$ denote a squarefree polynomial of degree $d>0$ and let $L_p(T)$ denote the denominator of the zeta function $Z_{f_p}(T)$. We know that the roots of $L_p(T)$ lie on the unit circle in the complex plane; show that in fact each is an nth root of unity for some $n\le d$. Then give a one-to-one correspondence between (i) cycle-types of degree-d permutations, (ii) possible factorization patterns of f_p in $\mathbf{F}_p[x]$, and (iii) the possible polynomials $L_p(T)$.

EXERCISE 1.4. Construct a monic squarefree quintic polynomial $f\in\mathbf{Z}[x]$ with no roots in $\mathbf{Q}$ such that $f_p(x)$ has a root in $\mathbf{F}_p$ for every prime p. Compute $c_0,\ldots,c_5$ and G_f.

EXERCISE 1.5. Let X be the arithmetic scheme $\operatorname{Spec}\mathbf{Z}[x,y]/(f,g)$, where

$$f(x,y):=y^2-2x^3+2x^2-2x-2,\qquad g(x,y):=4x^2-2xy+y^2-2.$$

By computing $Z_{X_p}(T)=L_p(T)^{-1}$ for sufficiently many small primes p, construct a list of the polynomials $L_p\in\mathbf{Z}[T]$ that you believe occur infinitely often, and estimate their relative frequencies. Use this data to derive a candidate for the matrix group $G_X:=\rho_X(\mathrm{Gal}(\overline{\mathbf{Q}}/\mathbf{Q}))$, where ρ_X is the Galois representation defined by the action of $\mathrm{Gal}(\overline{\mathbf{Q}}/\mathbf{Q})$ on $X(\overline{\mathbf{Q}})$. You may wish to use of computer algebra system such as Sage [**67**] or Magma or [**11**] to facilitate these calculations.

[10]This gives yet another way to define the Catalan numbers, one that does not appear to be among the 214 listed in [**84**].

2. Equidistribution, L-functions, and the Sato-Tate conjecture for elliptic curves

In this section we introduce the notion of equidistribution in compact groups G and relate it to analytic properties of L-functions of representations of G. We then explain (following Tate) why the Sato-Tate conjecture for elliptic curves follows from the holomorphicity and non-vanishing of a certain sequence of L-functions that one can associate to an elliptic curve over $\mathbf{Q}$ (or any number field).

2.1. Equidistribution. We now formally define the notion of equidistribution, following [**71**, §1A]. For a compact Hausdorff space X, we use $C(X)$ to denote the Banach space of complex-valued continuous functions $f\colon X \to \mathbf{C}$ equipped with the sup-norm $\|f\| := \sup_{x\in X} |f(x)|$. The space $C(X)$ is closed under pointwise addition and multiplication and contains all constant functions; it is thus a commutative $\mathbf{C}$-algebra with unit $\mathbb{1}_X$ (the function $x \mapsto 1$).[11] For any $\mathbf{C}$-valued functions f and g (continuous or not), we write $f \leq g$ whenever f and g are both $\mathbf{R}$-valued and $f(x) \leq g(x)$ for all $x \in X$; in particular, $f \geq 0$ means $\operatorname{im}(f) \subseteq \mathbf{R}_{\geq 0}$. The subset of $\mathbf{R}$-valued functions in $C(X)$ form a distributive lattice under this order relation.

DEFINITION 2.1. A (positive normalized Radon) *measure* on a compact Hausdorff space X is a continuous $\mathbf{C}$-linear map $\mu\colon C(X) \to \mathbf{C}$ that satisfies $\mu(f) \geq 0$ for all $f \geq 0$ and $\mu(\mathbb{1}_X) = 1$.

EXAMPLE 2.2. For each point $x \in X$ the map $f \mapsto f(x)$ defines the *Dirac measure* δ_x.

The value of μ on $f \in C(X)$ is often denoted using integral notation

$$\int_X f\mu := \mu(f),$$

and we shall use the two interchangeably.[12]

Having defined the measure μ as a function on $C(X)$, we would like to use it to assign values to (at least some) subsets of X. It is tempting to define the measure of a set $S \subseteq X$ as the measure of its indicator function $\mathbb{1}_S$, but in general the function $\mathbb{1}_S$ will not lie in $C(X)$; this occurs if and only if S is both open and closed (which we note applies to $S = X$). Instead, for each open set $S \subseteq X$ we define

$$\mu(S) = \sup\left\{\mu(f) : 0 \leq f \leq \mathbb{1}_S,\ f \in C(X)\right\} \in [0,1],$$

and for each closed set $S \subseteq X$ we define

$$\mu(S) = 1 - \mu(X - S) \in [0,1].$$

If $S \subseteq X$ has the property that for every $\epsilon > 0$ there exists an open set $U \supseteq S$ of measure $\mu(U) \leq \epsilon$, then we define $\mu(S) = 0$ and say that S has *measure zero*. If the boundary $\partial S := \overline{S} - S^0$ of a set S has measure zero, then we necessarily have $\mu(S^0) = \mu(\overline{S})$ and define $\mu(S)$ to be this common value; such sets are said to be *μ-quarrable*.

[11] In fact, it is a commutative C^*-algebra with complex conjugation as its involution, but we will not make use of this.

[12] Note that this is a definition; with a measure-theoretic approach one avoids the need to develop an integration theory.

For the purpose of studying equidistribution, we shall restrict our attention to μ-quarrable sets S. This typically does not include all measurable sets in the usual sense, by which we mean elements of the Borel σ-algebra Σ of X generated by the open sets under complements and countable unions and intersections (see Exercise 2.1). However, if we are given a regular Borel measure μ on X of total mass 1, by which we mean a countably additive function $\mu\colon \Sigma \to \mathbf{R}_{\geq 0}$ for which $\mu(S) = \inf\{\,\mu(U) : S \subseteq U,\ U \text{ open}\}$ and $\mu(X) = 1$, it is easy to check that defining $\mu(f) := \int_X f\mu$ for $f \in C(X)$ yields a measure under Definition 2.1; see [**41**, §1] for details. This measure is determined by the values μ takes on μ-quarrable sets [**99**]. In particular, if X is a compact group then its Haar measure induces a measure on X in the sense of Definition 2.1.

DEFINITION 2.3. A sequence $(x_1, x_2, x_3, \ldots)$ in X is said to be *equidistributed with respect to* μ, or simply μ*-equidistributed*, if for every $f \in C(X)$ we have

$$\mu(f) = \lim_{n\to\infty} \frac{1}{n}\sum_{i=1}^{n} f(x_i).$$

REMARK 2.4. When we speak of equidistribution, note that we are talking about a *sequence* (x_i) of elements of X in a particular order; it does not make sense to say that a *set* is equidistributed. For example, suppose we took the set of odd primes and arranged them in the sequence $(5, 13, 3, 17, 29, 7, \ldots)$ where we list two primes congruent to 1 modulo 4 followed by one prime congruent to 3 modulo 4. The sequence obtained by reducing this sequence modulo 4 is not equidistributed with respect to the uniform measure on $(\mathbf{Z}/4\mathbf{Z})^\times$, even though the sequence of odd primes in their usual order is (by Dirichlet's theorem on primes in arithmetic progressions). However, local rearrangements that change the index of an element by no more than a bounded amount do not change its equidistribution properties. This applies, in particular, to sequences indexed by primes of a number field ordered by norm; the equidistribution properties of such a sequence do not depend on how we order primes of the same norm.

If (x_i) is a sequence in X, for each real-valued function $f \in C(X)$ we define the kth-*moment* of the sequence $(f(x_i))$ by

$$\mathrm{M}_k[(f(x_i)] := \lim_{n\to\infty} \frac{1}{n}\sum_{i=1}^{n} f(x_i)^k.$$

If these limits exist for all $k \geq 0$, we then define the *moment sequence*

$$\mathrm{M}[f(x_i)] := (\mathrm{M}_0[(f(x_i)], \mathrm{M}_1[(f(x_i)], \mathrm{M}_2[(f(x_i)], \ldots).$$

If (x_i) is μ-equidistributed, then $\mathrm{M}_k[f(x_i)] = \mu(f^k)$ and the moment sequence

$$\mathrm{M}[f(x_i)] = (\mu(f^0), \mu(f^1), \mu(f^2), \ldots) \tag{2.1}$$

is independent of the sequence (x_i); it depends only on the function f and the measure μ.

REMARK 2.5. There is a partial converse that is relevant to some of our applications. To simplify matters, let us momentarily restrict our attention to real-valued functions; for the purposes of this remark, let $C(X)$ denote the Banach algebra of real-valued functions on X and replace $\mathbf{C}$ with $\mathbf{R}$ in Definition 2.1. Let (x_i) be a sequence in X and let $f \in C(X)$. Then $f(X)$ is a compact subset of $\mathbf{R}$, and we may view $(f(x_i))$ as a sequence in $f(X)$. If the moments $\mathrm{M}_k[f(x_i)]$ exist for all $k \geq 0$, then there is a *unique* measure on $f(X)$ with respect to which the sequence $(f(x_i))$ is equidistributed; this follows from the Stone-Weierstrass theorem. If μ is a measure on $C(X)$, we define the pushforward measure $\mu_f(g) := \mu(g \circ f)$ on $C(f(X)$ and see that the sequence $(f(x_i))$ is μ_f-equidistributed if and only if (2.1) holds. This gives a necessary (but in general not sufficient) condition for (x_i) to be μ-equidistributed that can be checked by comparing moment sequences. If we have a collection of functions $f_j \in C(X)$ such that the pushforward measures μ_{f_j} uniquely determine μ, we obtain a necessary and sufficient condition involving the moment sequences of the f_j with respect to μ. One can generalize this remark to complex-valued functions using the theory of C^*-algebras.

More generally, we have the following lemma.

LEMMA 2.6. *Let (f_j) be a family of functions whose linear combinations are dense in $C(X)$. If (x_i) is a sequence in X for which the limit $\lim_{n\to\infty} \frac{1}{n}\sum_{i=1}^n f_j(x_i)$ converges for every f_j, then there is a unique measure μ on X for which (x_i) is μ-equidistributed.*

PROOF. See [**71**, Lemma A.1, p. I-19]. □

PROPOSITION 2.7. *If (x_i) is a μ-equidistributed sequence in X and S is a μ-quarrable set in X then*

$$\mu(S) = \lim_{n\to\infty} \frac{\#\{x_i \in S : i \leq n\}}{n}.$$

PROOF. See Exercise 2.2. □

EXAMPLE 2.8. If $X = [0,1]$ and μ is the Lebesgue measure then a sequence (x_i) is μ-equidistributed if and only if for every $0 \leq a < b \leq 1$ we have

$$\lim_{n\to\infty} \frac{\#\{x_i \in [a,b] : i \leq n\}}{n} = b - a.$$

More generally, if X is a compact subset of $\mathbf{R}^n$ and μ is the normalized Lebesgue measure, then (x_i) is μ-equidistributed if and only if for every μ-quarrable $S \subseteq X$ we have $\lim_{n\to\infty} \frac{1}{n}\#\{x_i \in S : i \leq n\} = \mu(S)$.

2.2. Equidistribution in compact groups. We now specialize to the case where $X := \operatorname{conj}(G)$ is the space of conjugacy classes of a compact group G, obtained by taking the quotient of G as a topological space under the equivalence relation defined by conjugacy; let $\pi\colon G \to X$ denote the quotient map. We then equip X with the pushforward of the Haar measure μ on G (normalized so that $\mu(G) = 1$), which we also denote μ. Explicitly, π induces a map of Banach spaces

$$\begin{aligned} C(X) &\to C(G)\\ f &\mapsto f \circ \pi, \end{aligned}$$

and the value of μ on $C(X)$ is defined by

$$\mu(f) := \mu(f \circ \pi).$$

We say that a sequence (x_i) in X or a sequence (g_i) in G is *equidistributed* if it is μ-equidistributed (when we speak of equidistribution in a compact group without explicitly mentioning a measure, we always mean the Haar measure).

PROPOSITION 2.9. *Let G be a compact group with Haar measure μ, and let $X := \operatorname{conj}(G)$. A sequence (x_i) in X is μ-equidistributed if and only if for every irreducible character χ of G we have*

$$\lim_{n\to\infty} \frac{1}{n}\sum_{i=1}^{n} \chi(x_i) = \mu(\chi).$$

PROOF. As explained in [**71**, Prop. A.2], this follows from Lemma 2.6 and the Peter-Weyl theorem, since the irreducible characters χ of G generate a dense subset of $C(X)$. □

COROLLARY 2.10. *Let G be a compact group with Haar measure μ, and let $X := \operatorname{conj}(G)$. A sequence (x_i) in X is μ-equidistributed if and only if for every nontrivial irreducible character χ of G we have*

$$\lim_{n\to\infty} \frac{1}{n}\sum_{i=1}^{n} \chi(x_i) = 0.$$

PROOF. For the trivial character we have $\mu(1) = \mu(G) = 1$, and for any non-trivial irreducible character χ we must have $\mu(\chi) = \int_G \chi\mu = \int_G 1\cdot\chi\mu = 0$, by Schur orthogonality; the corollary follows. □

To illustrate these results, we now use Corollary 2.10 to prove an equidistribution result for elliptic curves over finite fields that will be useful later. We first recall some basic facts. Let E be an elliptic curve over a finite field $\mathbf{F}_q$; without loss of generality, assume $E/\mathbf{F}_q$ is given by a projective plane model. The *Frobenius endomorphism* $\pi_E : E \to E$ is defined by the rational map

$$(x : y : z) \mapsto (x^q : y^q : z^q).$$

Like all endomorphisms of elliptic curves, π_E has a characteristic polynomial of the form

$$T^2 - (\operatorname{tr}\pi_E)T + \deg\pi_E$$

satisfied by both π_E and its dual $\hat{\pi}_E$, where $\operatorname{tr}\pi_E = \pi_E + \hat{\pi}_E$ and $q = \deg\pi_E = \pi_E\hat{\pi}_E$ are both integers.[13] The set $E(\mathbf{F}_q)$ is, by definition, the subset of $E(\overline{\mathbf{F}}_q)$ fixed by π_E, equivalently, the kernel of the endomorphism $\pi_E - 1$. One can show that $\pi_E - 1$ is a separable, and therefore

$$\begin{aligned}\#E(\mathbf{F}_q) &= \#\ker(\pi_E - 1) = \deg(\pi_E - 1) = (\hat{\pi}_E - 1)(\pi_E - 1)\\ &= \hat{\pi}_E\pi_E + 1 - (\hat{\pi}_E + \pi_E)\\ &= q + 1 - \operatorname{tr}\pi_E.\end{aligned}$$

It follows that $t_q := q + 1 - \#E(\mathbf{F}_q)$ is the *trace of Frobenius* $\operatorname{tr}\pi_E$. As we showed in Section 1.6 for the case $q = p$, the zeta function of E can be written as

$$Z_E(T) = \frac{qT^2 - t_qT + 1}{(1-T)(1-qT)},$$

where the complex roots of $qT^2 - t_qT + 1$ have absolute value $q^{-1/2}$. This implies that we can write $t_q = \alpha + \bar{\alpha}$ for some $\alpha \in \mathbf{C}$ with $|\alpha| = q^{1/2}$, and we have $\#E(\mathbf{F}_q) = q + 1 - (\alpha + \bar{\alpha})$.

---

[13] By the *dual* of an endomorphism of a polarized abelian variety we mean the Rosati dual (see [**54**, §13]), which for elliptic curves we may identify with the dual isogeny.

We now observe that for any integer $r \geq 1$, the set $E(\mathbf{F}_{q^r})$ is the subset of $E(\overline{\mathbf{F}}_q)$ fixed by π_E^r, which corresponds to the q^r-power Frobenius automorphism; it follows that

$$\#E(\mathbf{F}_{q^r}) = q^r + 1 - (\alpha^r + \bar{\alpha}^r),$$

and therefore $\alpha^r + \bar{\alpha}^r$ is the trace t_{q^r} of the Frobenius endomorphism of the base change of E to $\mathbf{F}_{q^r}$.

As an application of Corollary 2.10, we now prove the following result, taken from [**24**, Prop 2.2]. Recall that $E/\mathbf{F}_q$ is said to be *ordinary* if t_q is not zero modulo the characteristic of $\mathbf{F}_q$.

PROPOSITION 2.11. *Let $E/\mathbf{F}_q$ be an ordinary elliptic curve and for integers $r \geq 1$, let $t_{q^r} := q^r + 1 - \#E(\mathbf{F}_{q^r})$ and define*

$$x_r := t_{q^r}/q^{r/2}.$$

The sequence (x_r) is equidistributed in $[-2, 2]$ with respect to the measure

$$\mu := \frac{1}{\pi}\frac{dz}{\sqrt{4-z^2}},$$

where dz is the Lebesgue measure on $[-2, 2]$.

PROOF. Let α be as above, with $|\alpha| = q^{1/2}$ and $\operatorname{tr}\pi_E = \alpha + \bar{\alpha}$. We then have $x_r = (\alpha^r + \bar{\alpha}^r)/q^{r/2}$ for all $r \geq 1$. Let $\mathrm{U}(1) := \{u \in \mathbf{C}^\times : u\bar{u} = 1\}$ be the unitary group. For $u = e^{i\theta}$, the Haar measure on $\mathrm{U}(1)$ corresponds to the uniform measure on $\theta \in [-\pi, \pi]$, this follows immediately from the translation invariance of the Haar measure. Let us compute the pushforward of the Haar measure of $\mathrm{U}(1)$ to $[-2, 2]$ via the map $u \mapsto z := u + \bar{u} = 2\cos\theta$. We have $dz = 2\sin\theta d\theta$, and see that the pushforward is precisely μ.

The nontrivial irreducible characters $\mathrm{U}(1) \to \mathbf{C}^\times$ all have the form $\phi_a(u) = u^a$ for some nonzero $a \in \mathbf{Z}$. For each such ϕ_a we have

$$\begin{aligned}\lim_{n\to\infty}\frac{1}{n}\sum_{r=1}^{n}\phi_a(\alpha^r/q^{r/2}) &= \lim_{n\to\infty}\frac{1}{n}\sum_{r=1}^{n}(\alpha/q^{1/2})^{ra}\\ &= \lim_{n\to\infty}\frac{1}{n}\frac{(\alpha/q^{1/2})^{a(n+1)} - (\alpha/q^{1/2})^a}{(\alpha/q^{1/2})^a - 1} = 0.\end{aligned}$$

The hypothesis that E is ordinary guarantees that $\alpha/q^{1/2}$ is not a root of unity (see Exercise 2.3), thus $(\alpha/q^{1/2})^a - 1$ is nonzero for all nonzero $a \in \mathbf{Z}$. Corollary 2.10 implies that $(\alpha^r/q^{r/2})$ is equidistributed in $\mathrm{U}(1)$, and therefore (x_r) is μ-equidistributed. □

See [**2**] for a generalization to smooth projective curves $C/\mathbf{F}_q$ of genus $g \geq 1$.

2.3. Equidistribution for L-functions. As above, let G be a compact group and let $X := \operatorname{conj}(G)$. Let K be a number field, and let $P := (\mathfrak{p}_1, \mathfrak{p}_2, \mathfrak{p}_3, \ldots)$ be a sequence consisting of all but finitely many primes $\mathfrak{p}$ of K ordered by norm; this means that $N(\mathfrak{p}_i) \leq N(\mathfrak{p}_j)$ for all $i \leq j$. Let $(x_\mathfrak{p})$ be a sequence in X indexed by P, and for each irreducible representation $\rho\colon G \to \mathrm{GL}_d(\mathbf{C})$, define the L-function

$$L(\rho, s) := \prod_{\mathfrak{p}\in P}\det(1 - \rho(x_\mathfrak{p})N(\mathfrak{p})^{-s})^{-1},$$

for $s \in \mathbf{C}$ with $\mathrm{Re}(s) > 1$.

THEOREM 2.12. *Let G and $(x_{\mathfrak{p}})$ be as above, and suppose $L(\rho, s)$ is meromorphic on* $\mathrm{Re}(s) \geq 1$ *with no zeros or poles except possibly at $s = 1$, for every irreducible representation ρ of G. The sequence $(x_{\mathfrak{p}})$ is equidistributed if and only if for each $\rho \neq 1$, the L-function $L(\rho, s)$ extends analytically to a function that is holomorphic and nonvanishing on* $\mathrm{Re}(s) \geq 1$.

PROOF. See the corollary to [**71**, Thm. A.2], or see [**24**, Thm. 2.3]. □

A notable case in which the hypothesis of Theorem 2.12 is known to hold is when $L(\rho, s)$ corresponds to an Artin L-function. As in Section 1.1, to each prime $\mathfrak{p}$ in K we associate an absolute Frobenius element $\mathrm{Frob}_{\mathfrak{p}} \in \mathrm{Gal}(\overline{K}/K)$, and for each finite Galois extension L/K we use $\mathrm{conj}_L(\mathrm{Frob}_{\mathfrak{p}})$ to denote the conjugacy class in $\mathrm{Gal}(L/K)$ of the restriction of $\mathrm{Frob}_{\mathfrak{p}}$ to L.

COROLLARY 2.13. *Let L/K be a finite Galois extension with $G := \mathrm{Gal}(L/K)$ and let P be the sequence of unramified primes of K ordered by norm (break ties arbitrarily). The sequence $(\mathrm{conj}_L(\mathrm{Frob}_{\mathfrak{p}}))_{\mathfrak{p} \in P}$ is equidistributed in $\mathrm{conj}(G)$; in particular, the Chebotarev density theorem (Theorem* 1.1*) holds.*

PROOF. For the trivial representation, the L-function $L(1, s)$ agrees with the Dedekind zeta function $\zeta_K(s)$ up to a finite number of holomorphic nonvanishing factors, and, as originally proved by Hecke, $\zeta_K(s)$ is holomorphic and nonvanishing on $\mathrm{Re}(s) \geq 1$ except for a simple pole at $s = 1$; see [**62**, Cor. VII.5.11], for example. For every nontrivial irreducible representation $\rho\colon G \to \mathrm{GL}_d(\mathbf{C})$, the L-function $L(\rho, s)$ agrees with the corresponding Artin L-function for ρ, up to a finite number of holomorphic nonvanishing factors, and, as originally proved by Artin, $L(\rho, s)$ is holomorphic and nonvanishing on $\mathrm{Re}(s) \geq 1$; see [**14**, p.225], for example. The corollary then follows from Theorem 2.12. □

2.4. Sato–Tate for CM elliptic curves. As a second application of Theorem 2.12, let us prove an equidistribution result for CM elliptic curves. To do so we need to introduce Hecke characters, which we will view as (quasi-)characters of the idèle class group of a number field.

DEFINITION 2.14. Let K be a number field and let $\mathbf{I}_K$ denote its idèle group. A *Hecke character* is a continuous homomorphism

$$\psi\colon \mathbf{I}_K \to \mathbf{C}^\times$$

whose kernel contains $K^\times$. The *conductor* of ψ is the $\mathbf{Z}_K$-ideal $\mathfrak{f} := \prod_{\mathfrak{p}} \mathfrak{p}^{e_{\mathfrak{p}}}$ in which each $e_{\mathfrak{p}}$ is the minimal nonnegative integer for which $1 + \hat{\mathfrak{p}}^{e_{\mathfrak{p}}} \subseteq \mathbf{Z}_{K_{\mathfrak{p}}}^\times \hookrightarrow \mathbf{I}_K$ lies in the kernel of ψ (all but finitely many $e_{\mathfrak{p}}$ are zero because ψ is continuous); here $\hat{\mathfrak{p}}$ denotes the maximal ideal of the valuation ring $\mathbf{Z}_{K_{\mathfrak{p}}}$ of $K_{\mathfrak{p}}$, the completion of K with respect to its $\mathfrak{p}$-adic absolute value.

Each Hecke character ψ has an associated *Hecke L-function*

$$L(\psi, s) := \prod_{\mathfrak{p} \nmid \mathfrak{f}} (1 - \psi(\mathfrak{p}) N(\mathfrak{p})^{-s})^{-1},$$

where $\psi(\mathfrak{p}) := \psi(\pi_{\hat{\mathfrak{p}}})$ for any uniformizer $\pi_{\hat{\mathfrak{p}}}$ of $\hat{\mathfrak{p}}$ (we have omitted the gamma factors at archimedean places). We now want to consider the sequence of unitarized values

$$x_{\mathfrak{p}} := \frac{\psi(\mathfrak{p})}{|\psi(\mathfrak{p})|} \in \mathrm{U}(1)$$

indexed by primes $\mathfrak{p} \nmid \mathfrak{f}$ ordered by norm.

LEMMA 2.15. *The sequence* $(x_\mathfrak{p})$ *is equidistributed in* $\mathrm{U}(1)$.

PROOF. As in the proof of Proposition 2.11, the nontrivial irreducible characters of $\mathrm{U}(1)$ are those of the form $\phi_a(z) = z^a$ with $a \in \mathbf{Z}$ nonzero, and in each case the corresponding L-function is a Hecke L-function (if ψ is a Hecke character, so is ψ^a and its unitarized version). If ψ is trivial then, as in the proof of Corollary 2.13, $L(1, s)$ is holomorphic and nonvanishing on $\mathrm{Re}(s) \geq 1$ except for a simple pole at $s = 1$, since the same is true of $\zeta_K(s)$. Hecke proved [**40**] that when ψ is nontrivial $L(\psi, s)$ is holomorphic and nonvanishing on $\mathrm{Re}(s) \geq 1$, and the lemma then follows from Theorem 2.12. □

As an application of Lemma 2.15, we can now prove the Sato-Tate conjecture for CM elliptic curves. Les us first consider the case where K is an imaginary quadratic field and E/K is an elliptic curve with CM by K (so $K \simeq \mathrm{End}(E) \otimes_\mathbf{Z} \mathbf{Q}$). As explained below, the general case (including $K = \mathbf{Q}$) follows easily.

Let $\mathfrak{f}$ be the *conductor* of E; this is a $\mathbf{Z}_K$-ideal divisible only by the primes of bad reduction for E; see [**81**, §IV.10] for a definition. A classical result of Deuring [**81**, Thm. II.10.5] implies the existence of a Hecke character ψ_E of K of conductor $\mathfrak{f}$ such that for each prime $\mathfrak{p} \nmid \mathfrak{f}$ we have $|\psi_E(\mathfrak{p})| = N(\mathfrak{p})^{1/2}$ and

$$\psi_E(\mathfrak{p}) + \overline{\psi_E(\mathfrak{p})} = t_\mathfrak{p},$$

where $t_\mathfrak{p} := \mathrm{tr}\, \pi_E = N(\mathfrak{p}) + 1 - \#E_\mathfrak{p}(\mathbf{F}_\mathfrak{p}) \in \mathbf{Z}$ is the trace of Frobenius of the reduction of E modulo $\mathfrak{p}$.

PROPOSITION 2.16. *Let K be an imaginary quadratic field and let E/K be an elliptic curve of conductor $\mathfrak{f}$ with CM by K. Let P be the sequence of primes of K that do not divide $\mathfrak{f}$ ordered by norm (break ties arbitrarily), and for $\mathfrak{p} \in P$ let $x_\mathfrak{p} := t_\mathfrak{p}/N(\mathfrak{p})^{1/2} \in [-2, 2]$ be the normalized Frobenius trace of $E_\mathfrak{p}$. The sequence $(x_\mathfrak{p})$ is equidistributed on $[-2, 2]$ with respect to the measure*

$$\mu_{\mathrm{cm}} := \frac{1}{\pi} \frac{dz}{\sqrt{4 - z^2}}.$$

PROOF. By the previous lemma, the sequence $(\psi_E(\mathfrak{p})/N(\mathfrak{p})^{1/2})_{\mathfrak{p} \in P}$ is equidistributed in $\mathrm{U}(1)$. As shown in the proof of Proposition 2.11, the measure μ_{cm} is the pushforward of the Haar measure on $\mathrm{U}(1)$ to $[-2, 2]$ under the map $u \mapsto u + \bar{u}$. For each $\mathfrak{p} \in P$ the image of $\psi_E(\mathfrak{p})/N(\mathfrak{p})^{1/2}$ under this map is

$$\frac{\psi_E(\mathfrak{p})}{N(\mathfrak{p})^{1/2}} + \frac{\overline{\psi_E(\mathfrak{p})}}{N(\mathfrak{p})^{1/2}} = \frac{t_p}{N(\mathfrak{p})^{1/2}} = x_\mathfrak{p}.$$ □

Figure 2 shows a trace histogram for the CM elliptic curve $y^2 = x^3 + 1$ over its CM field $\mathbf{Q}(\sqrt{-3})$.

Let us now consider the case of an elliptic curve $E/\mathbf{Q}$ with CM by F. For primes p of good reduction that are inert in F, the endomorphism algebra

$$\mathrm{End}(E_p)_\mathbf{Q} := \mathrm{End}(E_p) \otimes_\mathbf{Z} \mathbf{Q}$$

of the reduced curve E_p contains two distinct imaginary quadratic fields, one corresponding to the CM field $F \simeq \mathrm{End}(E)_\mathbf{Q}$ and the other generated by the Frobenius endomorphism (the two cannot coincide because p is inert in F but the Frobenius endomorphism has norm p in $\mathrm{End}(E_p)_\mathbf{Q}$). It follows that $\mathrm{End}(E_p)_\mathbf{Q}$ must be a quaternion algebra, E_p is supersingular, and for $p > 3$ we must have $t_p = 0$, since $t_p \equiv 0 \bmod p$ and $|t_p| \leq 2\sqrt{p}$; see [**82**, III,9,V.3] for these and other facts about endomorphism rings of elliptic curves.

At split primes $p = \mathfrak{p}\bar{\mathfrak{p}}$ the reduced curve E_p will be isomorphic to the reduction modulo $\mathfrak{p}$ of its base change to F (both of which are elliptic curves over $\mathbf{F}_p = \mathbf{F}_\mathfrak{p}$), and will have the same trace of Frobenius $t_p = t_\mathfrak{p}$. By the Chebotarev density theorem, the split and inert primes both have density 1/2, and it follows that the sequence of normalized Frobenius traces $x_\mathfrak{p} := t_p/\sqrt{p} \in [-2, 2]$ is equidistributed with respect to the measure $\frac{1}{2}\delta_0 + \frac{1}{2}\mu_{\text{cm}}$, where we use the Dirac measure δ_0 to put half the mass at 0 to account for the inert primes. This can be seen in Figure 3, which shows a trace histogram for the CM elliptic curve $y^2 = x^3 + 1$ over $\mathbf{Q}$; the thin spike in the middle of the histogram at zero has area 1/2 (one can also see that the nontrivial moments are half what they were in Figure 2).

A similar argument applies when E is defined over a number field K that does not contain the CM field F. For the sake of proving an equidistribution result we can restrict our attention to the degree-1 primes $\mathfrak{p}$ of K, those for which $N(\mathfrak{p}) = p$ is prime. Half of these primes $\mathfrak{p}$ will split in the compositum KF, and the subsequence of normalized traces $x_\mathfrak{p}$ at these primes will be equidistributed with respect to the measure μ_{cm}, and half will be inert in KF, in which case $x_\mathfrak{p} = t_\mathfrak{p} = 0$.

2.5. Sato–Tate for non-CM elliptic curves. We can now state the Sato-Tate conjecture in the form originally given by Tate, following [**71**, §1A]. Tate's seminal paper [**88**] describes what is now known as the *Tate conjecture*, which comes in two conjecturally equivalent forms **T1** and **T2**, the latter of which is

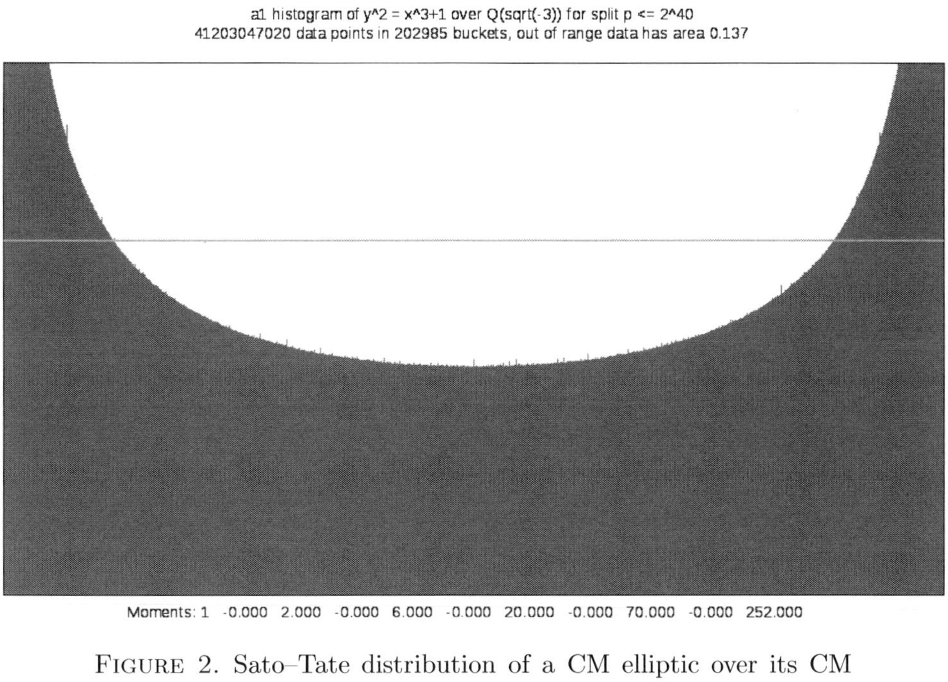

FIGURE 2. Sato–Tate distribution of a CM elliptic over its CM field (visit `http://math.mit.edu/~drew/g1_D2_a1f.gif` to see an animated version).

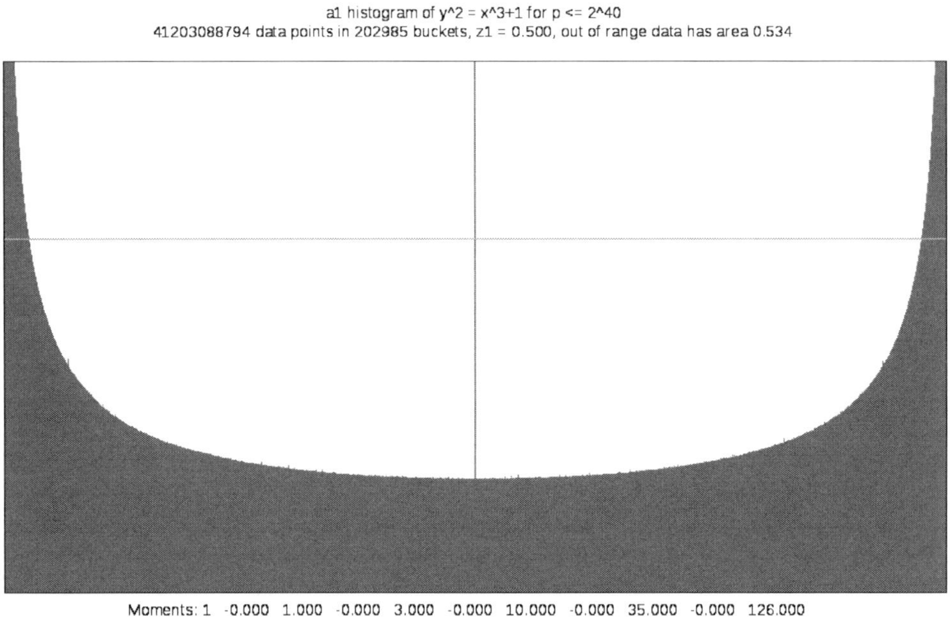

FIGURE 3. Sato–Tate distribution of a CM elliptic curve over $\mathbf{Q}$ (or visit `http://math.mit.edu/~drew/g1_D2_a1f.gif` to see an animated version).

stated in terms of L-functions. The Sato-Tate conjecture is obtained by applying **T2** to all powers of a fixed elliptic curve $E/\mathbf{Q}$ (as products of abelian varieties); see [**66**] for an introduction to the Tate conjecture and an explanation of how the Sato-Tate conjecture fits within it.

Let G be the compact group $\mathrm{SU}(2)$ of 2×2 unitary matrices with determinant 1. The irreducible representations of G are the mth symmetric powers ρ_m of the natural representation ρ_1 of degree 2 given by the inclusion $\mathrm{SU}(2) \subseteq \mathrm{GL}_2(\mathbf{C})$. Each element of $X := \operatorname{conj}(G)$ can be uniquely represented by a matrix of the form

$$\begin{pmatrix} e^{i\theta} & 0 \\ 0 & e^{-i\theta} \end{pmatrix},$$

where $\theta \in [0, \pi]$ is the eigenangle of the conjugacy class. It follows that each $f \in C(X)$ can be viewed as a continuous function $f(\theta)$ on the compact set $[0, \pi]$.

The pushforward of the Haar measure of G to X is given by

$$\mu = \frac{2}{\pi} \sin^2 \theta \, d\theta \tag{2.2}$$

(see Exercise 2.4), which means that for each $f \in C(X)$ we have

$$\mu(f) = \frac{2}{\pi} \int_0^\pi f(\theta) \sin^2 \theta \, d\theta.$$

Let $E/\mathbf{Q}$ be an elliptic curve without CM, let $P := (p)$ be the sequence of primes that do not divide the conductor N of E, in order, and for each $p \in P$ let $x_p \in X$ to be the element of X corresponding to the unique $\theta_p \in [0, \pi]$ for which

$2\cos\theta_p\sqrt{p} = t_p := p+1-\#E_p(\mathbf{F}_p)$ is the trace of Frobenius of the reduced curve E_p.

We are now in the setting of §2.3. We have a compact group $G := \mathrm{SU}(2)$, its space of conjugacy classes $X := \mathrm{conj}(G)$, a number field $K = \mathbf{Q}$, a sequence P containing all but finitely many primes of K ordered by norm, a sequence (x_p) in X indexed by P, and for each integer $m \geq 1$, an irreducible representation $\rho_m\colon G \to \mathrm{GL}_{m+1}(\mathbf{C})$. The L-function corresponding to ρ_m is given by

$$L(\rho_m, s) := \prod_{p\nmid N} \det(1-\rho_m(x_p)p^{-s})^{-1} = \prod_{p\nmid N}\prod_{k=0}^{m}(1-e^{i(m-2k)\theta_p}p^{-s})^{-1}.$$

For each $p\nmid N$, let α_p and $\bar{\alpha}_p$ be the roots of $T^2 - t_pT + p$, so that $\alpha_p = e^{i\theta_p}p^{1/2}$. If we now define

$$L^1_m(s) := \prod_{p\nmid N}\prod_{r=0}^{m}(1-\alpha_p^{m-r}\bar{\alpha}_p^r p^{-s})^{-1},$$

then for $m \geq 1$ we have

$$L(\rho_m, s) = L^1_m(s - m/2).$$

Tate conjectured in [**88**] that $L^1_m(s)$ is holomorphic and nonvanishing on the right half plane $\mathrm{Re}(s) \geq 1 + m/2$, which implies that each $L(\rho_m, s)$ is holomorphic and nonvanishing on $\mathrm{Re}(s) \geq 1$. Assuming this is true, Theorem 2.12 implies that the sequence (x_p) is μ-equidistributed, which is equivalent to the Sato-Tate conjecture.

We now recall the *modularity theorem* for elliptic curves over $\mathbf{Q}$, which states that there is a one-to-one correspondence between isogeny classes of elliptic curves $E/\mathbf{Q}$ of conductor N and modular forms

$$f(z) = \sum_{n\geq 1} a_n e^{2\pi i n z} \in S_2(\Gamma_0(N))^{\mathrm{new}} \qquad (a_n \in \mathbf{Z} \text{ with } a_1 = 1)$$

that are eigenforms for the action of the Hecke algebra on the space $S_2(\Gamma_0(N))$ of cuspforms of weight 2 and level N and *new* at level N, meaning not contained in $S_2(\Gamma_0(M))$ for any positive integer M properly dividing N. Such modular forms f are called (normalized) *newforms*, of weight 2 and level N, with rational coefficients. The modularity theorem was proved for squarefree N by Taylor and Wiles [**91, 98**], and extended to all conductors N by Breuil, Conrad, Diamond, and Taylor [**12**].

The modular form f is a simultaneous eigenform for all the Hecke operators T_n, and the normalization $a_1 = 1$ ensures that for each prime $p \nmid N$, the coefficient a_p is the eigenvalue of f for T_p. Under the correspondence given by the modularity theorem, the eigenvalue a_p is equal to the trace of Frobenius t_p of the reduced curve E_p, where E is any representative of the corresponding isogeny class. Here we are using the fact that if E and E' are isogenous elliptic curves over $\mathbf{Q}$ they necessarily have the same conductor N and the same trace of Frobenius t_p at ever $p \nmid N$.

There is an L-function $L(f, s)$ associated to the modular form f, and the modularity theorem guarantees that it coincides with the L-function $L(E, s)$ of E. So not only does $a_p = t_p$ for all $p \nmid N$, the Euler factors at the bad primes $p|N$ also agree. We need not concern ourselves with Euler factors at these primes, other than to note that they are holomorphic and nonvanishing on $\mathrm{Re}(s) \geq 3/2$. After removing the Euler factors at bad primes, the L-functions $L(E, s)$ and $L(f, s)$ both have the form

$$\prod_{p\nmid N}(1-a_pp^{-s}+p^{1-2s})^{-1}=\prod_{p\nmid N}\prod_{r=0}^{1}(1-\alpha_p^{1-r}\bar{\alpha}_p^{r}p^{-s})^{-1}=L_1^1(s),$$

where α_p and $\bar{\alpha}_p$ are the roots of $T^2-a_pT+p=T^2-t_pT+p$.

The L-function $L(f,s)$ is holomorphic and nonvanishing on $\mathrm{Re}(s)\geq 3/2$; see [**21**, Prop. 5.9.1]. The modularity theorem tells us that the same is true of $L(E,s)$, and therefore of $L_1^1(s)$. Thus the modularity theorem proves that Tate's conjecture regarding $L_m^1(s)$ holds when $m=1$. To prove the Sato-Tate conjecture one needs to show that this holds for all $m\geq 1$.

THEOREM 2.17. *Let* $f(z):=\sum_{n\geq 1}a_ne^{2\pi izn}\in S_2(\Gamma_0(N)^{\mathrm{new}}$ *be a normalized newform without CM. For each prime* $p\nmid N$ *let* $\alpha_p,\bar{\alpha}_p$ *be the roots of* T^2-a_pT+p. *Then*

$$\prod_{p\nmid N}\prod_{r=0}^{m}(1-\alpha_p^{m-r}\bar{\alpha}_p^{r}p^{-s})^{-1}=L_m^1(s)$$

is holomorphic and nonvanishing on $\mathrm{Re}(s)\geq 1+m/2$.

PROOF. Apply [**7**, Theorem B.2] with weight $k=2$, trivial nebentypus $\psi=1$, and trivial character $\chi=1$ (as noted in [**7**], this special case was already addressed in [**32**]). □

COROLLARY 2.18. *The Sato-Tate conjecture (Theorem* 1.9*) holds.*

REMARK 2.19. The Sato-Tate conjecture is also known to hold for elliptic curves over totally real fields, and over CM fields (imaginary quadratic extensions of totally real fields). The totally real case was initially proved for elliptic curves with potentially multiplicative reduction at some prime in [**32, 90**]; it was later shown this technical assumption can be removed (see the introduction of [**6**]). The generalization to CM fields was obtained at a recent IAS workshop [**3**] and still in the process of being written up in detail. As a consequence of this result the Sato-Tate conjecture for elliptic curves is now known for all number fields of degree 1 or 2 (but not for any higher degrees).

2.6. Exercises.

EXERCISE 2.1. Let X be a compact Hausdorff space. Show that a set $S\subseteq X$ is μ-quarrable for every measure μ on X if and only if the set S is both open and closed.

EXERCISE 2.2. Prove Proposition 2.7.

EXERCISE 2.3. Let E an elliptic curve over $\mathbf{F}_q$ and let α be a root of the characteristic polynomial of the Frobenius endomorphism π_E. Prove that $\alpha/\sqrt{q}$ is a root of unity if and only if E is supersingular.

EXERCISE 2.4. Show that the set of conjugacy classes of SU(2) is in bijection with the set of eigenangles $\theta\in[0,\pi]$. Then prove that the pushforward of the Haar measure of SU(2) onto $[0,\pi]$ is given by $\mu:=\frac{2}{\pi}\sin^2\theta\,d\theta$ (hint: show that SU(2) is isomorphic to the 3-sphere S^3 and use this isomorphism together with the translation invariance of the Haar measure to determine μ)

EXERCISE 2.5. Compute the trace moment sequence for SU(2) (that is, prove (1.6)). Embed U(1) in SU(2) via the map $u \mapsto \left(\begin{smallmatrix} u & 0 \\ 0 & \bar{u} \end{smallmatrix}\right)$ and compute its trace moment sequence (compare to Figure 2). Now determine the normalizer $N(\mathrm{U}(1))$ of U(1) in SU(2) and compute its trace moment sequence (compare to Figure 3).

3. Sato-Tate groups

In the previous section we showed that there are three distinct Sato-Tate distributions that arise for elliptic curves E over number fields K (only two of which occur when $K = \mathbf{Q}$). All three distributions can be associated to the Haar measure of a compact subgroup $G \subseteq \mathrm{SU}(2)$, in which we embed U(1) via the map $u \mapsto \left(\begin{smallmatrix} u & 0 \\ 0 & \bar{u} \end{smallmatrix}\right)$. We are interested in the pushforward μ of the Haar measure onto $\operatorname{conj}(G)$, which can be expressed in terms of the eigenangle $\theta \in [0, \pi]$. The three possibilities for G are listed below.

- U(1): we have $\mu(\theta) = \frac{1}{\pi} d\theta$ and trace moments:
$$(1, 0, 2, 0, 6, 0, 20, 0, 70, 0, 252, \ldots).$$
This case arises for CM elliptic curves over fields that contain the CM field.
- $N(\mathrm{U}(1))$: we have $\mu(\theta) = \frac{1}{2\pi} d\theta + \frac{1}{2}\delta_{\pi/2}$ and trace moments:
$$(1, 0, 1, 0, 3, 0, 10, 0, 35, 0, 126, \ldots).$$
This case arises for CM elliptic curves over fields that do not contain the CM field.
- SU(2): we have $\mu(\theta) = \frac{2}{\pi} \sin^2\theta \, d\theta$ and trace moments:
$$(1, 0, 1, 0, 2, 0, 5, 0, 14, 0, 42, \ldots).$$
This case arises for all non-CM elliptic curves (conjecturally so when K is not totally real or a CM field).

We have written μ in terms of θ, but we may view it as a linear function on the Banach space $C(X)$, where we identify $X := \operatorname{conj}(G)$ with $[0, \pi]$, by defining $\mu(f) := \int_0^\pi f(\theta)\mu(\theta)$, as in §2.1. In particular, μ assigns a value to the trace function $\operatorname{tr}\colon X \to [-2, 2]$, where $\operatorname{tr}(\theta) = 2\cos\theta$, and to its powers tr^n, which allows us to compute the trace moment sequence $(\mu(\operatorname{tr}^n))_{n\geq 0}$.

Our goal in this section is to define the compact group G as an invariant of the elliptic curve E, the *Sato-Tate group* of E, and to then generalize this definition to abelian varieties of arbitrary dimension. This will allow us to state the Sato-Tate conjecture for abelian varieties as an equidistribution statement with respect to the Haar measure of the Sato-Tate group.

3.1. The Sato-Tate group of an elliptic curve. Thus far the link between the elliptic curve E and the compact group G whose Haar measure is claimed (and in many cases proved) to govern the distribution of Frobenius traces has been made via the measure μ. That is, we have an equidistribution claim for the sequence (x_p) of normalized Frobenius traces associated to E that is phrased in terms of a measure μ that happens to be induced by the Haar measure of a compact group G. We now want to establish a direct relationship between E and G that defines G as an arithmetic invariant of E, *without assuming the Sato-Tate conjecture.*

In Section 1.1 we considered Galois representations $\rho_f\colon \operatorname{Gal}(\overline{\mathbf{Q}}/\mathbf{Q}) \to \mathrm{GL}_d(\mathbf{C})$ defined by the action of $\operatorname{Gal}(\overline{\mathbf{Q}}/\mathbf{Q})$ on the roots of a squarefree polynomial $f \in \mathbf{Z}[x]$. We thereby obtained a compact group G_f and a map that sends each prime p of good reduction for f to an element of $\operatorname{conj}(G_f)$ (namely, the map $p \mapsto \rho_f(\operatorname{Frob}_p)$). We were then able to relate the image of p under this map to the quantity $N_f(p)$ of interest, via (1.1). This construction did not involve any discussion of equidistribution, but we could then prove, via the Chebotarev density theorem, that the conjugacy classes $\rho_f(p)$ are equidistributed with respect to the pushforward of the Haar measure to $\operatorname{conj}(G_f)$.

We take a similar approach here. To each elliptic curve E over a number field K we will associate a compact group G that is constructed via a Galois representation attached to E, equipped with a map that sends each prime $\mathfrak{p}$ of good reduction for E to an element $x_\mathfrak{p}$ of $\operatorname{conj}(G)$ that we can directly relate to the quantity $N_E(\mathfrak{p}) := p + 1 - t_\mathfrak{p}$ whose distribution we wish to study. We may then conjecture (and prove, when E has CM or K is a totally real or CM field), that the sequence $(x_\mathfrak{p})$ is equidistributed in $X := \operatorname{conj}(G)$ (with respect to the pushforward of the Haar measure of G).

The group G is the *Sato–Tate* group of E, and will be denoted $\operatorname{ST}(E)$. It is a compact subgroup of $\operatorname{SU}(2)$, and our construction will make it easy to show that $\operatorname{ST}(E)$ is always one of the three groups $\operatorname{U}(1)$, $N(\operatorname{U}(1))$, $\operatorname{SU}(2)$ listed above, depending on whether E has CM or not, and if so, whether the CM field is contained in the ground field or not. None of this depends on any equidistribution results. This construction will be our prototype for the definition of the Sato-Tate group of an abelian variety of arbitrary dimension g, so we will work out the $g = 1$ case in some detail.

In order to associate a Galois representation to E/K, we need a set on which $\operatorname{Gal}(\overline{K}/K)$ can act. For each integer $n \geq 1$, let $E[n] := E(\overline{K})[n]$ denote the n-torsion subgroup of $E(\overline{K})$, a free $\mathbf{Z}/n\mathbf{Z}$-module of rank 2 (see [**82**, Cor. III.6.4]). The group $\operatorname{Gal}(\overline{K}/K)$ acts on points in $E(\overline{K})$ coordinate-wise, and $E[n]$ is invariant under this action because it is the kernel of the multiplication-by-n map $[n]$, an endomorphism of E that is defined over K; one can concretely define $E[n]$ as the zero locus of the n-division polynomials, which have coefficients in K. The action of $\operatorname{Gal}(\overline{K}/K)$ on $E[n]$ induces the *mod-n Galois representation*

$$\operatorname{Gal}(\overline{K}/K) \to \operatorname{Aut}(E[n]) \simeq \operatorname{GL}_2(\mathbf{Z}/n\mathbf{Z}).$$

This Galois representation is insufficient for our purposes, because the image $M_\mathfrak{p}$ of $\operatorname{Frob}_\mathfrak{p}$ in $\operatorname{GL}_2(\mathbf{Z}/n\mathbf{Z})$ does not determine $t_\mathfrak{p}$, we only have $t_\mathfrak{p} \equiv \operatorname{tr} M_\mathfrak{p} \bmod n$; we need to let $\operatorname{Gal}(\overline{K}/K)$ act on a bigger set.

So let us fix a prime ℓ (any prime will do), and consider the inverse system

$$\cdots \xrightarrow{[\ell]} E[\ell^3] \xrightarrow{[\ell]} E[\ell^2] \xrightarrow{[\ell]} E[\ell].$$

The inverse limit

$$T_\ell := \varprojlim_n E[\ell^n]$$

is the ℓ-adic *Tate-module* of E; it is a free $\mathbf{Z}_\ell$-module of rank 2. The group $\operatorname{Gal}(\overline{K}/K)$ acts on T_ℓ via its action on the groups $E[\ell^n]$, and this action is compatible with the multiplication-by-ℓ map $[\ell]$ because this map is defined over K (it can be written as a rational map with coefficients in K). This yields the *ℓ-adic Galois representation*

$$\rho_{E,\ell}\colon \operatorname{Gal}(\overline{K}/K) \to \operatorname{Aut}(T_\ell) \simeq \operatorname{GL}_2(\mathbf{Z}_\ell).$$

The representation $\rho_{E,\ell}$ enjoys the following property: for every prime $\mathfrak{p} \nmid \ell$ of good reduction for E the image of $\mathrm{Frob}_\mathfrak{p}$ is a matrix $M_\mathfrak{p} \in \mathrm{GL}_2(\mathbf{Z}_\ell)$ that has the same characteristic polynomial as the Frobenius endomorphism of $E_\mathfrak{p}$, namely, $T^2 - t_\mathfrak{p}T + N(\mathfrak{p})$, where $t_\mathfrak{p} := \mathrm{tr}\,\pi_{E_\mathfrak{p}}$. Note that the matrix $M_\mathfrak{p}$ is determined only up to conjugacy; there is ambiguity both in our choice of $\mathrm{Frob}_\mathfrak{p}$ (see §1.1) and in our choice of a basis for T_ℓ, which fixes the isomorphism $\mathrm{Aut}(T_\ell) \simeq \mathrm{GL}_2(\mathbf{Z}_\ell)$. We should thus think of $\rho_{E,\ell}(\mathrm{Frob}_\mathfrak{p})$ as representing a conjugacy class in $\mathrm{GL}_2(\mathbf{Z}_\ell)$.

We prefer to work over the field $\mathbf{Q}_\ell$, rather than its ring of integers $\mathbf{Z}_\ell$, so let us define the *rational Tate module*

$$V_\ell := T_\ell \otimes_\mathbf{Z} \mathbf{Q},$$

which is a 2-dimensional $\mathbf{Q}_\ell$-vector space equipped with an action of $\mathrm{Gal}(\overline{K}/K)$. We may then view the Galois representation $\rho_{E,\ell}$ as having image $G_\ell \subseteq \mathrm{GL}_2(\mathbf{Q}_\ell)$. We also prefer to work with an algebraic group, so let us define G_ℓ^{zar} to be the $\mathbf{Q}_\ell$-algebraic group obtained by taking the Zariski closure of G_ℓ in $\mathrm{GL}_2(\mathbf{Q}_\ell)$. This means that G_ℓ^{zar} is the affine variety defined by the ideal of $\mathbf{Q}_\ell$-polynomials that vanish on the set G_ℓ; it is a subvariety of $\mathrm{GL}_2/\mathbf{Q}_\ell$ that is closed under the group operation and thus an algebraic group over $\mathbf{Q}_\ell$. The algebraic group G_ℓ^{zar} is the *ℓ-adic monodromy group* of E (it is also denoted G_ℓ^{alg}).

Background 3.1 (Algebraic groups). An affine (or linear) *algebraic group* over a field k is a group object in the category of (not necessarily irreducible) affine varieties over k. The only projective algebraic groups we shall consider are smooth and connected, hence abelian varieties, so when we use the term algebraic group without qualification, we mean an affine algebraic group.[14] The canonical example is GL_n, which can be defined as an affine variety in $\mathbf{A}^{n^2+1}$ (over any field) by the equation $t \det M = 1$ (here $\det M$ denotes the determinant polynomial in n^2 variables M_{ij}), with morphisms $m\colon \mathrm{GL}_n \times \mathrm{GL}_n \to \mathrm{GL}_n$ and $i\colon \mathrm{GL}_n \to \mathrm{GL}_n$ defined by polynomial maps corresponding to matrix multiplication and inversion (one uses t as the inverse of $\det A$ when defining i). The classical groups SL_n, Sp_{2n},U_n, SU_n, O_n, SO_n are all affine algebraic groups (assume $\mathrm{char}(k) \neq 2$ for O_n and SO_n), as are the groups $\mathrm{USp}_{2n} := \mathrm{Sp}_{2n} \cap \mathrm{U}_{2n}$ and GSp_{2n} that are of particular interest to us; the $\mathbf{R}$ and $\mathbf{C}$ points of these groups are *Lie groups* (differentiable manifolds with a group structure). If G is an affine algebraic group over k and L/k is a field extension, the Zariski closure of any subgroup $H \subseteq G(L)$ of the L-points of G is equal to the set of rational points of an affine variety defined over L that is also an algebraic group via the morphisms m and i defining G. Thus every subgroup $H \subseteq G(L)$ uniquely determines an algebraic group over L whose rational points coincide with the Zariski closure of H; as an abuse of terminology we may refer to this algebraic group as the Zariski closure of H in $G(L)$ (or in G_L, the base change of G to L). The connected and irreducible components of an algebraic group G coincide, and are necessarily finite in number. The connected component G^0 of the identity is itself an algebraic group, a normal subgroup of G compatible with base change. For more on algebraic groups see any of the classic texts [**10**, **42**, **83**], or see [**55**] for a more modern treatment.

[14]There are interesting algebraic groups (group schemes of finite type over a field) that are neither affine nor projective (even if we restrict our attention to those that are smooth and connected), but we shall not consider them here.

Having defined the $\mathbf{Q}_\ell$-algebraic group G_ℓ^{zar}, we now restrict our attention to the subgroup $G_\ell^{1,\mathrm{zar}}$ obtained by imposing the symplectic constraint

$$M^t\Omega M = \Omega, \qquad\qquad \Omega := \left(\begin{smallmatrix} 0 & -1 \\ 1 & 0 \end{smallmatrix}\right),$$

which corresponds to putting a *symplectic form* (a nondegenerate bilinear alternating pairing) on the vector space V_ℓ (we could of course choose any Ω that defines such a form). This condition can clearly be expressed by a polynomial (a quadratic form in fact), thus $G_\ell^{1,\mathrm{zar}}$ is an algebraic group over $\mathbf{Q}_\ell$ contained in Sp_2. We remark that $\mathrm{Sp}_2 = \mathrm{SL}_2$, so we could have just required $\det M = 1$, but this is an accident of low dimension: the inclusion $\mathrm{Sp}_{2n} \subseteq \mathrm{SL}_{2n}$ is strict for all $n > 1$.

Finally, let us choose an embedding $\iota\colon \mathbf{Q}_\ell \to \mathbf{C}$, and let $G_{\ell,\iota}^{1,\mathrm{zar}}$ be the $\mathbf{C}$-algebraic group obtained from $G_\ell^{1,\mathrm{zar}}$ by base change to $\mathbf{C}$ (via ι). The group $G_{\ell,\iota}^{1,\mathrm{zar}}(\mathbf{C})$ is a subgroup of $\mathrm{Sp}_2(\mathbf{C})$ that we may view as a Lie group with finitely many connected components. It therefore contains a maximal compact subgroup that is unique up to conjugacy [**63**, Thm. IV.3.5], and we take this as the *Sato–Tate group* $\mathrm{ST}(E)$ of E (which is thus defined only up to conjugacy). It is a compact subgroup of $\mathrm{USp}(2) = \mathrm{SU}(2)$ (this equality is another accident of low dimension).

For each prime $\mathfrak{p} \nmid \ell$ of good reduction for E, let $M_\mathfrak{p}$ denote the image of $\mathrm{Frob}_\mathfrak{p}$ under the maps

$$\mathrm{Gal}(\overline{K}/K) \xrightarrow{\rho_{E,\ell}} G_\ell \hookrightarrow G_\ell^{\mathrm{zar}}(\mathbf{Q}_\ell) \hookrightarrow G_{\ell,\iota}^{\mathrm{zar}}(\mathbf{C}),$$

where the middle map is inclusion and we use the embedding $\iota\colon \mathbf{Q}_\ell \to \mathbf{C}$ to obtain the last injection. We now consider the normalized Frobenius image

$$\bar{M}_\mathfrak{p} := N(\mathfrak{p})^{-1/2} M_\mathfrak{p};$$

it is a matrix with trace $t_\mathfrak{p}/N(\mathfrak{p})^{-1/2} \in [-2, 2]$ and determinant 1, and its eigenvalues $e^{\pm i\theta_\mathfrak{p}}$ lie on the unit circle.[15] The eigenangle $\theta_\mathfrak{p}$ determines a unique conjugacy class in $\mathrm{ST}(E)$, which we take as $x_\mathfrak{p}$. The characteristic polynomial of $x_\mathfrak{p}$ is the normalized L-polynomial $\bar{L}_\mathfrak{p}(T) := L_p(N(\mathfrak{p})^{-1/2}T)$, where $L_\mathfrak{p}(T)$ is the numerator of the zeta function of $E_\mathfrak{p}$, and $L_\mathfrak{p}(N(\mathfrak{p})^{-s})$ is the Euler factor at $\mathfrak{p}$ in the L-series $L(E, s)$.

The Sato–Tate conjecture then amounts to the statement that the sequence $(x_\mathfrak{p})$ in $X := \mathrm{conj}(\mathrm{ST}(E))$ is equidistributed. Notice that the statement is the same in both the CM and non-CM cases, but the measure on X is different, because $\mathrm{ST}(E)$ is different. Indeed, there are three possibilities for $\mathrm{ST}(E)$, corresponding to the three distributions that we noted at the beginning of this section.

THEOREM 3.2. *Let E be an elliptic curve over a number field K. Up to conjugacy in* $\mathrm{SU}(2)$ *we have*

$$\mathrm{ST}(E) = \begin{cases} \mathrm{U}(1) & \text{if } E \text{ has CM defined over } K,\\ N(\mathrm{U}(1)) & \text{if } E \text{ has CM not defined over } K,\\ \mathrm{SU}(2) & \text{if } E \text{ does not have CM,} \end{cases}$$

where $\mathrm{U}(1)$ *is embedded in* $\mathrm{SU}(2)$ *via* $u \mapsto \left(\begin{smallmatrix} u & 0 \\ 0 & \bar{u} \end{smallmatrix}\right)$.

[15] Note that we embed $G_\ell^{\mathrm{zar}}(\mathbf{Q}_\ell)$ in $G_{\ell,\iota}^{\mathrm{zar}}(\mathbf{C})$ *before* normalizing by $N(\mathfrak{p})^{-1/2}$; as pointed out by Serre [**77**, p. 131], we want to take the square root in $\mathbf{C}$ where it is unambiguously defined.

PROOF. If E has CM defined over K then G_ℓ is abelian, because the action of $\mathrm{Gal}(\overline{K}/K)$ on V_ℓ factors through the abelian group $\mathrm{Gal}(L/K)$, where the field $L := K(E[\ell^\infty])$ is obtained by adjoining the coordinates of the ℓ-power torsion points of E; this follows from [**81**, Thm. II.2.3]. Therefore G_ℓ lies in a *Cartan subgroup* of $\mathrm{GL}_2(\mathbf{Q}_\ell)$ (a maximal abelian subgroup), which necessarily splits when we pass to $G_{\ell,\iota}^{\mathrm{zar}}(\mathbf{C})$, where it is conjugate to the group of diagonal matrices. This implies that $\mathrm{ST}(E)$ is conjugate to $\mathrm{U}(1)$, the subgroup of diagonal matrices in $\mathrm{SU}(2)$.

If E has CM not defined over K, then G_ℓ lies in the normalizer of a Cartan subgroup of $\mathrm{GL}_2(\mathbf{Q}_\ell)$, but not in the Cartan itself, and $\mathrm{ST}(E)$ is conjugate to the normalizer $N(\mathrm{U}(1))$ of $\mathrm{U}(1)$ in $\mathrm{SU}(2)$; the argument is as above, but now the action of $\mathrm{Gal}(\overline{K}/K)$ factors through $\mathrm{Gal}(FL/K)$, where F is the CM field and $\mathrm{Gal}(FL/K)$ contains the abelian subgroup $\mathrm{Gal}(FL/FK)$ with index 2.

If E does not have CM then Serre's open image theorem (see [**71**, §IV.3] and [**72**]) implies that G_ℓ is a finite index subgroup of $\mathrm{GL}_2(\mathbf{Z}_\ell)$; we therefore have $G_\ell^{1,\mathrm{zar}} = \mathrm{SL}_2$, which implies $\mathrm{ST}(E) = \mathrm{SU}(2)$. □

It follows from Theorem 3.2 that (up to conjugacy), the Sato–Tate group $\mathrm{ST}(E)$ does not depend on our choice of the prime ℓ or the embedding $\iota\colon \mathbf{Q}_\ell \to \mathbf{C}$ that we used. We should also note that $\mathrm{ST}(E)$ depends only on the isogeny class of E; this follows from the fact that we used the rational Tate module V_ℓ to define it (indeed, two abelian varieties over a number field are isogenous if and only if their rational Tate modules are isomorphic as Galois modules, by Faltings' isogeny theorem [**23**], but we are only using the easy direction of this equivalence here).

3.2. The Sato–Tate group of an abelian variety. We now wish to generalize our definition of the Sato–Tate group of an elliptic curve to abelian varieties. Recall that an *abelian variety* is a smooth connected projective variety that is also an algebraic group, where the group operations are now given by morphisms of projective varieties; on any affine patch they can be defined by a polynomial map. Remarkably, the fact that abelian varieties are commutative algebraic groups is not part of the definition, it is a consequence; see [**54**, Cor. 1.4]. We also recall that an *isogeny* of abelian varieties is simply an isogeny of algebraic groups, a surjective morphism with finite kernel.

Abelian varieties of dimension g may arise as the Jacobian $\mathrm{Jac}(C)$ of a smooth projective curve C/k of genus g. If C has a k-rational point (as when C is an elliptic curve), one can functorially identify $\mathrm{Jac}(C)$ with the *divisor class group* $\mathrm{Pic}^0(C)$, the group of degree-zero divisors modulo principal divisors, but one can unambiguously define the abelian variety $\mathrm{Jac}(C)$ in any case; see [**54**, Ch. III] for details.

If C is a smooth projective curve over a number field K and $A := \mathrm{Jac}(C)$ is its Jacobian, then for every prime $\mathfrak{p}$ of good reduction for C, the abelian variety A also has good reduction at $\mathfrak{p}$,[16] and the L-polynomial $L_\mathfrak{p}(T)$ appearing in the numerator of the zeta function $Z_{C_\mathfrak{p}}(T)$ is reciprocal to the characteristic polynomial $\chi_\mathfrak{p}(T)$ of the Frobenius endomorphism $\pi_{A_\mathfrak{p}}$ of $A_\mathfrak{p}$, which acts on points of A via the $N(\mathfrak{p})$-power Frobenius automorphism (coordinate-wise). In particular, we have the

[16]For $g > 1$ the converse does not hold (in general); this impacts only finitely many primes $\mathfrak{p}$ and will not concern us.

identity

$$L_{\mathfrak{p}}(T) = T^{2g}\chi_{\mathfrak{p}}(T^{-1}), \tag{3.1}$$

in which both sides are integer polynomials of degree $2g$ whose complex roots have absolute value $N(\mathfrak{p})^{-1/2}$. As with elliptic curves, one can consider the L-function $L(A,s)$ attached to A, which is defined as an Euler product with factors $L_{\mathfrak{p}}(N(\mathfrak{p})^{-s})$ at each prime $\mathfrak{p}$ where A has good reduction.[17] Studying the distribution of the normalized L-polynomials $\bar{L}_{\mathfrak{p}}(T)$ associated to C is thus equivalent to studying the distribution of the normalized characteristic polynomials of $\pi_{A_{\mathfrak{p}}}$, and also equivalent to studying the distribution of the normalized Euler factors of $L(A,s)$.

REMARK 3.3. Each of these three perspectives is successively more general than the previous, the last vastly so. There are abelian varieties over K that are not the Jacobian of any curve defined over K, and L-functions that can be written as Euler products over primes of K that are not the L-function of any abelian variety. One can more generally consider the distribution of normalized Euler factors of *motivic L-functions*, which we also expect to be governed by the Haar measure of a Sato-Tate group associated to the underlying motive, as defined in [**76**, **77**]; see [**26**] for some concrete examples in weight 3.

The recipe for defining the Sato-Tate group $\mathrm{ST}(A)$ of an abelian variety A/K of genus g is a direct generalization of the $g=1$ case. We proceed as follows:

1. Pick a prime ℓ, define the Tate module $T_\ell := \varprojlim_n A[\ell^n]$, a free $\mathbf{Z}_\ell$-module of rank $2g$, and the rational Tate module $V_\ell := T_\ell \otimes_{\mathbf{Z}} \mathbf{Q}$, a $\mathbf{Q}_\ell$-vector space of dimension $2g$.
2. Use the Galois representation $\rho_{A,\ell}\colon \mathrm{Gal}(\overline{K}/K) \to \mathrm{Aut}(V_\ell) \simeq \mathrm{GL}_{2g}(\mathbf{Q}_\ell)$ to define $G_\ell := \operatorname{im}\rho_{A,\ell}$.
3. Let G_ℓ^{zar} be the Zariski closure of G_ℓ in $\mathrm{GL}_{2g}(\mathbf{Q}_\ell)$ (as an algebraic group), and define $G_\ell^{1,\mathrm{zar}}$ by adding the symplectic constraint $M^t\Omega M = \Omega$, so that $G_\ell^{1,\mathrm{zar}}$ is a $\mathbf{Q}_\ell$-algebraic subgroup of Sp_{2g}.
4. Pick an embedding $\iota\colon \mathbf{Q}_\ell \to \mathbf{C}$ and use it to define $G_{\ell,\iota}^{1,\mathrm{zar}}$ as the base-change of $G_\ell^{1,\mathrm{zar}}$ to $\mathbf{C}$.
5. Define $\mathrm{ST}(A) \subseteq \mathrm{USp}(2g)$ as a maximal compact subgroup of $G_{\ell,\iota}^{1,\mathrm{zar}}(\mathbf{C})$, unique up to conjugacy.
6. For each good prime $\mathfrak{p} \nmid \ell$, let $M_{\mathfrak{p}}$ be the image of $\mathrm{Frob}_{\mathfrak{p}}$ in $G_{\ell,\iota}^{\mathrm{zar}}(\mathbf{C})$ and define $x_{\mathfrak{p}} \in \mathrm{conj}(\mathrm{ST}(A))$ to be the conjugacy class of $\overline{M}_{\mathfrak{p}} := N(\mathfrak{p})^{-1/2}M_{\mathfrak{p}}$, in $\mathrm{ST}(A)$.

Step 6 requires some justification; it is not obvious why $\overline{M}_{\mathfrak{p}}$ should necessarily be conjugate to an element of $\mathrm{ST}(A)$. Here we are relying on two key facts.

First, the image G_ℓ of $\rho_{A,\ell}$ in $\mathrm{GL}_{2g}(\mathbf{Q}_\ell)$ actually lies in $\mathrm{GSp}_{2g}(\mathbf{Q}_\ell)$, the group of *symplectic similitudes*. The algebraic group GSp_{2g} is defined by imposing the constraint

$$M^t\Omega M = \lambda\Omega, \qquad \Omega := \begin{pmatrix} 0 & -I_g \\ I_g & 0 \end{pmatrix},$$

where λ is necessarily an element of the multiplicative group $\mathbf{G}_m := \mathrm{GL}_1$, since M is invertible. The morphism $\mathrm{GSp}_{2g} \to \mathbf{G}_m$ defined by λ is the *similitude character*,

[17] Explicitly determining the Euler factors at bad primes is difficult when $\dim A > 1$. Practical methods are known only in special cases, such as when A is the Jacobian of a hyperelliptic curve (even in this case there is still room for improvement).

and we have an exact sequence of algebraic groups

$$1 \to \mathrm{Sp}_{2g} \hookrightarrow \mathrm{GSp}_{2g} \xrightarrow{\lambda} \mathbf{G}_m \to 1.$$

The action of $\mathrm{Gal}(\overline{K}/K)$ on the Tate module is compatible with the Weil pairing, and this forces the image G_ℓ of $\rho_{E,\ell}$ to lie in $\mathrm{GSp}_{2g}(\mathbf{Q}_\ell)$; see Exercise 3.1. By fixing a symplectic basis for V_ℓ in step 1 we can view $\rho_{A,\ell}$ as a continuous homomorphism

$$\rho_{A,\ell}\colon \mathrm{Gal}(\overline{K}/K) \to \mathrm{GSp}_{2g}(\mathbf{Q}_\ell) \subseteq \mathrm{GL}_{2g}(\mathbf{Q}_\ell)$$

For $g = 1$ we have $\mathrm{GL}_2 = \mathrm{GSp}_2$, but for $g > 1$ the algebraic group GSp_{2g} is properly contained in GL_{2g}.

Second, we are relying on the fact that $M_\mathfrak{p}$, and therefore $\overline{M}_\mathfrak{p}$, is *semisimple* (diagonalizable, since we are working over $\mathbf{C}$). This follows from Tate's proof of the Tate conjecture for abelian varieties over finite fields (combine the main theorem and part (a) of Theorem 2 in [**89**]). The matrix $\overline{M}_\mathfrak{p}$ is thus diagonalizable and has eigenvalues of absolute value 1; it therefore lies in a compact subgroup of $G_{\ell,\iota}^{1,\mathrm{zar}}(\mathbf{C})$ (take the closure of the group it generates). This compact group is necessarily conjugate to a subgroup of the maximal compact subgroup $\mathrm{ST}(A)$, which must contain an element conjugate to $\overline{M}_\mathfrak{p}$.

REMARK 3.4. When defining the Sato-Tate group in more general settings one instead uses the semisimple component of the (multiplicative) Jordan decomposition (see [**10**, Thm. I.4.4]) of $\overline{M}_\mathfrak{p}$ to define $x_\mathfrak{p}$, as in [**77**, §8.3.3]. This avoids the need to assume the conjectured *semisimplicity of Frobenius*, which is known for abelian varieties but not in general.

BACKGROUND 3.5 (Weil pairing). If A is an abelian variety over a field k and $A^\vee$ is its dual abelian variety (see [**54**, §I.8]), then for each $n \geq 1$ prime to the characteristic of k, the *Weil pairing* is a nondegenerate bilinear map

$$A[n] \times A^\vee[n] \to \mu_n(\overline{k})$$

that commutes with the action of $\mathrm{Gal}(\overline{k}/k)$; here μ_n denotes the group of nth roots of unity (the algebraic group defined by $x^n = 1$). Letting n vary over powers of a prime $\ell \neq \mathrm{char}(k)$ and taking inverse limits yields a bilinear map on the corresponding Tate modules:

$$e_\ell\colon T_\ell \times T_\ell^\vee \to \mu_{\ell^\infty}(\overline{k}) := \varprojlim_n \mu_{\ell^n}(\overline{k}).$$

Given a *polarization*, an isogeny $\phi\colon A \to A^\vee$, we can use it to define a bilinear pairing

$$\begin{aligned} e_\ell^\phi\colon T_\ell \times T_\ell &\to \mu_{\ell^\infty}(\overline{k}) \\ (x,y) &\mapsto e_\ell(x,\phi(y)) \end{aligned}$$

that is also compatible with the action of $\mathrm{Gal}(\overline{k}/k)$. One can always choose a polarization ϕ so that the pairing e_ℓ^ϕ is nondegenerate and skew symmetric, meaning that $e_\ell^\phi(a,b) = e_\ell^\phi(b,a)^{-1}$ for all $a, b \in T_\ell$; see [**54**, Prop. I.13.2]. When A is the Jacobian of a curve it is naturally equipped with a *principal polarization* ϕ, an isomorphism $A \xrightarrow{\sim} A^\vee$, for which this automatically holds; in this situation it is common to simply identify e_ℓ with e_ℓ^ϕ without mentioning ϕ explicitly.

We should note that our definition of the Sato-Tate group $\mathrm{ST}(A)$ required us to choose a prime ℓ and an embedding $\iota : \mathbf{Q}_\ell \to \mathbf{C}$. Up to conjugacy in $\mathrm{USp}(2g)$ one expects the Sato-Tate group to be independent of these choices; this is known for $g \le 3$ (see [**4**]), but open in general. We shall nevertheless refer to $\mathrm{ST}(A)$ as "the" Sato-Tate group of A, with the understanding that we are fixing once and for all a prime ℓ and an embedding $\iota : \mathbf{Q}_\ell \to \mathbf{C}$ (note that these choices do not depend on A or even its dimension g).

3.3. The Sato-Tate conjecture for abelian varieties. Having defined the Sato-Tate group of an abelian variety over a number field we can now state the Sato-Tate conjecture for abelian varieties.

CONJECTURE 3.6. *Let A be an abelian variety over a number field K, let* $\mathrm{ST}(A)$ *denote its Sato-Tate group, and let $(x_\mathfrak{p})$ be the sequence of conjugacy classes of normalized images of Frobenius elements in* $\mathrm{ST}(A)$ *at primes $\mathfrak{p}$ of good reduction for A, ordered by norm (break ties arbitrarily). Then the sequence $(x_\mathfrak{p})$ is equidistributed (with respect to the pushforward of the Haar measure of* $\mathrm{ST}(A)$ *to its space of conjugacy classes).*

3.4. The identity component of the Sato-Tate group. There are two algebraic groups that one can associate to an abelian variety A over a number field K that are closely related to its Sato–Tate group, the *Mumford–Tate group* and the *Hodge group*, both of which conjecturally determine the identity component of the Sato–Tate group (provably so whenever the Mumford–Tate conjecture is known, which includes all abelian varieties of dimension $g \le 3$, as shown in [**4**]). In order to define these groups we need to recall some facts about complex abelian varieties and their associated Hodge structures.

BACKGROUND 3.7 (complex abelian varieties). Let A be an abelian variety of dimension g over $\mathbf{C}$. Then $A(\mathbf{C})$ is a connected compact Lie group and therefore isomorphic to a torus V/Λ, where $V \simeq \mathbf{C}^g$ is a complex vector space of dimension g and $\Lambda \simeq \mathbf{Z}^{2g}$ is a full lattice in V that we view as a free $\mathbf{Z}$-module; one can obtain Λ as the kernel of the exponential map $\exp\colon T_0(A(\mathbf{C})) \to A(\mathbf{C})$, where $T_0(A(\mathbf{C}))$ denotes the tangent space at the identity. While every complex abelian variety corresponds to a complex torus, the converse is true only when $g = 1$. The complex tori $X := V/\Lambda$ that correspond to abelian varieties are those that admit a *polarization* (or *Riemann form*), a positive definite Hermitian form $H\colon V \times V \to \mathbf{C}$ with $\operatorname{Im} H(\Lambda,\Lambda) = \mathbf{Z}$ (here Im means imaginary part). Given a polarization H on X, the map $v \mapsto H(v,\cdot)$ defines an isogeny to the *dual torus* $X^\vee := V^*/\Lambda^*$, where

$$V^* := \{f\colon V \to \mathbf{C} : f(\alpha v) = \bar{\alpha} f(v) \text{ and } f(v_1 + v_2) = f(v_1) + f(v_2)\},$$

and $\Lambda^* := \{f \in V^* : \operatorname{Im} f(\Lambda) \subseteq \mathbf{Z}\}$. This isogeny is a polarization of X as an abelian variety; conversely, any polarization on A (one always exists) can be used to define a polarization on the complex torus $A(\mathbf{C})$. One can then show that the map $A \mapsto A(\mathbf{C})$ defines an equivalence of categories between complex abelian varieties and polarizable complex tori. For more background on complex abelian varieties, see the overviews in [**54**, §1] or [**59**, §1], or see [**8**] for a comprehensive treatment.

Now let A be an abelian variety over a number field K, fix an embedding $K \hookrightarrow \mathbf{C}$, and let $\mathbf{C}^g/\Lambda$ be the complex torus corresponding to $A(\mathbf{C})$. We may identify Λ with the singular homology group $H_1(A(\mathbf{C}), \mathbf{Z})$, and we similarly have $\Lambda_R := \Lambda \otimes_{\mathbf{Z}} R \simeq H_1(A(\mathbf{C}), R)$ for any ring R.

The isomorphisms $A(\mathbf{C}) \simeq \mathbf{C}^g/\Lambda$ and $A(\mathbf{C}) \simeq \mathbf{R}^{2g}/\Lambda$ of complex and real Lie groups allow us to view

$$\Lambda_{\mathbf{R}} \simeq H_1(A(\mathbf{C}), \mathbf{R})$$

as a real vector space of dimension $2g$ equipped with a *complex structure*, by which we mean an $\mathbf{R}$-algebra homomorphism $h\colon \mathbf{C} \to \operatorname{End}(\Lambda_{\mathbf{R}})$. In the language of Hodge theory, this amounts to the statement that (Λ, h) is an *integral Hodge structure* (pure of weight -1).

We can also view h as morphism of $\mathbf{R}$-algebraic groups $h\colon \mathbf{S} \to \mathrm{GL}_{\Lambda_{\mathbf{R}}}$. Here $\mathbf{S}$ denotes the *Deligne torus* (also known as the *Serre torus*), obtained by viewing $\mathbf{C}^\times$ as an $\mathbf{R}$-algebraic group (this amounts to taking the restriction of scalars of $\mathbf{G}_m := \mathrm{GL}_1$ from $\mathbf{C}$ to $\mathbf{R}$; see Exercise 3.2). The morphism h can be defined over $\mathbf{R}$ because $\mathbf{C}^g/\Lambda$ is a polarizable torus, since it comes from an abelian variety (in general this need not hold). The real Lie group $\mathbf{S}(\mathbf{R}) \simeq \mathbf{C}^\times$ is generated by $\mathbf{R}^\times$ and $\mathrm{U}(1) = \{z \in \mathbf{C}^\times : z\bar{z} = 1\}$, which intersect in $\{\pm 1\}$; taking Zariski closures yields $\mathbf{R}$-algebraic subgroups $\mathbf{G}_m$ and U_1 of $\mathbf{S}$ that intersect in μ_2. Restricting h to $\mathrm{U}_1 \subseteq \mathbf{S}$ yields a morphism $\mathrm{U}_1 \to \mathrm{GL}_{\Lambda_{\mathbf{R}}}$ with the following property: the image of each $u \in \mathrm{U}_1(\mathbf{R}) = \mathrm{U}(1)$ has eigenvalues u, u^{-1} with multiplicity g; see [**8**, Prop. 17.1.1]. The image of such a map is known as a *Hodge circle.*

The *rational Hodge structure* $(\Lambda_{\mathbf{Q}}, h)$ is obtained by replacing the lattice Λ with $\Lambda_{\mathbf{Q}} := \Lambda \otimes_{\mathbf{Z}} \mathbf{Q}$ and can be used to define the Mumford-Tate group.

DEFINITION 3.8. The *Mumford–Tate group* $\mathrm{MT}(A)$ is the smallest $\mathbf{Q}$-algebraic group G in $\mathrm{GL}_{\Lambda_{\mathbf{Q}}}$ for which $h(\mathbf{S}) \subseteq G(\mathbf{R})$; equivalently, it is the $\mathbf{Q}$-Zariski closure of $h(\mathbf{S}(\mathbf{R}))$ in $\mathrm{GL}_{\Lambda_{\mathbf{R}}}$. The *Hodge group* $\mathrm{Hg}(A)$ is similarly defined as the $\mathbf{Q}$-Zariski closure of $h(\mathrm{U}(1))$ in $\mathrm{GL}_{\Lambda_{\mathbf{R}}}$.

As defined above, the Mumford–Tate group $\mathrm{MT}(A)$ is a $\mathbf{Q}$-algebraic subgroup of GL_{2g}. But the complex torus $\mathbf{C}^g/\Lambda$ is polarizable, which means that we can put a symplectic form on Λ_R that is compatible with h, and this implies that in fact $\mathrm{MT}(A)$ is a $\mathbf{Q}$-algebraic subgroup of GSp_{2g}. Similarly, the Hodge group $\mathrm{Hg}(A)$ is a $\mathbf{Q}$-algebraic subgroup of Sp_{2g}, and in fact $\mathrm{Hg}(A) = \mathrm{MT}(A) \cap \mathrm{Sp}_{2g}$; this is sometimes used as an alternative definition of $\mathrm{Hg}(A)$. Much of the interest in the Hodge group arises from the fact that it gives us an isomorphism of $\mathbf{Q}$-algebras

$$\operatorname{End}(A_{\mathbf{C}})_{\mathbf{Q}} \simeq \operatorname{End}(\Lambda_{\mathbf{Q}})^{\mathrm{Hg}(A)},$$

where $\operatorname{End}(A_{\mathbf{C}})_{\mathbf{Q}} := \operatorname{End}(A_{\mathbf{C}}) \otimes_{\mathbf{Z}} \mathbf{Q}$ and $\mathrm{Hg}(A)$ acts on $\operatorname{End}(\Lambda_{\mathbf{Q}})$ by conjugation; see [**8**, Prop. 17.3.4]. To see why this isomorphism is useful, let us note one application.

THEOREM 3.9. *For an abelian variety A of dimension g over a number field K, the Hodge group* $\mathrm{Hg}(A)$ *is commutative if and only if the endomorphism algebra* $\operatorname{End}(A_{\overline{K}})_{\mathbf{Q}}$ *contains a commutative semisimple* $\mathbf{Q}$*-algebra of dimension $2g$.*

PROOF. See [**8**, Prop. 17.3.5]. □

For $g = 1$ the abelian varieties A that satisfy the two equivalent properties of Theorem 3.9 are CM elliptic curves. More generally, such abelian varieties are said to be of *CM-type*. For abelian varieties of general type one has the opposite extreme: $\operatorname{End}(A_{\overline{K}})_{\mathbf{Q}} = \mathbf{Q}$ and $\mathrm{Hg}(A) = \mathrm{Sp}_{2g}$; see [**8**, Prop. 17.4.2].

In the previous section we defined two $\mathbf{Q}_\ell$-algebraic groups $G_\ell^{\mathrm{zar}} \subseteq \mathrm{GSp}_{2g}$ and $G_\ell^{1,\mathrm{zar}} \subseteq \mathrm{Sp}_{2g}$ associated to A. It is reasonable to ask how they are related to the $\mathbf{Q}$-algebraic groups $\mathrm{MT}(A)$ and $\mathrm{Hg}(A)$. Unlike the groups G_ℓ^{zar} and $G_\ell^{1,\mathrm{zar}}$, the algebraic groups $\mathrm{MT}(A)$ and $\mathrm{Hg}(A)$ are necessarily connected (by construction).[18] Deligne proved that the identity component of G_ℓ^{zar} is always a subgroup of $\mathrm{MT}(A) \otimes_{\mathbf{Q}} \mathbf{Q}_\ell$, equivalently, that the identity component of $G_\ell^{1,\mathrm{zar}}$ is a subgroup of $\mathrm{Hg}(A) \otimes_{\mathbf{Q}} \mathbf{Q}_\ell$); see [**20**]. It is conjectured that these inclusions are in fact equalities.

Conjecture 3.10 (Mumford–Tate Conjecture). *The identity component of G_ℓ^{zar} is equal to $\mathrm{MT}(A) \otimes_{\mathbf{Q}} \mathbf{Q}_\ell$; equivalently, the identity component of $G_\ell^{1,\mathrm{zar}}$ is equal to $\mathrm{Hg}(A) \otimes_{\mathbf{Q}} \mathbf{Q}_\ell$.*

This conjecture is known to hold for abelian varieties of dimension $g \le 3$; see [**4**, Th. 6.11] where it is shown that this follows from [**57**]. When it holds, the Mumford–Tate group (and the Hodge group) uniquely determines the identity component of the Sato–Tate group, up to conjugation in $\mathrm{USp}(2g)$; see [**25**, Lemma 2.8]. Neither the Mumford–Tate group nor the Hodge group tell us anything about the component groups of G_ℓ^{zar}, $G_\ell^{1,\mathrm{zar}}$, $\mathrm{ST}(A)$ (the three are isomorphic; see [**77**, §8.3.4]), but there is a closely related $\mathbf{Q}$-algebraic group that conjecturally does.

Conjecture 3.11 (Algebraic Sato–Tate Conjecture). *There exists a $\mathbf{Q}$-algebraic subgroup $\mathrm{AST}(A)$ of Sp_{2g} such that $G_\ell^{1,\mathrm{zar}} = \mathrm{AST}(A) \otimes_{\mathbf{Q}} \mathbf{Q}_\ell$.*

Banaszak and Kedlaya [**4**] have shown that this conjecture holds for $g \le 3$ via an explicit description of $\mathrm{AST}(A)$ using *twisted Lefschetz groups*.

3.5. The component group of the Sato-Tate group. We have seen that the Mumford–Tate group conjecturally determines the identity component $\mathrm{ST}(A)^0$ of the Sato–Tate group $\mathrm{ST}(A)$ of an abelian variety A over a number field K (provably so in dimension $g \le 3$). The identity component $\mathrm{ST}(A)^0$ is a normal finite index subgroup of $\mathrm{ST}(A)$, and we now want to consider the component group $\mathrm{ST}(A)/\mathrm{ST}(A)^0$. As above, for any field extension L/K, we use A_L to denote the base change of A to L.

Theorem 3.12. *Let A be an abelian variety over a number field K. There is a unique finite Galois extension L/K with the property that $\mathrm{ST}(A_L)$ is connected and $\mathrm{Gal}(L/K) \simeq \mathrm{ST}(A)/\mathrm{ST}(A)^0$. The extension L/K is unramified outside the primes of bad reduction for A, and for every subextension F/K of L/K we have $\mathrm{Gal}(L/F) \simeq \mathrm{ST}(A_F)/\mathrm{ST}(A_F)^0$.*

Proof. As explained in [**77**, §8.3.4], the component groups of G_ℓ^{zar} and $\mathrm{ST}(A)$ are isomorphic. Let Γ be the Galois group of the maximal subextension K_{S_ℓ} of $\mathrm{Gal}(\overline{K}/K)$ that is unramified away from the set S_ℓ consisting of the primes of bad reduction for A and the primes of K lying above ℓ. The ℓ-adic Galois representation $\rho_{A,\ell}\colon \mathrm{Gal}(\overline{K}/K) \to \mathrm{Aut}(V_\ell)$ induces a continuous surjective homomorphism

$$\Gamma \to G_\ell^{\mathrm{zar}}/(G_\ell^{\mathrm{zar}})^0,$$

[18]This is true more generally for all motives of odd weight. For motives of even weight the situation is more delicate; complications arise from the fact that we are then working with orthogonal groups rather than symplectic groups; see [**4**, **5**].

whose kernel is a normal open subgroup Γ_0 of Γ. The corresponding fixed field L is a finite Galois extension of K, and it is the minimal Galois extension of K for which $\mathrm{ST}(A_L)$ is connected. It is clearly uniquely determined and unramified outside S_ℓ, and we have isomorphisms

$$\mathrm{Gal}(L/K) \simeq \Gamma/\Gamma_0 \simeq G_\ell^{\mathrm{zar}}/(G_\ell^{\mathrm{zar}})^0 \simeq \mathrm{ST}(A)/\mathrm{ST}(A)^0.$$

As shown by Serre [**75**], the component group of G_ℓ^{zar}, and therefore of $\mathrm{ST}(A)$, is independent of ℓ, and the above argument applies to any choice of ℓ. Thus L/K can be ramified only at primes of bad reduction for A. For any subextension F/K of L/K, replacing A by A_F in the argument above yields the same field L, with $\mathrm{Gal}(L/F) \simeq \mathrm{ST}(A_F)/\mathrm{ST}(A_F)^0$. □

3.6. Exercises.

EXERCISE 3.1. Let A be an abelian variety of dimension g over a number field K. Show that one can choose a basis for $V_\ell = T_\ell \otimes_{\mathbf{Z}} \mathbf{Q}$ so that the matrix M describing the action of any $\sigma \in \mathrm{Gal}(\overline{K}/K)$ on V_ℓ satisfies $M^t \Omega M = \lambda \Omega$ for some $\lambda \in \mathbf{Q}_\ell^\times$, where $\Omega := \left(\begin{smallmatrix} 0 & -I \\ I & 0 \end{smallmatrix}\right)$. Conclude that the image of the corresponding Galois representation lies in $\mathrm{GSp}_{2g}(\mathbf{Q}_\ell)$ and describe the map $\mathrm{Gal}(\overline{K}/K) \to \mathbf{Q}_\ell^\times$ induced by the similitude character λ.

EXERCISE 3.2. Define the Deligne torus $\mathbf{S}$ as an $\mathbf{R}$-algebraic group in $\mathbf{A}^4$ (give equations that define it as an affine variety and polynomial maps for the group operations), and then express the $\mathbf{R}$-algebraic groups $\mathbf{G}_m$ and U_1 as subgroups of $\mathbf{S}$ that intersect in μ_2. Prove that $\mathbf{S}(\mathbf{R})$ and $\mathbf{C}^\times$ are isomorphic as real Lie groups (give explicit maps in both directions).

EXERCISE 3.3. Let L/K be a finite separable extension of degree d, written as $L = K(\alpha)$. Given an affine L-variety Y defined by polynomials $P_k \in L[y_1, \ldots, y_n]$, we can construct an affine K-variety $\mathrm{Res}_{L/K}(Y)$ by writing each $y_i = \sum_{j=0}^{d-1} x_{ij}\alpha^j$ in terms of the K-basis $\{1, \alpha, \ldots, \alpha^{d-1}\}$ for L and using the minimal polynomial of α to replace each $P_k(y_1, \ldots, y_n)$ by a polynomial in $K[x_{11}, \ldots, x_{1d}, \ldots, x_{n1} \ldots, x_{nd}]$. The K-variety $\mathrm{Res}_{L/K}(Y)$ is the *Weil restriction* (or *restriction of scalars*) of Y. Prove that the $\mathbf{R}$-algebraic group $\mathbf{S}$ (the Deligne torus) is the Weil restriction of the $\mathbf{C}$-algebraic group $\mathbf{G}_m$, that is, $\mathbf{S} = \mathrm{Res}_{\mathbf{C}/\mathbf{R}}(\mathbf{G}_m)$.

4. Sato–Tate axioms and Galois endomorphism types

In this section we present the Sato-Tate axioms and consider the problem of classifying Sato-Tate groups of abelian varieties of a given dimension g. We then compute trace moment sequences of all connected Sato-Tate groups of abelian varieties of dimension $g \le 3$ and present formulas for the trace moment sequence of $\mathrm{USp}(2g)$ (the generic case) that apply to all g,

4.1. Sato–Tate axioms. In [**77**, §8.2] Serre gives a set of axioms that any Sato–Tate group is expected to satisfy. Serre considers Sato–Tate groups in a more general context than we do here, so we will state the axioms as they apply to Sato–Tate groups of abelian varieties. As in §3.4, for a Lie group G we define a *Hodge circle* to be a subgroup H of G that is the image of a continuous homomorphism $\theta\colon \mathrm{U}(1) \to G^0$ whose elements $\theta(u)$ have eigenvalues u and u^{-1} with multiplicity g (note that H necessarily lies in the identity component G^0 of G).

DEFINITION 4.1. A group G satisfies the *Sato–Tate axioms* (for abelian varieties of dimension $g \geq 1$) if and only if the following hold:

(ST1) (Lie condition) G is a closed subgroup of $\mathrm{USp}(2g)$.

(ST2) (Hodge condition) The Hodge circles in G generate a dense non-trivial subgroup of G^0.[19]

(ST3) (rationality condition) For each component H of G and irreducible character χ of $\mathrm{GL}_{2g}(\mathbf{C})$, we have $\int_H \chi\mu \in \mathbf{Z}$, where μ is the Haar measure on G normalized so that $\mu(\mathbb{1}_H) = 1$.

REMARK 4.2. Definition 4.1 generalizes easily to self-dual motives with rational coefficients. Given an integer weight $w \geq 0$ and Hodge numbers $h^{p,q} \in \mathbf{Z}_{\geq 0}$ indexed by $p, q \in \mathbf{Z}_{\geq 0}$ with $p + q = w$ such that $h^{p,q} = h^{q,p}$ when w is odd, let $d := \sum h^{p,q}$. For abelian varieties we have $w = 1$ and $h^{1,0} = h^{0,1} = g$. In axiom (ST1) we require G to be a closed subgroup of $\mathrm{USp}(d)$ (resp. $\mathrm{O}(d)$) when w is odd (resp. even), and in axiom (ST2) we require elements $\theta(u)$ of a Hodge circle to have eigenvalues u^{p-q} with multiplicity $h^{p,q}$; axiom (ST3) is unchanged.

Axiom (ST1) implies that G is a compact Lie group, and (ST2) rules out finite groups, since G must contain at least one Hodge circle and therefore contains a subgroup isomorphic to $\mathrm{U}(1)$. When G is connected, (ST3) holds automatically and only (ST1) and (ST2) need to be checked; this is an easy application of representation theory, see [**49**, Prop. 2]. Axiom (ST3) plays no role when $g = 1$ (see the proof of Proposition 4.4 below), but for $g > 1$ it is crucial. When $g = 2$, for example, for every integer $n \geq 1$ we can diagonally embed $\mathrm{U}(1) \times \mathrm{U}(1)[n]$ in $\mathrm{USp}(4)$ to get infinitely many non-conjugate closed groups $G \subseteq \mathrm{USp}(4)$ whose identity component is a Hodge circle. All of these groups satisfy (ST1) and (ST2), but only finitely many satisfy (ST3). Indeed, if we take χ and let C be a component on which the projection to $\mathrm{U}(1)[n]$ has order n, we have

$$\int_C \chi\mu = \zeta_n + \bar{\zeta}_n \in \mathbf{Z}$$

only for $n \in \{2, 3, 4, 6\}$. More generally, we have the following theorem.

THEOREM 4.3. *Up to conjugacy, for any fixed dimension $g \geq 1$ the number of subgroups of* $\mathrm{USp}(2g)$ *that satisfy the Sato–Tate axioms is finite.*

PROOF. See [**25**, Rem. 3.3] □

Theorem 4.3 motivates the following *classification problem*: given an integer $g \geq 1$, determine the subgroups of $\mathrm{USp}(2g)$ that satisfy the Sato–Tate axioms. The case $g = 1$ is easy.

PROPOSITION 4.4. *For $g = 1$ the three groups* $\mathrm{U}(1)$, $N(\mathrm{U}(1)$ *and* $\mathrm{SU}(2)$ *listed in Theorem* 3.2 *are the only groups that satisfy the Sato–Tate axioms (up to conjugacy).*

PROOF. Suppose G satisfies the Sato–Tate axioms. Then G^0 contains a conjugate of $\mathrm{U}(1)$ embedded in $\mathrm{USp}(2)$ via $u \mapsto \left(\begin{smallmatrix} u & 0 \\ 0 & \bar{u} \end{smallmatrix}\right)$, as in Theorem 3.2, and it must be a compact connected Lie group. The only nontrivial compact connected Lie

[19]The statement of (ST2) in [**25**] inadvertently omits the requirement that the Hodge circles generate a dense subgroup.

groups in $\mathrm{USp}(2) = \mathrm{SU}(2)$ are $\mathrm{U}(1)$ and $\mathrm{SU}(2)$ itself (this follows from the classification of compact connected Lie groups but is easy to see directly). Thus either $G^0 = \mathrm{SU}(2)$, in which case $G = \mathrm{SU}(2)$, or G^0 is conjugate to $\mathrm{U}(1)$ and must be a normal subgroup of G (the identity component of a compact Lie group is always a normal subgroup of finite index). The group $\mathrm{U}(1)$ has index 2 in its normalizer, so $\mathrm{U}(1)$ and $N(\mathrm{U}(1))$ are the only possibilities for G when $G^0 = \mathrm{U}(1)$. □

COROLLARY 4.5. *For $g = 1$ a group G satisfies the Sato–Tate axioms if and only if it is the Sato–Tate group of an elliptic curve over a number field.*

The classification problem for $g = 2$ is more difficult, but it has been solved.

THEOREM 4.6. *Up to conjugacy in* $\mathrm{USp}(4)$ *there are* 55 *groups that satisfy the Sato–Tate axioms for $g = 2$. Of these* 55, *the following* 6 *are connected:*

$$\mathrm{U}(1)_2, \quad \mathrm{SU}(2)_2, \quad \mathrm{U}(1)\times\mathrm{U}(1), \quad \mathrm{U}(1)\times\mathrm{SU}(2), \quad \mathrm{SU}(2)\times\mathrm{SU}(2), \quad \mathrm{USp}(4),$$

were $\mathrm{U}(1)_2$ *denotes* $\mathrm{U}(1) = \left\{\left(\begin{smallmatrix} u & 0 \\ 0 & \bar{u}\end{smallmatrix}\right) : u \in \mathbf{C}^\times\right\}$ *diagonally embedded in* $\mathrm{USp}(4)$, *and similarly for* $\mathrm{SU}(2)_2$.

PROOF. See [**25**, Thm. 3.4], which gives an explicit description of the 55 groups. □

REMARK 4.7. Those familiar with the classification of connected compact Lie groups may notice that the group $\mathrm{U}(2)$, which can be embedded in $\mathrm{USp}(4)$, is missing from Theorem 4.6. This is because it fails to satisfy the Hodge condition (ST2); it contains subgroups isomorphic to $\mathrm{U}(1)$, but there is no way to embed $\mathrm{U}(1) \hookrightarrow \mathrm{U}(2) \hookrightarrow \mathrm{USp}(4)$ and get eigenvalues u and u^{-1} with multiplicity 2; see [**26**, Rem. 2.3]. However, for motives of weight 3 and Hodge numbers $h^{3,0} = h^{2,1} = h^{1,2} = h^{0,3} = 1$ the modified Hodge condition noted in Remark 4.2 is satisfied by a subgroup of $\mathrm{USp}(4)$ isomorphic to $\mathrm{U}(2)$; see [**26**] for details, including two examples of weight 3 motives with Sato-Tate group $\mathrm{U}(2)$.

Corollary 4.5 does not hold for $g = 2$.

THEOREM 4.8. *Of the* 55 *groups appearing in Theorem* 4.6, *only* 52 *arise as the Sato–Tate group of an abelian surface over a number field. Of these,* 34 *arise for abelian surfaces defined over* $\mathbf{Q}$.

PROOF. See [**25**, Thm. 1.5]. □

The three subgroups of $\mathrm{USp}(4)$ that satisfy the Sato–Tate axioms but are not the Sato–Tate group of any abelian surface over a number field are the normalizer of $\mathrm{U}(1)\times\mathrm{U}(1)$ in $\mathrm{USp}(4)$, whose component group is the dihedral group of order 8, and two of its subgroups, one of index 2 and one of index 4. The proof that these three groups do not occur is obtained by first establishing a bijection between *Galois endomorphism types* (see Definition 4.10 below) and Sato–Tate groups, and then showing that there are only 52 Galois endomorphism types of abelian surfaces. Explicit examples of genus 2 curves whose Jacobians realize these 52 possibilities can be found in [**25**, Table 11], and animated histograms of their Sato–Tate distributions are available at

`http://math.mit.edu/~drew/g2SatoTateDistributions.html`.

The classification problem for $g = 3$ remains open, but the connected cases have been determined (see Table 2 in the next section). Before leaving our discussion of the Sato–Tate axioms, it is reasonable to ask whether Sato–Tate groups necessarily satisfy them. Of course we expect this to be the case, but it is difficult to prove in general. However, it can be proved to hold in all cases where the Mumford–Tate conjecture is known, including all cases with $g \leq 3$.

PROPOSITION 4.9. *Let A be an abelian variety of dimension g over a number field K for which the Mumford–Tate conjecture holds. Then* $\mathrm{ST}(A)$ *satisfies the Sato–Tate axioms.*

PROOF. See [**25**, Prop. 3.2]. □

4.2. Galois endomorphism types. We will work in the abstract category $\mathcal{C}$ whose objects are pairs (G, E) of a finite group G and an $\mathbf{R}$-algebra E equipped with an $\mathbf{R}$-linear action of G, and whose morphisms $\Phi\colon (G, E) \to (G', E')$ are pairs (ϕ_G, ϕ_E), where $\phi_G\colon G \to G'$ is a morphism of groups, and $\phi_E\colon E \to E'$ is an equivariant morphism of $\mathbf{R}$-algebras, meaning that

$$\phi_E(e^g) = \phi_E(e)^{\phi_G(g)} \qquad \text{for all } g \in G \text{ and } e \in E. \tag{4.1}$$

To each abelian variety A/K we now associate an isomorphism class $[G, E]$ in $\mathcal{C}$ as follows. The minimal extension L/K for which $\mathrm{End}(A_L) = \mathrm{End}(A_{\overline{K}})$ is a finite Galois extension of K; we shall take G to be $\mathrm{Gal}(L/K)$ and E to be the real endomorphism algebra $\mathrm{End}(A_L)_{\mathbf{R}} := \mathrm{End}(A_L) \otimes_{\mathbf{Z}} \mathbf{R}$. The Galois group $\mathrm{Gal}(L/K)$ acts on $\mathrm{End}(A_L)$ via its action on the coefficients of the rational maps defining each element of $\mathrm{End}(A_K)$; this induces an $\mathbf{R}$-linear action of $\mathrm{Gal}(L/K)$ on $\mathrm{End}(A_L)_{\mathbf{R}}$ via composition with the natural map $\mathrm{End}(A_L) \to \mathrm{End}(A_L)_{\mathbf{R}}$. The pair $(\mathrm{Gal}(L/K), \mathrm{End}(A_L)_{\mathbf{R}})$ is thus an object of $\mathcal{C}$.

DEFINITION 4.10. The *Galois endomorphism type* $\mathrm{GT}(A)$ of an abelian variety A/K is the isomorphism class of the pair $(\mathrm{Gal}(L/K), \mathrm{End}(A_L)_{\mathbf{R}})$ in the category $\mathcal{C}$, where L is the minimal extension of K for which $\mathrm{End}(A_L) = \mathrm{End}(A_{\overline{K}})$.

EXAMPLE 4.11. Let E be an elliptic curve over a number field K. If E does not have CM, or if it has CM defined over K, then its endomorphisms are all defined over $L = K$; otherwise, its endomorphisms are all defined over its CM field L, an imaginary quadratic extension of K. The real endomorphism algebra $\mathrm{End}(E_L)_{\mathbf{R}}$ is isomorphic to $\mathbf{R}$ when E does not have CM, and isomorphic to $\mathbf{C}$ when E does have CM. We therefore have

$$\mathrm{GT}(E) = \begin{cases} [\mathrm{C}_1, \mathbf{C}] & \text{if } E \text{ has CM defined over } K \\ [\mathrm{C}_2, \mathbf{C}] & \text{if } E \text{ has CM not defined over } K \\ [\mathrm{C}_1, \mathbf{R}] & \text{if } E \text{ does not have CM} \end{cases}$$

Here C_n denotes the cyclic group of order n; in the case $[\mathrm{C}_2, \mathbf{C}]$ the action of C_2 on $\mathbf{C}$ corresponds to complex conjugation.

The three Galois endomorphism types listed in Example 4.11 correspond to the three Sato-Tate groups listed in Theorem 3.2. Under this correspondence, the real endomorphism algebra $\mathrm{End}(E_L)_{\mathbf{R}}$ determines the identity component $\mathrm{ST}(E)^0$ (up to conjugacy), and the Galois group $\mathrm{Gal}(L/K)$ is isomorphic to the component group $\mathrm{ST}(E)/\mathrm{ST}(E)^0$. Moreover, the field L is precisely the field L given by Theorem 3.12.

THEOREM 4.12. *Let A be an abelian variety A of dimension $g \leq 3$ defined over a number field K and let L be the minimal field for which* $\mathrm{End}(A_L) = \mathrm{End}(A_{\overline{K}})$. *The conjugacy class of the Sato-Tate group* $\mathrm{ST}(A)$ *determines the Galois endomorphism type* $\mathrm{GT}(A)$*; moreover, the conjugacy class of the identity component* $\mathrm{ST}(A)^0$ *determines the isomorphism class of* $\mathrm{End}(A_L)_{\mathbf{R}}$ *and* $\mathrm{ST}(A)/\mathrm{ST}(A)^0 \simeq \mathrm{Gal}(L/K)$. *For $g \leq 2$ the converse holds: the Galois endomorphism type* $\mathrm{GT}(A)$ *determines the Sato–Tate group* $\mathrm{ST}(A)$ *up to conjugacy.*

PROOF. See Proposition 2.19 and Theorem 1.4 in [**25**]. □

It is expected that in fact the Sato–Tate group always determines the Galois endomorphism type, and that the converse holds for $g \leq 3$. For $g = 3$ we at least know that the real endomorphism algebra $\mathrm{End}(A_L)_{\mathbf{R}}$ determines the identity component $\mathrm{ST}(A)^0$ and that $\mathrm{Gal}(L/K) \simeq \mathrm{ST}(A)/\mathrm{ST}(A)^0$. At first glance it might seem that this should determine $\mathrm{ST}(A)$, but it does not, even when $g = 2$. One needs to also understand how $\mathrm{Gal}(L/K)$ acts on $\mathrm{End}(A_L)_{\mathbf{R}}$ and relate this to the Sato-Tate group $\mathrm{ST}(A)$. In [**25**] this is accomplished for $g = 2$ by looking at the lattice of $\mathbf{R}$-subalgebras of $\mathrm{End}(A_L)_{\mathbf{R}}$ fixed by subgroups of $\mathrm{Gal}(L/K)$ and showing that this is enough to uniquely determine $\mathrm{ST}(A)$; see [**25**, Thm. 4.3]. To apply the same approach when $g = 3$ we need a more detailed classification of the Galois endomorphism types and Sato–Tate groups for $g = 3$ than is currently available.

For $g = 4$ the Galois endomorphism type does not always determine the Sato–Tate group. This is due to an exceptional counterexample constructed by Mumford in [**58**], in which he proves the existence of an abelian four-fold A for which $\mathrm{End}(A_{\overline{K}}) = \mathbf{Z}$ but $\mathrm{MT}(A) \neq \mathrm{GSp}_8$. The fact that $\mathrm{MT}(A)$ is properly contained in GSp_8 implies that $\mathrm{ST}(A)$ must be properly contained in $\mathrm{USp}(8)$ (this does not depend on the Mumford–Tate conjecture, here we are only using the inclusion proved by Deligne). On the other hand, for an abelian variety of general type one has $\mathrm{End}(A_{\overline{K}}) = \mathbf{Z}$ and $\mathrm{ST}(A) = \mathrm{USp}(2g)$; see [**31, 100**] for an explicit criterion that applies to almost all Jacobians of hyperelliptic curves.

For $g > 4$ one can construct exceptional examples as a product of an abelian variety with one of Mumford's exceptional four-folds, so in general the Galois endomorphism type cannot determine the Sato–Tate group for any $g \geq 4$. However, such examples will not be simple and will have $\mathrm{End}(A) \neq \mathbf{Z}$. In [**74**] Serre proves an analog of his open image theorem for elliptic curves that applies to abelian varieties of dimension $g = 2, 6$ and g odd. For these values of g, if $\mathrm{End}(A_{\overline{K}}) = \mathbf{Z}$ then $\mathrm{ST}(A) = \mathrm{USp}(2g)$ and no direct analog of Mumford's construction exists.

REMARK 4.13. For $g \leq 3$, the field L in Theorem 3.12 (the minimal L for which $\mathrm{ST}(A_L)$ is connected) is the same as the field L in Theorem 4.12 (the minimal L for which $\mathrm{End}(A_L) = \mathrm{End}(A_{\overline{K}})$). In any case, the former always contains the latter: if $\mathrm{ST}(A_L)$ is connected then we necessarily have $\mathrm{End}(A_{\overline{K}}) = \mathrm{End}(A_L)$. This can be seen as a consequence of Bogomolov's theorem [**9**], which states that G_ℓ is open in $G_\ell^{\mathrm{zar}}(\mathbf{Q}_\ell)$, and Faltings' theorem [**23**] that $\mathrm{End}(A)_{\mathbf{Q}_\ell} \simeq \mathrm{End}(V_\ell(A))^{G_\ell}$. If $\mathrm{ST}(A)$ (and therefore G_ℓ^{zar}) is connected, then $\mathrm{End}(A)$ is invariant under base change (now apply this to $A = A_L$).

Tables 1 and 2 list the real endomorphism algebras and corresponding identity components of Sato-Tate groups that arise in dimensions $g = 2, 3$. A complete list of the 52 Galois endomorphism types and corresponding Sato-Tate groups for $g = 2$ can be found in [**25**, Thm. 4.3] and [**25**, Table 9].

TABLE 1. Real endomorphism algebras and Sato–Tate identity components for abelian surfaces.

geometric type of abelian surface	$\mathrm{End}(A_{\overline{K}})_{\mathbf{R}}$	$\mathrm{ST}(A)^0$
square of CM elliptic curve	$\mathrm{M}_2(\mathbf{C})$	$\mathrm{U}(1)_2$
QM abelian surface square of non-CM elliptic curve	$\mathrm{M}_2(\mathbf{R})$	$\mathrm{SU}(2)_2$
CM abelian surface product of CM elliptic curves	$\mathbf{C}\times\mathbf{C}$	$\mathrm{U}(1)\times\mathrm{U}(1)$
product of CM and non-CM elliptic curves	$\mathbf{C}\times\mathbf{R}$	$\mathrm{U}(1)\times\mathrm{SU}(2)$
RM abelian surface product of non-CM elliptic curves	$\mathbf{R}\times\mathbf{R}$	$\mathrm{SU}(2)\times\mathrm{SU}(2)$
abelian surface of general type	$\mathbf{R}$	$\mathrm{USp}(4)$

TABLE 2. Real endomorphism algebras and Sato–Tate identity components for abelian threefolds (EC=Elliptic curve, AS=Abelian surface).

geometric type of abelian three-fold	$\mathrm{End}(A_K)_{\mathbf{R}}$	$\mathrm{ST}(A)^0$
cube of a CM EC	$\mathrm{M}_3(\mathbf{C})$	$\mathrm{U}(1)_3$
cube of a non-CM EC	$\mathrm{M}_3(\mathbf{R})$	$\mathrm{SU}(2)_3$
product of CM EC and CM EC^2	$\mathbf{C}\times\mathrm{M}_2(\mathbf{C})$	$\mathrm{U}(1)\times\mathrm{U}(1)_2$
product of CM EC and QM abelian surface product of CM EC and non-CM EC^2	$\mathbf{C}\times\mathrm{M}_2(\mathbf{R})$	$\mathrm{U}(1)\times\mathrm{SU}(2)_2$
product of non-CM EC and CM EC^2	$\mathbf{R}\times\mathrm{M}_2(\mathbf{C})$	$\mathrm{SU}(2)\times\mathrm{U}(1)_2$
product of non-CM EC and QM AS product of non-CM EC and non-CM EC^2	$\mathbf{R}\times\mathrm{M}_2(\mathbf{R})$	$\mathrm{SU}(2)\times\mathrm{SU}(2)_2$
sextic CM abelian threefold product of CM EC and CM abelian surface product of three CM ECs	$\mathbf{C}^3$	$\mathrm{U}(1)^3$
product of non-CM EC and CM AS product of non-CM EC and two CM ECs	$\mathbf{C}^2\times\mathbf{R}$	$\mathrm{U}(1)^2\times\mathrm{SU}(2)$
product of CM EC and RM AS product of CM EC and two non-CM ECs	$\mathbf{C}\times\mathbf{R}^2$	$\mathrm{U}(1)\times\mathrm{SU}(2)^2$
RM abelian threefold product of non-CM EC and RM AS product of 3 non-CM ECs	$\mathbf{R}^3$	$\mathrm{SU}(2)^3$
product of CM EC and generic AS	$\mathbf{C}\times\mathbf{R}$	$\mathrm{U}(1)\times\mathrm{USp}(4)$
product of non-CM EC and generic AS	$\mathbf{R}\times\mathbf{R}$	$\mathrm{SU}(2)\times\mathrm{USp}(4)$
quadratic CM abelian threefold	$\mathbf{C}$	$\mathrm{U}(3)$
generic abelian threefold	$\mathbf{R}$	$\mathrm{USp}(6)$

As can be seen in Tables 1 and 2, the Sato–Tate group is in some respects a rather coarse invariant; for example, it cannot distinguish a product of non-CM elliptic curves from a geometrically simple abelian surface with real multiplication (RM). On the other hand, the Haar measures of the 52 Sato–Tate groups of abelian surfaces over number fields all give rise to distinct distributions of characteristic polynomials, which, under the Sato–Tate conjecture, match the distribution of normalized L-polynomials, and there are some rather fine distinctions among these distributions that the Sato–Tate group detects. For example, there are only 37 distinct trace distributions among the 52 groups, one needs to look at both the linear and quadratic coefficients of the characteristic polynomials in order to distinguish them.

It is possible for two non-conjugate Sato–Tate groups to be isomorphic as abstract groups yet give rise to distinct trace distributions. For example, the connected Sato-Tate groups $\mathrm{SU}(2) \times \mathrm{U}(1)_2$ and $\mathrm{U}(1) \times \mathrm{SU}(2)_2$ that appear in Table 2 are both abstractly isomorphic to the real Lie group $\mathrm{U}(1) \times \mathrm{SU}(2)$, but these two embeddings of $\mathrm{U}(1) \times \mathrm{SU}(2)$ in $\mathrm{USp}(6)$ have different trace distributions.

As shown by the example below, this phenomenon can also occur for disconnected Sato-Tate groups with the same identity component.

EXAMPLE 4.14. Consider the hyperelliptic curves

$$C_1\colon y^2 = x^6 + 3x^5 + 15x^4 - 20x^3 + 60x^2 - 60x + 28,$$
$$C_2\colon y^2 = x^6 + 6x^5 - 15x^4 + 20x^3 - 15x^2 + 6x - 1,$$

and let $A_1 := \mathrm{Jac}(C_1)$ and $A_2 := \mathrm{Jac}(C_2)$ denote their Jacobians. Over $\overline{\mathbf{Q}}$ both A_1 and A_2 are isogenous to the square of the elliptic curve $y^2 = x^3 + 1$, which has CM by $\mathbf{Q}(\sqrt{-3})$. We necessarily have $\mathrm{ST}(A_1)^0 = \mathrm{ST}(A_2)^0 = \mathrm{U}(1)_2$, and the component groups are both isomorphic to the dihedral group of order 12. However, their Sato–Tate groups are different: in terms of the labels used in [**25**], we have $\mathrm{ST}(A_1) = D_{6,1}$, while $\mathrm{ST}(A_2) = D_{6,2}$ (see [**25**, §3.4] for explicit descriptions of these groups in terms of generators), and their normalized trace distributions are quite different. For C_1 the density of zero traces is 3/4, whereas for C_2 it is 7/12 (these ratios represent the proportion of Sato–Tate group components on which the trace is identically zero), and their normalized trace moment sequences are

$$(1, 0, 1, 0, 9, 0, 110, 0, 1505, 0, 21546, \ldots),$$
$$(1, 0, 2, 0, 18, 0, 200, 0, 2450, 0, 31752, \ldots),$$

respectively. The Sato-Tate conjecture for these two curves was proved in [**27**], so this difference in Sato-Tate groups provably impacts the normalized trace distributions of A_1 and A_2.

4.3. Sato–Tate measures. Once we know the Sato–Tate group $\mathrm{ST}(A)$ of an abelian variety A, we are in a position to compute various statistic related to the distribution of its conjugacy classes, such as the moments of characteristic polynomial coefficients (or any other conjugacy class invariant). We can then test the Sato–Tate conjecture by comparing these to corresponding statistics obtained by computing normalized L-polynomials $\bar{L}_{\mathfrak{p}}(T)$ for all primes $\mathfrak{p}$ of good reduction for A up to some norm bound B.

The first step is to determine the Haar measure on $\mathrm{ST}(A)^0$. For $g = 1$ there are only two possibilities: either $\mathrm{ST}(A)^0 = \mathrm{U}(1)$ or $\mathrm{ST}(A)^0 = \mathrm{SU}(2)$, where, as usual we embed U(1) in SU(2) via $u \mapsto \left(\begin{smallmatrix} u & 0 \\ 0 & \bar{u} \end{smallmatrix}\right)$. In terms of the eigenangle θ, the pushforward measure on $\operatorname{conj}(\mathrm{ST}(A)^0)$ is one of

$$\mu_{\mathrm{U}(1)} := \tfrac{1}{\pi} d\theta,$$
$$\mu_{\mathrm{SU}(2)} := \tfrac{2}{\pi} \sin^2\theta \, d\theta,$$

with $0 \le \theta \le \pi$. This also addresses two of the possibilities for $\mathrm{ST}(A)^0$ that arise when $g = 2$, the groups $\mathrm{U}(1)_2$ and $\mathrm{SU}(1)_2$ listed in the first two rows of Table 1; these denote two identical copies of U(1) and SU(2) diagonally embedded in USp(4). When expressed in terms of the eigenangle θ, the measure $\mu_{\mathrm{U}(1)_2}$ is exactly the same as $\mu_{\mathrm{U}(1)}$ (and similarly for $\mu_{\mathrm{SU}(2)_2}$), but note that we will get a different distribution on characteristic polynomials (which now have degree 4 rather than degree 2), because each eigenvalue now occurs with multiplicity 2; in particular, the trace becomes $4\cos\theta$ rather than $2\cos\theta$.

For the groups $\mathrm{ST}(A)^0$ that appear in the next three rows of Table 1, the measure on $\operatorname{conj}(\mathrm{ST}(A)^0)$ is a product of measures that we already know:

$$\mu_{\mathrm{U}(1)\times\mathrm{U}(1)} := \tfrac{1}{\pi^2} d\theta_1 \, d\theta_2,$$
$$\mu_{\mathrm{U}(1)\times\mathrm{SU}(2)} := \tfrac{2}{\pi^2} \sin^2\theta_2 \, d\theta_1 \, d\theta_2,$$
$$\mu_{\mathrm{SU}(2)\times\mathrm{SU}(2)} := \tfrac{4}{\pi^2} \sin^2\theta_1 \sin^2\theta_2 \, d\theta_1 \, d\theta_2.$$

To obtain the measure for the generic case $\mathrm{ST}(A) = \mathrm{ST}(A)^0 = \mathrm{USp}(4)$, we use the Weyl integration formula for $\mathrm{USp}(2g)$ (which includes the case $\mathrm{USp}(2) = \mathrm{SU}(2)$ that we already know):

$$\mu_{\mathrm{USp}(2g)} := \frac{1}{g!}\left(\prod_{1\le j<k\le g} (2\cos\theta_j - 2\cos\theta_k)^2\right) \prod_{1\le j\le g} \left(\tfrac{2}{\pi}\sin^2\theta_j \, d\theta_j\right), \tag{4.2}$$

with $0 \le \theta_j \le \pi$, see [**97**, Thm. 7.8B] or [**45**, §5.0.4].

This covers all the Sato-Tate groups listed in Table 1 for $g = 2$. By taking appropriate products of measures we know and applying the Weyl integration formula with $g = 3$, we obtain all the $g = 3$ cases listed in Table 2 except for U(3), where we need the Weyl integration formula for $\mathrm{U}(g)$:

$$\mu_{\mathrm{U}(g)} := \frac{1}{g!}\left(\prod_{1\le j<k\le g} \left|e^{i\theta_j} - e^{i\theta_k}\right|\right) \prod_{1\le j\le g} \tfrac{1}{2\pi} d\theta_j, \tag{4.3}$$

with $0 \le \theta_j \le 2\pi$ (note the 2π); see [**97**, Thm. 7.4B] or [**45**, §5.0.3].

With the measure $\mu_{\mathrm{ST}(A)^0}$ in hand, for any continuous class function f on $\mathrm{ST}(A)$, we can compute

$$\mu_{\mathrm{ST}(A)}(f) := \int_{\mathrm{ST}(A)} f(x)\mu_{\mathrm{ST}(A)}(x) = \sum_x \int_{\mathrm{ST}(A)^0} f(xy)\mu_{\mathrm{ST}(A)^0}(y),$$

as a finite sum over left coset representatives $x\,\mathrm{ST}(A)^0$ of $\mathrm{ST}(A)/\mathrm{ST}(A)^0$; we refer the reader to [**25**, §5.1.1] for further details and explicit results in the case $g = 2$.

4.4. Trace moment sequences. Having determined Haar measures for various Sato–Tate groups $\mathrm{ST}(A)$, we now consider the problem of computing the trace moment sequence of a connected Sato–Tate group; so assume $\mathrm{ST}(A) = \mathrm{ST}(A)^0$. For each integer $n \geq 0$ we wish to compute the nth moment

$$\mathrm{E}_{\mathrm{ST}(A)}[\mathrm{tr}^n] = \int_0^\pi \cdots \int_0^\pi \left(\sum_{j=1}^g 2\cos\theta_j\right)^n \mu_{\mathrm{ST}(A)}(\theta_1,\ldots,\theta_g).$$

We have already done this computation for the groups U(1) and SU(2) that arise in dimension $g = 1$. For U(1) we have

$$\mathrm{E}_{\mathrm{U}(1)}[\mathrm{tr}^n] = \frac{1}{\pi}\int_0^\pi (2\cos\theta)^n\, d\theta = b_n := \binom{n}{n/2},$$

where we adopt the convention that $\binom{n}{n/2} = 0$ when n is odd, and for SU(2) we have

$$\mathrm{E}_{\mathrm{SU}(2)}[\mathrm{tr}^n] = \frac{2}{\pi}\int_0^\pi (2\cos\theta)^n \sin^2\theta\, d\theta = c_n := \frac{2}{n+2}\binom{n}{n/2}.$$

We thus obtain the moment sequences

$$\mathrm{M}_{\mathrm{U}(1)}[\mathrm{tr}] = (1,\, 0,\, 2,\, 0,\, 6,\, 0,\, 20,\, 0,\, 70,\, 0,\, 252,\ldots),$$
$$\mathrm{M}_{\mathrm{SU}(2)}[\mathrm{tr}] = (1,\, 0,\, 1,\, 0,\, 2,\, 0,\, 5,\, 0,\, 14,\, 0,\, 42,\ldots).$$

For $g = 2$, observe that for 5 of the 6 connected Sato–Tate groups listed in Table 1 we can compute their trace moment sequences directly from the trace moment sequences for U(1) and SU(2); no integration is required. For $\mathrm{U}(1)_2$ and $\mathrm{SU}(2)_2$ we simply have

$$E_{\mathrm{U}(1)_2}[\mathrm{tr}^n] = E_{\mathrm{U}(1)}[2^n\,\mathrm{tr}^n] = 2^n b_n,$$
$$E_{\mathrm{SU}(2)_2}[\mathrm{tr}^n] = E_{\mathrm{SU}(2)}[2^n\,\mathrm{tr}^n] = 2^n c_n,$$

and for $\mathrm{U}(1)\times\mathrm{U}(1)$, $\mathrm{U}(1)\times\mathrm{SU}(2)$, $\mathrm{SU}(2)\times\mathrm{SU}(2)$ we take binomial convolutions to obtain[20]

(4.4)
$$E_{\mathrm{U}(1)\times\mathrm{U}(1)}[\mathrm{tr}^n] = \sum_{r=0}^n \binom{n}{r} E_{\mathrm{U}(1)}[\mathrm{tr}^r]E_{\mathrm{U}(1)}[\mathrm{tr}^{n-r}] = \sum_{r=0}^n \binom{n}{r} b_r b_{n-r} = b_n^2,$$

(4.5)
$$E_{\mathrm{U}(1)\times\mathrm{SU}(2)}[\mathrm{tr}^n] = \sum_{r=0}^n \binom{n}{r} E_{\mathrm{U}(1)}[\mathrm{tr}^r]E_{\mathrm{SU}(2)}[\mathrm{tr}^{n-r}] = \sum_{r=0}^n \binom{n}{r} b_r c_{n-r} = \tfrac{1}{2}c_n b_{n+2},$$

(4.6)
$$E_{\mathrm{SU}(2)\times\mathrm{SU}(2)}[\mathrm{tr}^n] = \sum_{r=0}^n \binom{n}{r} E_{\mathrm{SU}(2)}[\mathrm{tr}^r]E_{\mathrm{SU}(2)}[\mathrm{tr}^{n-r}] = \sum_{r=0}^n \binom{n}{r} c_r c_{n-r} = c_n c_{n+2}.$$

For the generic case USp(4) we apply (4.2) with $g = 2$ to obtain

$$\begin{aligned} E_{\mathrm{USp}(4)}[\mathrm{tr}^n] &= \tfrac{2^{n+3}}{\pi^2}\int_0^\pi\int_0^\pi (\cos\theta_1+\cos\theta_2)^n(\cos\theta_1-\cos\theta_2)^2\sin^2\theta_1\sin^2\theta_2\, d\theta_1 d\theta_2 \\ &= c_n c_{n+4} - c_{n+2}^2. \end{aligned}$$

[20] It is at this point we see the utility of starting our moment sequences at M_0.

Here we have applied the general determinantal formula from [**49**, Thm. 1] that allows one to compute the moment generating function of the kth eigenvalue power-sum in $\mathrm{USp}(2g)$. Recall that the *moment generating function* of a moment sequence $(m_0, m_1, m_2, \ldots)$ is the exponential generating function

$$\mathcal{M}(z) := \sum_{n=0}^{\infty} m_n \frac{z^n}{n!}.$$

One uses exponential generating functions so that products of moment generating functions correspond to binomial convolutions of moment sequences; this means that if $\mathcal{M}_1(z)$ and $\mathcal{M}_2(z)$ are the moment generating functions of two independent random variable X_1 and X_2, then the moment generating function of $X_1 + X_2$ is simply $\mathcal{M}_1(z)\mathcal{M}_2(z)$.

The determinantal formula for the first eigenvalue power-sum (the trace) is

$$\mathcal{M}_{\mathrm{USp}(2g)}[\mathrm{tr}] = \det_{g\times g}\left(\mathcal{C}^{i+j-2}\right)_{ij},$$

where $\mathcal{C}^m$ is the moment generating function defined by

$$\mathcal{C}^m(z) := \sum_{r=0}^{m} \binom{n}{r} \left(\mathcal{B}_{2r-n} - \mathcal{B}_{2r-n+2}\right), \qquad \mathcal{B}_s(z) := \sum_{n=0}^{\infty} \frac{z^{2n+s}}{s!(n+s)!}.$$

The function $\mathcal{B}_s(z)$ is related to a hyperbolic Bessel function of the first kind; see [**49**, p. 13] for details.

For the connected Sato–Tate groups that arise in dimension $g = 2$ we obtain

$$\begin{aligned}
\mathrm{M}_{\mathrm{U}(1)_2}[\mathrm{tr}] &= (1,\ 0,\ 8,\ 0,\ 96,\ 0,\ 1280,\ 0,\ 17920,\ 0,\ 258048,\ \ldots),\\
\mathrm{M}_{\mathrm{SU}(2)_2}[\mathrm{tr}] &= (1,\ 0,\ 4,\ 0,\ 32,\ 0,\ 320,\ 0,\ 3584,\ 0,\ 43008,\ \ldots),\\
\mathrm{M}_{\mathrm{U}(1)\times\mathrm{U}(1)}[\mathrm{tr}] &= (1,\ 0,\ 4,\ 0,\ 36,\ 0,\ 400,\ 0,\ 4900,\ 0,\ 63504,\ \ldots),\\
\mathrm{M}_{\mathrm{U}(1)\times\mathrm{SU}(2)}[\mathrm{tr}] &= (1,\ 0,\ 3,\ 0,\ 20,\ 0,\ 175,\ 0,\ 1764,\ 0,\ 19404,\ \ldots),\\
\mathrm{M}_{\mathrm{SU}(2)\times\mathrm{SU}(2)}[\mathrm{tr}] &= (1,\ 0,\ 2,\ 0,\ 10,\ 0,\ 70,\ 0,\ 588,\ 0,\ 5544,\ \ldots),\\
\mathrm{M}_{\mathrm{USp}(4)}[\mathrm{tr}] &= (1,\ 0,\ 1,\ 0,\ 3,\ 0,\ 14,\ 0,\ 84,\ 0,\ 594,\ \ldots),
\end{aligned}$$

and for $g = 3$ we have

$$\begin{aligned}
\mathrm{M}_{\mathrm{U}(1)_3}[\mathrm{tr}] &= (1,\ 0,\ 18,\ 0,\ 486,\ 0,\ 14580,\ 0,\ 459270,\ 0,\ 14880348,\ \ldots),\\
\mathrm{M}_{\mathrm{SU}(2)_3}[\mathrm{tr}] &= (1,\ 0,\ 9,\ 0,\ 162,\ 0,\ 3645,\ 0,\ 91854,\ 0,\ 2480058,\ \ldots),\\
\mathrm{M}_{\mathrm{U}(1)\times\mathrm{U}(1)_2}[\mathrm{tr}] &= (1,\ 0,\ 10,\ 0,\ 198,\ 0,\ 4900,\ 0,\ 134470,\ 0,\ 3912300,\ \ldots),\\
\mathrm{M}_{\mathrm{U}(1)\times\mathrm{SU}(2)_2}[\mathrm{tr}] &= (1,\ 0,\ 6,\ 0,\ 86,\ 0,\ 1660,\ 0,\ 37254,\ 0,\ 916020,\ \ldots),\\
\mathrm{M}_{\mathrm{SU}(2)\times\mathrm{U}(1)_2}[\mathrm{tr}] &= (1,\ 0,\ 9,\ 0,\ 146,\ 0,\ 2965,\ 0,\ 68334,\ 0,\ 1707930,\ \ldots),\\
\mathrm{M}_{\mathrm{SU}(2)\times\mathrm{SU}(2)_2}[\mathrm{tr}] &= (1,\ 0,\ 5,\ 0,\ 58,\ 0,\ 925,\ 0,\ 17598,\ 0,\ 374850,\ \ldots),\\
\mathrm{M}_{\mathrm{U}(1)\times\mathrm{U}(1)\times\mathrm{U}(1)}[\mathrm{tr}] &= (1,\ 0,\ 6,\ 0,\ 90,\ 0,\ 1860,\ 0,\ 44730,\ 0,\ 1172556,\ \ldots),\\
\mathrm{M}_{\mathrm{U}(1)\times\mathrm{U}(1)\times\mathrm{SU}(2)}[\mathrm{tr}] &= (1,\ 0,\ 5,\ 0,\ 62,\ 0,\ 1065,\ 0,\ 21714,\ 0,\ 492366,\ \ldots),\\
\mathrm{M}_{\mathrm{U}(1)\times\mathrm{SU}(2)\times\mathrm{SU}(2)}[\mathrm{tr}] &= (1,\ 0,\ 4,\ 0,\ 40,\ 0,\ 570,\ 0,\ 9898,\ 0,\ 19521,\ \ldots),\\
\mathrm{M}_{\mathrm{SU}(2)\times\mathrm{SU}(2)\times\mathrm{SU}(2)}[\mathrm{tr}] &= (1,\ 0,\ 3,\ 0,\ 24,\ 0,\ 285,\ 0,\ 4242,\ 0,\ 73206,\ \ldots),\\
\mathrm{M}_{\mathrm{U}(1)\times\mathrm{USp}(4)}[\mathrm{tr}] &= (1,\ 0,\ 3,\ 0,\ 21,\ 0,\ 214,\ 0,\ 2758,\ 0,\ 41796,\ \ldots),\\
\mathrm{M}_{\mathrm{SU}(2)\times\mathrm{USp}(4)}[\mathrm{tr}] &= (1,\ 0,\ 2,\ 0,\ 11,\ 0,\ 94,\ 0,\ 1050,\ 0,\ 14076,\ \ldots),\\
\mathrm{M}_{\mathrm{U}(3)}[\mathrm{tr}] &= (1,\ 0,\ 2,\ 0,\ 12,\ 0,\ 120,\ 0,\ 1610,\ 0,\ 25956,\ \ldots),\\
\mathrm{M}_{\mathrm{USp}(6)}[\mathrm{tr}] &= (1,\ 0,\ 1,\ 0,\ 3,\ 0,\ 15,\ 0,\ 104,\ 0,\ 909,\ \ldots),
\end{aligned}$$

Recall that for $g = 1$ the trace moment sequence $(1, 0, 1, 0, 2, 0, 5, 0, 14, 0, 42, \ldots)$ of the generic Sato–Tate group SU(2) corresponds to the sequence of Catalan numbers with 0's inserted at the odd moments. There is a standard combinatorial interpretation of this sequence: the nth moment counts the number of returning walks of length n on a 1-dimensional integer lattice that stay to the right of the origin (there are no such walks when n is odd, hence the odd moments are zero).

This combinatorial interpretation generalizes to higher genus. For $g = 2$ the trace moment sequence for the generic Sato–Tate group USp(4) counts returning walks on a 2-dimensional integer lattice that satisfy $x_1 \geq x_2 \geq 0$ (so now there are 3 walks of length 4, not just 2). In general, for any $g \geq 1$ the trace moment sequence for the generic Sato–Tate group USp($2g$) counts returning walks on a g-dimensional integer lattice that satisfy $x_1 \geq \ldots \geq x_g \geq 0$; this follows from a general result of Grabiner and Magyar [**30**] that relates the decomposition of tensor powers of certain representations of classical Lie groups to lattice paths that are constrained to lie in the closure of the fundamental *Weyl chamber* of the corresponding Lie algebra (which can be defined as an intersection of hyperplanes orthogonal to elements of a basis for the root system).

This combinatorial feature has an interesting asymptotic consequence. For any integers $g' \geq g > 0$, the moment sequences $\mathrm{M}_{\mathrm{USp}(2g')}[\mathrm{tr}]$ and $\mathrm{M}_{\mathrm{USp}(2g)}[\mathrm{tr}]$ must agree up to the $2g$th moment; see Exercise 4.3. Thus the moments sequences $\mathrm{M}_{\mathrm{USp}(2g)}[\mathrm{tr}]$ converge to a limiting sequence as $g \to \infty$:

$$\begin{aligned}
\mathrm{M}_{\mathrm{USp}(2)}[\mathrm{tr}] &= (1,\, 0,\, 1,\, 0,\, 2,\, 0,\, 5,\, 0,\, 14,\, 0,\, 42,\, \ldots),\\
\mathrm{M}_{\mathrm{USp}(4)}[\mathrm{tr}] &= (1,\, 0,\, 1,\, 0,\, 3,\, 0,\, 14,\, 0,\, 84,\, 0,\, 594,\, \ldots),\\
\mathrm{M}_{\mathrm{USp}(6)}[\mathrm{tr}] &= (1,\, 0,\, 1,\, 0,\, 3,\, 0,\, 15,\, 0,\, 104,\, 0,\, 909,\, \ldots),\\
\mathrm{M}_{\mathrm{USp}(8)}[\mathrm{tr}] &= (1,\, 0,\, 1,\, 0,\, 3,\, 0,\, 15,\, 0,\, 105,\, 0,\, 944\, \ldots).\\
&\ \ \vdots\\
\mathrm{M}_{\mathrm{USp}(\infty)}[\mathrm{tr}] &= (1,\, 0,\, 1,\, 0,\, 3,\, 0,\, 15,\, 0,\, 105,\, 0,\, 945,\, \ldots).
\end{aligned}$$

The limiting sequence $\mathrm{M}_{\mathrm{USp}(\infty)}[\mathrm{tr}]$ is precisely the moment sequence of the standard normal distribution (mean 0 and variance 1); the nth moment is zero if n is odd, and for even n it is given by

$$(n-1)!! := n(n-2)(n-4)\cdots 3\cdot 1.$$

Figure 4 shows the a_1-distributions for $g = 1, 2, 3, 4$, normalized to the same scale, which illustrates convergence to the standard normal distribution.

4.5. Exercises.

EXERCISE 4.1. Give combinatorial proofs of the identities used in (4.4), (4.5), (4.6).

EXERCISE 4.2. Using the combinatorial interpretation of the trace moment sequence $\mathrm{M}_{\mathrm{USp}(2g)}[\mathrm{tr}]$, prove that for $g' > g$ the moment sequences $\mathrm{M}_{\mathrm{USp}(2g')}[\mathrm{tr}]$ and $\mathrm{M}_{\mathrm{USp}(2g)}[\mathrm{tr}]$ agree up to the $2g$th moment but disagree at the $(2g+2)$th moment. Then show that the limiting trace moment sequence $\mathrm{M}_{\mathrm{USp}(\infty)}[\mathrm{tr}]$ is equal to the moment sequence of the standard normal distribution.

EXERCISE 4.3. Characterize each of the 6 trace moment sequences that arise for connected Sato–Tate groups in dimension $g = 2$ by showing that each sequence counts returning walks on an 2-dimensional integer lattice that are constrained to a certain region of the plane.

EXERCISE 4.4. Similarly characterize the 14 trace moment sequences that arise for connected Sato–Tate groups in dimension $g = 3$ in terms of returning walks on a 3-dimensional integer lattice.

EXERCISE 4.5. For each of the 5 non-generic connected Sato–Tate groups that arise in dimension $g = 2$ compute the moment sequence for a_2, the quadratic coefficient of the characteristic polynomial.

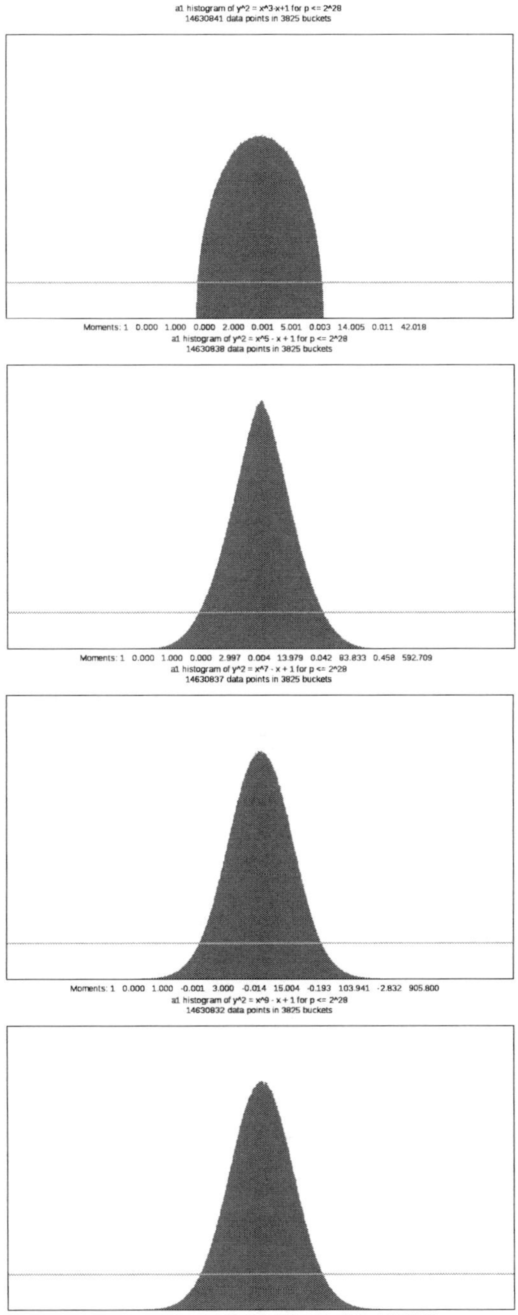

FIGURE 4. Generic trace distributions for $g = 1, 2, 3, 4$ (shown with the same vertical scale).

References

[1] J. D. Achter and J. Holden, *Notes on an analogue of the Fontaine-Mazur conjecture* (English, with English and French summaries), J. Théor. Nombres Bordeaux **15** (2003), no. 3, 627–637. MR2142226

[2] O. Ahmadi and I. E. Shparlinski, *On the distribution of the number of points on algebraic curves in extensions of finite fields*, Math. Res. Lett. **17** (2010), no. 4, 689–699, DOI 10.4310/MRL.2010.v17.n4.a9. MR2661173

[3] P. Allen, F. Calegari, A. Caraiani, T. Gee, D. Helm, B. Le-Hung, J. Newton, P. Scholze, R. Taylor, J. Thorne, `http://www.math.ias.edu/files/Emerging1016report.pdf`, *Applications to modularity of recent progress on the cohomology of Shimura Varieties*, IAS working group report, 2016 (paper to appear).

[4] G. Banaszak and K. S. Kedlaya, *An algebraic Sato-Tate group and Sato-Tate conjecture*, Indiana Univ. Math. J. **64** (2015), no. 1, 245–274, DOI 10.1512/iumj.2015.64.5438. MR3320526

[5] G. Banaszak and K. S. Kedlaya, *Motivic Serre group, algebraic Sato-Tate group and Sato-Tate conjecture*, Frobenius distributions: Lang-Trotter and Sato-Tate conjectures, Contemp. Math., vol. 663, Amer. Math. Soc., Providence, RI, 2016, pp. 11–44, DOI 10.1090/conm/663/13348. MR3502937

[6] T. Barnet-Lamb, T. Gee, and D. Geraghty, *The Sato-Tate conjecture for Hilbert modular forms*, J. Amer. Math. Soc. **24** (2011), no. 2, 411–469, DOI 10.1090/S0894-0347-2010-00689-3. MR2748398

[7] T. Barnet-Lamb, D. Geraghty, M. Harris, and R. Taylor, *A family of Calabi-Yau varieties and potential automorphy II*, Publ. Res. Inst. Math. Sci. **47** (2011), no. 1, 29–98, DOI 10.2977/PRIMS/31. MR2827723

[8] C. Birkenhake and H. Lange, *Complex abelian varieties*, 2nd ed., Grundlehren der Mathematischen Wissenschaften [Fundamental Principles of Mathematical Sciences], vol. 302, Springer-Verlag, Berlin, 2004. MR2062673

[9] F. A. Bogomolov, *Sur l'algébricité des représentations l-adiques* (French, with English summary), C. R. Acad. Sci. Paris Sér. A-B **290** (1980), no. 15, A701–A703. MR574307

[10] A. Borel, *Linear algebraic groups*, 2nd ed., Graduate Texts in Mathematics, vol. 126, Springer-Verlag, New York, 1991. MR1102012

[11] W. Bosma, J. Cannon, and C. Playoust, *The Magma algebra system. I. The user language*, J. Symbolic Comput. **24** (1997), no. 3-4, 235–265, DOI 10.1006/jsco.1996.0125. MR1484478

[12] C. Breuil, B. Conrad, F. Diamond, and R. Taylor, *On the modularity of elliptic curves over* $\mathbf{Q}$*: wild 3-adic exercises*, J. Amer. Math. Soc. **14** (2001), no. 4, 843–939, DOI 10.1090/S0894-0347-01-00370-8. MR1839918

[13] A. Bucur and K. S. Kedlaya, *An application of the effective Sato-Tate conjecture*, Frobenius distributions: Lang-Trotter and Sato-Tate conjectures, Contemp. Math., vol. 663, Amer. Math. Soc., Providence, RI, 2016, pp. 45–56, DOI 10.1090/conm/663/13349. MR3502938

[14] J. W. S. Cassels and A. Fröhlich (eds.), *Algebraic number theory*, 2nd ed., London Mathematical Society, London, 2010. MR3618860

[15] W. Castryck, A. Folsom, H. Hubrechts, and A. V. Sutherland, *The probability that the number of points on the Jacobian of a genus 2 curve is prime*, Proc. Lond. Math. Soc. (3) **104** (2012), no. 6, 1235–1270, DOI 10.1112/plms/pdr063. MR2946086

[16] L. Clozel, M. Harris, and R. Taylor, *Automorphy for some l-adic lifts of automorphic mod l Galois representations*, Publ. Math. Inst. Hautes Études Sci. **108** (2008), 1–181, DOI 10.1007/s10240-008-0016-1. MR2470687

[17] A. C. Cojocaru, R. Davis, A. Silverberg, and K. E. Stange, *Arithmetic properties of the Frobenius traces defined by a rational abelian variety (with two appendices by J-P. Serre)*, Int. Math. Res. Not. IMRN **12** (2017), 3557–3602. MR3693659

[18] P. Deligne, *La conjecture de Weil. I* (French), Inst. Hautes Études Sci. Publ. Math. **43** (1974), 273–307. MR0340258

[19] P. Deligne, *La conjecture de Weil. II* (French), Inst. Hautes Études Sci. Publ. Math. **52** (1980), 137–252. MR601520

[20] P. Deligne, J. S. Milne, A. Ogus, and K.-y. Shih, *Hodge cycles, motives, and Shimura varieties*, Lecture Notes in Mathematics, vol. 900, Springer-Verlag, Berlin-New York, 1982. MR654325

[21] F. Diamond and J. Shurman, *A first course in modular forms*, Graduate Texts in Mathematics, vol. 228, Springer-Verlag, New York, 2005. MR2112196

[22] J. Diestel and A. Spalsbury, *The joys of Haar measure*, Graduate Studies in Mathematics, vol. 150, American Mathematical Society, Providence, RI, 2014. MR3186070

[23] G. Faltings, *Endlichkeitssätze für abelsche Varietäten über Zahlkörpern* (German), Invent. Math. **73** (1983), no. 3, 349–366, DOI 10.1007/BF01388432. MR718935

[24] F. Fité, *Equidistribution, L-functions, and Sato-Tate groups*, Trends in number theory, Contemp. Math., vol. 649, Amer. Math. Soc., Providence, RI, 2015, pp. 63–88, DOI 10.1090/conm/649/13020. MR3415267

[25] F. Fité, K. S. Kedlaya, V. Rotger, and A. V. Sutherland, *Sato-Tate distributions and Galois endomorphism modules in genus 2*, Compos. Math. **148** (2012), no. 5, 1390–1442, DOI 10.1112/S0010437X12000279. MR2982436

[26] F. Fité, K. S. Kedlaya, and A. V. Sutherland, *Sato-Tate groups of some weight 3 motives*, Frobenius Distributions: Lang-Trotter and Sato-Tate conjectures, Contemp. Math., vol. 663, Amer. Math. Soc., Providence, RI, 2016, pp. 57–101, DOI 10.1090/conm/663/13350. MR3502939

[27] F. Fité and A. V. Sutherland, *Sato-Tate distributions of twists of* $y^2 = x^5 - x$ *and* $y^2 = x^6 + 1$, Algebra Number Theory **8** (2014), no. 3, 543–585, DOI 10.2140/ant.2014.8.543. MR3218802

[28] F. Fité and A. V. Sutherland, *Sato-Tate groups of* $y^2 = x^8 + c$ *and* $y^2 = x^7 - cx$, Frobenius distributions: Lang-Trotter and Sato-Tate conjectures, Contemp. Math., vol. 663, Amer. Math. Soc., Providence, RI, 2016, pp. 103–126, DOI 10.1090/conm/663/13351. MR3502940

[29] J. von zur Gathen and J. Gerhard, *Modern computer algebra*, 3rd ed., Cambridge University Press, Cambridge, 2013. MR3087522

[30] D. J. Grabiner and P. Magyar, *Random walks in Weyl chambers and the decomposition of tensor powers*, J. Algebraic Combin. **2** (1993), no. 3, 239–260, DOI 10.1023/A:1022499531492. MR1235279

[31] C. Hall, *An open-image theorem for a general class of abelian varieties*, Bull. Lond. Math. Soc. **43** (2011), no. 4, 703–711, DOI 10.1112/blms/bdr004. MR2820155

[32] M. Harris, N. Shepherd-Barron, and R. Taylor, *A family of Calabi-Yau varieties and potential automorphy*, Ann. of Math. (2) **171** (2010), no. 2, 779–813, DOI 10.4007/annals.2010.171.779. MR2630056

[33] W. B. Hart, `http://link.springer.com/chapter/10.1007/978-3-642-15582-6_18`, *Fast Library for Number Theory: An introduction*, in `http://link.springer.com/book/10.1007/978-3-642-15582-6`, *Proceedings of the Third International Congress on Mathematical Software (ICMS 2010)*, LNCS 6327, Springer, 2010, 88–91.

[34] W. B. Hart, F. Johansson and S. Pancratz, *Fast Library for Number Theory*, version 2.5.2, `http://flintlib.org`, 2015.

[35] D. Harvey, *Kedlaya's algorithm in larger characteristic*, Int. Math. Res. Not. IMRN **22** (2007), DOI 10.1093/imrn/rnm095. MR2376210

[36] D. Harvey, *Counting points on hyperelliptic curves in average polynomial time*, Ann. of Math. (2) **179** (2014), no. 2, 783–803, DOI 10.4007/annals.2014.179.2.7. MR3152945

[37] D. Harvey, *Computing zeta functions of arithmetic schemes*, Proc. Lond. Math. Soc. (3) **111** (2015), no. 6, 1379–1401, DOI 10.1112/plms/pdv056. MR3447797

[38] D. Harvey and A. V. Sutherland, *Computing Hasse-Witt matrices of hyperelliptic curves in average polynomial time*, LMS J. Comput. Math. **17** (2014), no. suppl. A, 257–273, DOI 10.1112/S1461157014000187. MR3240808

[39] D. Harvey and A. V. Sutherland, *Computing Hasse-Witt matrices of hyperelliptic curves in average polynomial time, II*, Frobenius distributions: Lang-Trotter and Sato-Tate conjectures, Contemp. Math., vol. 663, Amer. Math. Soc., Providence, RI, 2016, pp. 127–147, DOI 10.1090/conm/663/13352. MR3502941

[40] E. Hecke, *Eine neue Art von Zetafunktionen und ihre Beziehungen zur Verteilung der Primzahlen* (German), Math. Z. **6** (1920), no. 1-2, 11–51, DOI 10.1007/BF01202991. MR1544392

[41] H. Heyer, *Probability measures on locally compact groups*, Springer-Verlag, Berlin-New York, 1977. MR0501241

[42] J. E. Humphreys, *Linear algebraic groups*, Graduate Texts in Mathematics, vol. 21, Springer-Verlag, New York-Heidelberg, 1975. MR0396773

[43] C. Johansson, *On the Sato-Tate conjecture for non-generic abelian surfaces*, Trans. Amer. Math. Soc. **369** (2017), no. 9, 6303–6325, DOI 10.1090/tran/6847. MR3660222

[44] I. Kaplansky, http://projecteuclid.org/euclid.bams/1183510807, *Lattices of continuous functions*, Bulletin of the Amer. Math. Soc. **6** (1947), 617–623. DOI 10.1090/S0002-9904-1947-08856-X. MR20715

[45] N. M. Katz and P. Sarnak, *Random matrices, Frobenius eigenvalues, and monodromy*, American Mathematical Society Colloquium Publications, vol. 45, American Mathematical Society, Providence, RI, 1999. MR1659828

[46] K. S. Kedlaya, *Counting points on hyperelliptic curves using Monsky-Washnitzer cohomology*, J. Ramanujan Math. Soc. **16** (2001), no. 4, 323–338. MR1877805

[47] K. S. Kedlaya, *Computing zeta functions via p-adic cohomology*, Algorithmic number theory, Lecture Notes in Comput. Sci., vol. 3076, Springer, Berlin, 2004, pp. 1–17, DOI 10.1007/978-3-540-24847-7_1. MR2137340

[48] K. S. Kedlaya and A. V. Sutherland, *Computing L-series of hyperelliptic curves*, Algorithmic Number Theory, Lecture Notes in Comput. Sci., vol. 5011, Springer, Berlin, 2008, pp. 312–326, DOI 10.1007/978-3-540-79456-1_21. MR2467855

[49] K. S. Kedlaya and A. V. Sutherland, *Hyperelliptic curves, L-polynomials, and random matrices*, Arithmetic, geometry, cryptography and coding theory, Contemp. Math., vol. 487, Amer. Math. Soc., Providence, RI, 2009, pp. 119–162, DOI 10.1090/conm/487/09529. MR2555991

[50] K. S. Kedlaya and C. Umans, *Fast polynomial factorization and modular composition*, SIAM J. Comput. **40** (2011), no. 6, 1767–1802, DOI 10.1137/08073408X. MR2863194

[51] J. Klüners and G. Malle, *A database for field extensions of the rationals*, LMS J. Comput. Math. **4** (2001), 182–196, DOI 10.1112/S1461157000000851. MR1901356

[52] P. Koosis, *The logarithmic integral. I*, Cambridge Studies in Advanced Mathematics, vol. 12, Cambridge University Press, Cambridge, 1998. Corrected reprint of the 1988 original. MR1670244

[53] S. Lang and H. Trotter, *Frobenius distributions in* GL_2*-extensions*, Lecture Notes in Mathematics, Vol. 504, Springer-Verlag, Berlin-New York, 1976. MR0568299

[54] J. S. Milne, http://www.jmilne.org/math/CourseNotes/av.html*Abelian varieties*, v2.00, 2008.

[55] J. S. Milne, http://www.jmilne.org/math/CourseNotes/iAG200.pdf, *Algebraic groups: An introduction to the theory of algebraic group schemes over fields*, v2.00, 2015.

[56] B. Moonen, *An introduction to Mumford–Tate groups*, Monte Verita Lecture Notes, available at http://www.math.ru.nl/~bmoonen/Lecturenotes/MTGps.pdf, 2004.

[57] B. J. J. Moonen and Yu. G. Zarhin, *Hodge classes on abelian varieties of low dimension*, Math. Ann. **315** (1999), no. 4, 711–733, DOI 10.1007/s002080050333. MR1731466

[58] D. Mumford, *A note of Shimura's paper "Discontinuous groups and abelian varieties"*, Math. Ann. **181** (1969), 345–351, DOI 10.1007/BF01350672. MR248146

[59] D. Mumford, *Abelian varieties*, Tata Institute of Fundamental Research Studies in Mathematics, vol. 5, Published for the Tata Institute of Fundamental Research, Bombay; by Hindustan Book Agency, New Delhi, 2008. With appendices by C. P. Ramanujam and Yuri Manin; Corrected reprint of the second (1974) edition. MR2514037

[60] M. R. Murty and V. K. Murty, *Non-vanishing of L-functions and applications*, Modern Birkhäuser Classics, Birkhäuser/Springer Basel AG, Basel, 1997. [2011 reprint of the 1997 original] [MR1482805]. MR3025442

[61] V. K. Murty, *Explicit formulae and the Lang-Trotter conjecture*, Rocky Mountain J. Math. **15** (1985), no. 2, 535–551, DOI 10.1216/RMJ-1985-15-2-535. MR823264

[62] J. Neukirch, *Algebraic number theory*, Grundlehren der Mathematischen Wissenschaften [Fundamental Principles of Mathematical Sciences], vol. 322, Springer-Verlag, Berlin, 1999. Translated from the 1992 German original. MR1697859

[63] A. L. Onishchik and A. L. Vinberg (eds.), http://www.springer.com/us/book/9783540546832, *Lie groups and Lie algebras III: Structure of Lie groups and Lie algebras*, Springer, 1994.

[64] OEIS Foundation Inc., *The On-Line Encyclopedia of Integer Sequences*, online database at http://oeis.org, 2016.

[65] J. Pila, *Frobenius maps of abelian varieties and finding roots of unity in finite fields*, Math. Comp. **55** (1990), no. 192, 745–763, DOI 10.2307/2008445. MR1035941

[66] D. Ramakrishnan, *Remarks on the Tate Conjecture for beginners*, notes from the AIM Tate Conjecture Workshop, available at `http://www.aimath.org/WWN/tateconjecture/tateconjecture.pdf`, 2007.

[67] The Sage Developers, *Sage Mathematics Software*, Version 7.0, available at `http://www.sagemath.org`, 2016.

[68] W. F. Sawin, *Ordinary primes for Abelian surfaces* (English, with English and French summaries), C. R. Math. Acad. Sci. Paris **354** (2016), no. 6, 566–568, DOI 10.1016/j.crma.2016.01.025. MR3494322

[69] R. Schoof, *Elliptic curves over finite fields and the computation of square roots mod p*, Math. Comp. **44** (1985), no. 170, 483–494, DOI 10.2307/2007968. MR777280

[70] R. Schoof, *Counting points on elliptic curves over finite fields*, J. Théor. Nombres Bordeaux **7** (1995), no. 1, 219–254. MR1413578

[71] J.-P. Serre, *Abelian l-adic representations and elliptic curves*, Research Notes in Mathematics, vol. 7, A K Peters, Ltd., Wellesley, MA, 1998. Revised reprint of the 1968 original. MR1484415

[72] J.-P. Serre, *Propriétés galoisiennes des points d'ordre fini des courbes elliptiques* (French), Invent. Math. **15** (1972), no. 4, 259–331, DOI 10.1007/BF01405086. MR387283

[73] J.-P. Serre, *Résumé des cours de 1985-1986, Annuaire du Collège de France, 1986, 95–99*; in *Oeuvres – Collected Papers, Volume IV*, Springer, 2003, 33–37.

[74] J.-P. Serre, *Lettre à Marie-France Vignéras du 10/2/1986*, pages 38–55, in `http://www.springer.com/us/book/9783642398391`, *Oeuvres – Collected Papers, Volume IV*, Springer, 2000.

[75] J.-P. Serre, *Lettres à Ken Ribet du 1/1/1981 et du 29/1/1981*, pages 1–20 in `http://www.springer.com/us/book/9783642398391`, *Oeuvres – Collected Papers, Volume IV*, Springer, 2000.

[76] J.-P. Serre, *Propriétés conjecturales des groups de Galois motiviques et des représentations ℓ-adiques*, pages 377-400 in `http://bookstore.ams.org/pspum-55-1/`*Motives: Part I*, U. Jansen, S. Kleiman, and J.-P. Serre eds., AMS Proceedings of Symposia in Pure Mathematics **55**, 1994.

[77] J.-P. Serre, `http://www.crcnetbase.com/isbn/9781466501935`, *Lectures on $N_X(p)$*, Research Notes in Mathematics **11**, CRC Press, 2012.

[78] Y.-D. Shieh, `http://iml.univ-mrs.fr/~kohel/phd/thesis_shieh.pdf`, *Arithmetic aspects of point counting and Frobenius distributions*, Ph.D. thesis, Université d'Aix-Marseille, 2015.

[79] Y.-D. Shieh, *Character theory approach to Sato-Tate groups*, LMS J. Comput. Math. **19** (2016), no. suppl. A, 301–314, DOI 10.1112/S1461157016000279. MR3540962

[80] V. Shoup, *NTL: A Library for doing Number Theory*, version 9.6.4, available at `http://www.shoup.net/ntl/`, 2016.

[81] J. H. Silverman, *Advanced topics in the arithmetic of elliptic curves*, Graduate Texts in Mathematics, vol. 151, Springer-Verlag, New York, 1994. MR1312368

[82] J. H. Silverman, *The arithmetic of elliptic curves*, 2nd ed., Graduate Texts in Mathematics, vol. 106, Springer, Dordrecht, 2009. MR2514094

[83] T. A. Springer, *Linear algebraic groups*, 2nd ed., Modern Birkhäuser Classics, Birkhäuser Boston, Inc., Boston, MA, 2009. MR2458469

[84] R. P. Stanley, *Catalan numbers*, Cambridge University Press, New York, 2015. MR3467982

[85] A. V. Sutherland, `smalljac`, version 5.0, available at `http://math.mit.edu/~drew`, 2017.

[86] A. V. Sutherland, *Order computations in generic groups*, ProQuest LLC, Ann Arbor, MI, 2007. Thesis (Ph.D.)–Massachusetts Institute of Technology. MR2717420

[87] A. V. Sutherland, *Structure computation and discrete logarithms in finite abelian p-groups*, Math. Comp. **80** (2011), no. 273, 477–500, DOI 10.1090/S0025-5718-10-02356-2. MR2728991

[88] J. Tate, *Algebraic cycles and poles of zeta functions*, pp. 224–241 in `http://bookstore.ams.org/cworks-24-1/`, *Collected works of John Tate: Part I (1951–1975)*, B. Mazur and J.-P. Serre eds., American Mathematical Society, 2016.

[89] J. Tate, *Endomorphisms of abelian varieties over finite fields*, Invent. Math. **2** (1966), 134–144, DOI 10.1007/BF01404549. MR206004

[90] R. Taylor, *Automorphy for some l-adic lifts of automorphic mod l Galois representations. II*, Publ. Math. Inst. Hautes Études Sci. **108** (2008), 183–239, DOI 10.1007/s10240-008-0015-2. MR2470688

[91] R. Taylor and A. Wiles, *Ring-theoretic properties of certain Hecke algebras*, Ann. of Math. (2) **141** (1995), no. 3, 553–572, DOI 10.2307/2118560. MR1333036
[92] J. Thorner, *The error term in the Sato-Tate conjecture*, Arch. Math. (Basel) **103** (2014), no. 2, 147–156, DOI 10.1007/s00013-014-0673-x. MR3254358
[93] P. van Wamelen, *On the CM character of the curves* $y^2 = x^q - 1$, Journal of Number Theory **64** (1997), 59–83.
[94] A. Weil, *Sur les courbes algébriques et les variétés qui s'en déduisent* (French), Actualités Sci. Ind., no. 1041, Publ. Inst. Math. Univ. Strasbourg **7** (1945), Hermann et Cie., Paris, 1948. MR0027151
[95] A. Weil, *Variétés abéliennes et courbes algébriques* (French), Actualités Sci. Ind., no. 1064, Publ. Inst. Math. Univ. Strasbourg 8 (1946), Hermann & Cie., Paris, 1948. MR0029522
[96] A. Weil, *Numbers of solutions of equations in finite fields*, Bull. Amer. Math. Soc. **55** (1949), 497–508, DOI 10.1090/S0002-9904-1949-09219-4. MR29393
[97] H. Weyl, *The Classical Groups. Their Invariants and Representations*, Princeton University Press, Princeton, N.J., 1939. MR0000255
[98] A. Wiles, *Modular elliptic curves and Fermat's last theorem*, Ann. of Math. (2) **141** (1995), no. 3, 443–551, DOI 10.2307/2118559. MR1333035
[99] A. Wulfsohn, *A note on the vague topology for measures*, Proc. Cambridge Philos. Soc. **58** (1962), 421–422. MR137811
[100] Y. G. Zarhin, *Hyperelliptic Jacobians without complex multiplication*, Math. Res. Lett. **7** (2000), no. 1, 123–132, DOI 10.4310/MRL.2000.v7.n1.a11. MR1748293

Department of Mathematics, Massachusetts Institute of Technology, 77 Massachusetts Avenue, Cambridge, Massachusetts 02139
Email address: `drew@math.mit.edu`